11th ADVANCES IN RELIABILITY
TECHNOLOGY SYMPOSIUM

Proceedings of the 11th Advances in Reliability Technology Symposium held at the University of Liverpool, Liverpool, UK, 18–20 April 1990

Organised by
The National Centre of Systems Reliability and the University of Liverpool

In association with
UMIST and the University of Bradford

Support Organisations
British Computer Society
Centre for Software Reliability
European Safety & Reliability Association
Institute of Quality Assurance
Institution of Electrical Engineers
Safety and Reliability Society
Institution of Mechanical Engineers

Organising Committee

P. Comer (*Chairman, Secretary*)	National Centre of Systems Reliability
A. Z. Keller	University of Bradford
R. N. Allan	UMIST
R. F. de la Mare	University of Bradford
P. Martin	University of Liverpool
B. A. White	National Centre of Systems Reliability
M. Cottington	National Centre of Systems Reliability

Session Chairmen

A. Z. Keller	University of Bradford, UK
A. J. Bourne	former Head, National Centre of Systems Reliability, UK
R. N. Allan	UMIST, UK
P. Humphreys	National Centre of Systems Reliability, UK
C. J. Dale	Cranfield Information Technology Institute, UK
I. A. Watson	National Centre of Systems Reliability, UK
R. F. de la Mare	University of Bradford, UK
R. Billinton	University of Saskatchewan, Canada
P. Martin	University of Liverpool, UK

11th ADVANCES IN RELIABILITY TECHNOLOGY SYMPOSIUM

Edited by

PETER COMER

National Centre of Systems Reliability,
Culcheth, Warrington, Cheshire, UK

ELSEVIER APPLIED SCIENCE
LONDON and NEW YORK

ELSEVIER SCIENCE PUBLISHERS LTD
Crown House, Linton Road, Barking, Essex IG11 8JU, England

Sole Distributor in the USA and Canada
ELSEVIER SCIENCE PUBLISHING CO., INC.
655 Avenue of the Americas, New York, NY 10010, USA

WITH 59 TABLES AND 102 ILLUSTRATIONS

© 1990 ELSEVIER SCIENCE PUBLISHERS LTD
© 1990 GOVERNMENT OF CANADA—pp. 323–336
Softcover reprint of the hardcover 1st edition 1990

British Library Cataloguing in Publication Data

Advances in Reliability Technology Symposium (11th: 1990:
University of Liverpool)
1. Engineering equipment. Reliability
I. Title   II. Comer, Peter
620.00452

ISBN-13:978-94-010-6828-4   e-ISBN-13:978-94-009-0761-4
DOI:10.1007/978-94-009-0761-4

Library of Congress CIP data applied for

Preface

On behalf of the Organising Committee of the 11th ARTS I would like to welcome all the delegates, session chairpersons and authors. I particularly welcome new delegates, delegates from mainland Europe and from other countries.

At the time of the last symposium, our tenth anniversary, we looked back on the growth of the symposium and the support it had received from so many people. Not least was the support given by Mrs Ruth Campbell who, between this symposium and the last, has retired from the National Centre of Systems Reliability. The Organising Committee would hereby like to acknowledge a very special debt of gratitude, over many years, to Ruth.

Our gratitude also goes to Dr A. Z. Keller of the University of Bradford, the Organising Committee Chairman at the 10th Symposium, our President for 11th ARTS and, since the beginning, a staunch supporter of the ARTS. Our thanks go to Mme A. Carnino of Electricité de France for being our after-dinner speaker and to Mr A. J. Bourne for being our keynote speaker. Their speeches have been keenly anticipated.

Behind the symposium, the detailed and hard work of the administrative staff of the National Centre of Systems Reliability continues even after it has ended. Our thanks go to them, particularly, and to the Universities of Liverpool, Manchester Institute of Science and Technology and Bradford for their consistent support to the symposium.

Lastly, the most important part in the symposium has to be the contributors. Without the papers there would not be a symposium, so, as at previous symposia, our final thanks go to the authors.

Peter Comer
Organising Committee Chairman, 11th ARTS

Preface

Contents

Human Factors

Safety Assessment

Modelling

Dependency Modelling

Software Reliability

Electrical Systems

Late Submission

List of Contributors

R. N. Allan, Electrical Energy and Power Systems Group, UMIST, PO Box 88, Manchester M60 1QD, UK

J. I. Ansell, Department of Management Systems and Sciences, University of Hull, Hull HU6 7RX, UK

R. Billinton, Power System Research Group, University of Saskatchewan, Saskatoon, Sakatchewan S7N 0W0, Canada

J. C. C. Bueno, Petroleo Brasileiro SA—PETROBRAS, Serviço de Engenharia, Av. Chile 65, 11th Floor, Rio de Janeiro, RJ, Brazil

M. Carey, RM Consultants Ltd, Genesis Centre, Birchwood Science Park, Risley, Warrington WA3 7BH, UK

D. Chen, Department of Mechanical Engineering, Wuhan Iron and Steel University, Wuhan, People's Republic of China

S. J. Chester, Department of Mathematics, Statistics and Operational Research, Nottingham Polytechnic, Burton Street, Nottingham, UK

I. M. Clark, Canadian Electrical Association, Suite 580, One Westmount Square, Montreal, Quebec H3Z 2P9, Canada

C. Davies, Technology Group, Aerosystems International, Scientific House, 40–44 Coombe Road, New Malden, Surrey KT3 4QF, UK

M. Eid, Commissariat à l'Energie Atomique, Saclay, 91191 Gif sur Yvette Cedex, France

W. Gall, Human Factors Unit, Safety and Reliability Directorate, AEA Technology, Wigshaw Lane, Culcheth, Warrington WA3 4NE, UK

I. K. Gibson, Safety and Reliability Directorate, AEA Technology, Wigshaw Lane, Culcheth, Warrington WA3 4NE, UK

P. Helyer, Technology Group, Aerosystems International, Scientific House, 40–44 Coombe Road, New Malden, Surrey KT3 4QF, UK

T. O. Inga-Rojas, Electrical Energy and Power Systems Group, UMIST, PO Box 88, Manchester M60 IQD, UK

A. M. Jenkins, Human Factors Unit, Safety and Reliability Directorate, AEA Technology, Wigshaw Lane, Culcheth, Warrington WA3 4NE, UK

J. C. P. Kam, Department of Mechanical Engineering, University College London, Torrington Place, London WC1E 7JE, UK

A. Z. Keller, Industrial Technology School, University of Bradford, Bradford BD7 1DP, UK

M. Kersken, Gesellschaft für Reaktorsicherheit (GRS) mbH, Forschungsgelände, 8046 Garching, FRG

C. Labouchere, Optimis Ltd, 139 High Street, Wootton Bassett, Swindon, Wiltshire, UK

J. A. Lockett, Safety and Engineering Science Division, AEE Winfrith, Dorset, UK

P. Manasse, Department of Statistics and Computational Mathematics, University of Liverpool, PO Box 147, Liverpool L69 3BX, UK

J. M. Marshall, Department of Mathematics, Statistics and Operational Research, Nottingham Polytechnic, Burton Street, Nottingham, UK

P. Martin, Department of Mechanical Engineering, University of Liverpool, PO Box 147, Liverpool L69 3BX, UK

R. H. Matthews, Safety and Reliability Directorate, AEA Technology, Wigshaw Lane, Culcheth, Warrington WA3 4NE, UK

T. A. Mazzuchi, Department of Mathematics and Systems Engineering, Koninklijke/Shell-Laboratorium, PO Box 3003, 1003 AA Amsterdam, The Netherlands

M. Oprisan, Canadian Electrical Association, Suite 580, One Westmount Square, Montreal, Quebec H3Z 2P9, Canada

B. B. W. Ostrom, Atomic Energy of Canada Ltd, Chalk River Nuclear Laboratories, Chalk River, Ontario K0J 1J0, Canada

C. F. Pensom, Atomic Energy of Canada Ltd, Chalk River Nuclear Laboratories, Chalk River, Ontario K0J 1J0, Canada

U. D. Perera, IBM (UK) Ltd, Havant Plant, PO Box 6, Havant, Hampshire PO9 1SA, UK

M. J. Phillips, Department of Mathematics, University of Leicester, Leicester LE1 7RH, UK

E. N. Pistikopoulos, Department of Mathematics and Systems Engineering, Koninklijke/Shell-Laboratorium, PO Box 3003, 1003 AA Amsterdam, The Netherlands

P. Pretesacque, Compagnie Générale des Matières Nucléaires (COGEMA), BP 4, 78141 Vélizy, Villacoublay Cedex, France

I. C. Pyle, SD-Scicon plc, Abbey House, Farnborough, Hampshire GU14 7NA, UK

I. S. Qamber, EE & CS Department, College of Engineering, University of Bahrain, Bahrain

A. Rushton, Department of Chemical Engineering, Loughborough University of Technology, Loughborough, Leicestershire LE11 3TU, UK

K. Shen, Department of Machine Design, Lund Institute of Technology, Box 118, S-221 00 Lund, Sweden

D. J. Sherwin, School of Engineering Production, University of Birmingham, PO Box 363, Birmingham B15 2TT, UK

P. Thorpe, Department of Chemical Engineering, Loughborough University of Technology, Loughborough, Leicestershire LE11 3TU, UK

A. Veevers, Department of Statistics and Computational Mathematics, University of Liverpool, PO Box 147, Liverpool L69 3BX, UK

L. A. Walls, Department of Mathematics and Statistics, Paisley College of Technology, High Street, Paisley PA1 2BE, UK

S. Whalley, RM Consultants Ltd, Genesis Centre, Birchwood Science Park, Risley, Warrington WA3 7BH, UK

D. W. Wightman, Department of Mathematics, Statistics and Operational Research, Nottingham Polytechnic, Burton Street, Nottingham, UK

D. J. Winfield, Atomic Energy of Canada Ltd, Chalk River Nuclear Laboratories, Chalk River, Ontario K0J 1J0, Canada

J. Winterton, University of Bradford Management Centre, Emm Lane, Bradford, West Yorkshire BD9 4JL, UK

J. Woodhouse, Optimis Ltd, 139 High Street, Wootton Bassett, Swindon, Wiltshire, UK

M. Xie, Division of Quality Technology, Linköping University, S-581 83 Linköping, Sweden

M. Zachar, Nuclear Transport Ltd, Delancey Drive, Risley, Warrington WA3 6AS, UK

THE APPLICATION OF TWO RECENTLY DEVELOPED HUMAN RELIABILITY TECHNIQUES
TO COGNITIVE ERROR ANALYSIS

W GALL, BSc*
Human Factors Unit
Safety and Reliability Directorate
AEA Technology
Wigshaw Lane, Culcheth, Warrington WA3 4NE, UK
(Head of Unit P Humphreys)
*Author is now at H M Nuclear Installations Inspectorate.

ABSTRACT

Cognitive error can lead to catastrophic consequences for manned systems,
including those whose design renders them immune to the effects of physical
slips made by operators. Four such events, which occurred recently, were
analysed. The analysis identifies the factors which contributed to the
errors and suggests practical strategies for error recovery or prevention.
Two types of analysis were conducted: an unstructured analysis based on the
analyst's knowledge of psychological theory, and a structured analysis
using two recently-developed human reliability analysis techniques. In
general, the structured techniques required less effort to produce results
and these were comparable to those of the unstructured analysis.

INTRODUCTION

Human error analysts identify four broad categories of human error; slips,
lapses, mistakes and violations. A slip is an erroneous action in which the
operator formulates the correct intention but fails in its execution. A
lapse is a failing of memory. A mistake is where the operator
misinterprets a situation and formulates an incorrect plan of action. A
violation is an action taken by an operator in deliberate breach of
operating procedures, very often, as a calculated risk taken in order to
achieve a specific goal (references 1 & 2).

The study which forms the subject of the present paper examined human
errors involved in four recent incidents in nuclear power plants. The
incidents were initially analysed by a group of human reliability experts.
Group members independently examined selected material relating to the
incidents and later compared their analyses in order to reach a consensus
on causes and potential remedies. No formal methods of analysis were
determined in advance, each member simply read through the material and
drew conclusions based on their own experience and expertise. The author

additionally applied two recently developed human reliability analysis techniques, to the incidents, in order to determine their usefulness.

UNSTRUCTURED INCIDENT ANALYSIS

Incident 1 France PWR
The first incident occurred after a PWR had been shutdown for refuelling. Plant operating procedures were contravened when the reactor was restarted with the safety injection systems unavailable. An OECD Nuclear Energy Agency report on the incident highlighted two main sources of error: operating procedures were incomplete and poorly presented and, although alarms sounded, they were not attended to.

The NEA report indicates that aspects of the start up procedures were devised and checked by a shift team assisted by the Radiation Protection and Safety Engineer, but were implemented by the next shift. The procedures produced by the original shift failed to consider the unusual plant state during start up and consequently failed to specify clearly all of the necessary actions to ensure plant safety. It is known that certain phenomena associated with groups can produce and consolidate errors (reference 3). Examples of such group processes are over-confidence, reliance on the knowledge of others and shared, therefore individually diminished, responsibility (reference 4). In addition to group-oriented effects, as outlined above, individuals in both shifts failed to 'think in causal nets' and focused only on the linear progression of events towards the goal failing to consider contingencies or potential side effects (reference 5).

It is possible that the shifts deliberately contravened operating specifications in order to solve some of their planning problems or to achieve a high priority goal. Such violations of the operating specifications are not necessarily mistakes, but rather calculated risks. (It is a mistake if the 'calculation' is incorrect).

Incident 2 W Germany BWR
The second incident, comprising at least three errors, occurred when start up was initiated in a BWR with a residual heat removal system (RHRS) checkvalve left open. Despite failed attempts to close the checkvalve from the control room start up was continued and an attempt was made to close the valve by opening a second testline to create a pressure differential across the valve. This action failed causing primary coolant to be lost to the atmosphere. At this point start up was discontinued until the RHRS checkvalve was successfully closed. When start up was re-attempted, however, a third error was committed when a pressurised RHRS train was connected to the refuelling water storage tank causing coolant, again, to be lost to the atmosphere.

Expert analyses of this incident provided two explanations for start up being continued with the RHRS checkvalve left open. Firstly the operator may have violated operating procedures by not informing the shift supervisor of the incorrect valve position. If this were not the case it must be assumed the operator believed the checkvalve to be closed, despite an indication it was open. It is possible he intended to make pressure and temperature checks to confirm the position of the checkvalve but forget to do so (a lapse). Alternatively, he may have relied on an indication that a particular relief valve had opened to indicate the closure of the checkvalve. (This indication is known to have failed).

The attempt to blow closed the open checkvalve by releasing pressure to the testline was identified as a mistake or cognitive error. This action indicated a major flaw in the mental model of the operator as the action rather than closing the checkvalve could have caused a testline rupture outside of the containment.

Opening the RHRS train to the RWST was probably a lapse or slip. The operator, preoccupied with trying to close the checkvalve perhaps lost concentration on this task. He could have suffered a lapse of memory, that is, failed to recall the depressurisation step, or perhaps made a physical slip, that is, operated the wrong valve control.

Incident 3 US BWR

During a routine instrument test a maintenance technician erroneously opened a valve. A number of instruments were affected including the feedwater level monitor which registered high, and caused the feedwater pumps to reduce their flow. The technician realised his mistake and closed all variable and reference leg isolation valves, this action, however, caused a pressure pulse which in turn caused level instruments to register low.The two reactor recirculation (RR) pumps tripped and a half scram signal and alarm were initiated. A complex chain of events followed leading to severe power oscillations in the reactor. The operating team considered four possible strategies to solve the problem, but, before any of these could be put into effect, the reactor tripped.

The expert analysts concluded that the maintenance technician made a slip in selecting the wrong valve due to the close proximity and similarity of valves. He quickly diagnosed his error, probably as a result of some check or feedback from the system and isolated the reference from the variable leg in an attempt to rectify the problem. Available reports do not focus on the maintenance technician and it cannot be ascertained whether his attempted remedial actions resulted from a poor mental model of the system. He probably did not expect the subsequent power oscillations (no-one did) but simply reacted without realising the consequences.

The human factors analysts concluded that the operating team performed to the best of their ability under the circumstances. They diagnosed that an instrument malfunction had caused the high level alarm and that an actual ATWS (Anticipated Transient Without Scram) event was not taking place although RR pump flow had been lost in response to the ATWS signal. Similarly, the immediate actions undertaken were logical and resourceful. These actions included: attempting to establish feedwater reheating (having addressed and considered negligible the potential problem of introducing unheated water into the core), attempting to restart the pumps in accordance with the perceived most appropriate operating procedure, and planning the contingency to manually scram the reactor in the event of failing to restart the RR pumps. The diagnoses and actions were deemed to demonstrate a sophisticated knowledge of reactor behaviour and a clear understanding of the nature of the transient. The incident may, therefore, be considered a 'cognitive success', albeit a success in which the final action was taken out of the hands of the operators by automatic protection systems.

Incident 4 Sweden PWR

The fourth incident occurred in a PWR, here a night shift team performed routine 'local criticality' tests with the hydraulic scram system and other safety systems untested and inoperable. When the team handed over to the day shift criticality tests were discontinued when it was noticed that the hydraulic scram system was not available. The situation was reported to management who subsequently ordered an investigation. An operation

4

planner clarified the operational order, and tests recommenced with the scram system and other systems operable, but still untested.

The expert analysis of this incident revealed a number of errors by those responsible for the testing. Firstly, checks, that the prerequisites for the criticality tests had been fulfilled, were not carried out. This is either an example of a slip or a violation. Secondly, a lapse was made when an engineer missed a step in a procedure which would have ensured the scram and safety systems were tested. This lapse was caused by the engineer being distracted by the reactor physicist. Because of these errors the criticality tests were performed before the safety tests. This lead to confusion within the day shift resulting in them concentrating on the unavailability of the scram and safety systems. This is an example of bounded rationality where the day shift focused only on one piece of information, system unavailability, ignoring other information concerning testing of the system. It was also thought that diffusion of responsibility within the team played a part in the incident as it is unlikely that all personnel would not appreciate that tests of the safety and scram systems had not been carried out.

STRUCTURED INCIDENT ANALYSIS

A recent review (reference 6) suggested two techniques of particular value in certain aspects of human reliability analysis: CADA (reference 7) (the Critical Action and Decision Approach) for identifying cognitive errors, and HEART (reference 8) (the Human Error Assessment and Reduction Technique). These techniques were applied to the four incidents by the author. Certain ad hoc modifications were made to both techniques during the analysis, these are discussed later.

CADA Analysis
The full CADA analysis is extensive and cannot be reproduced in the current paper owing to space limitations. A summary of the error analysis, however, can be found in appendix 1. In general, CADA was found straight forward to apply. From the information contained in the incident reports, it was possible to identify quickly and easily, the critical decision making stages that could have lead to the incidents. Three exceptions were noted, these were :
1. Incident 3 maintainer closes valves to correct his original error. It was not possible to ascertain whether this was a considered decision or a simple reflex action or some combination of rule-based and knowledge-based behaviour. In this case, it was difficult to apply CADA because information was lacking, not because of a failing in the technique.
2. Incident 3, operators plan to restart RR pumps. CADA suggests that the operators progressed through all stages of the decision making process and arrived at an appropriate decision.
3. Incident 4, perform tests with scram system inoperable. It was difficult to determine whether the problem stemmed from an inadequate mental model, or from a lack of knowledge. It was difficult, therefore, to determine whether it was a cognitive error suitable for CADA analysis.

The questions and conditions suggested by CADA were, in general, pertinent and in a number of specific instances, identified the same operator errors as the original investigations into the incidents by the licensees. CADA questions and conditions do not, however, address such issues as equipment availability or assistance of ancillary personnel or other team members. Occasionally it was found necessary to look beyond the

obvious errors in applying CADA, for example in Incident 4. The external
error here may seem to be an Execution error on closer inspection,
however, the problem of testing without the scram system operable arose
because of an Observation/Data Collection error in the planning of the
tests. It could not be determined whether CADA would have identified all of
the same root causes as those disclosed by teams investigating the
incidents. In order to establish this, it would be necessary to apply
CADA, in parallel, to the same source data and then compare the results.

HEART Analysis
Again, the full HEART analysis cannot be presented in the space available,
however, a summary of the error analysis appears in appendix 2. More
difficulty was experienced in applying the HEART technique than in the CADA
technique. It was sometimes difficult to identify the task type using
HEART. For example, the maintenance technician's task in Incident 3 could
have been 'simple', 'complex', or 'highly practised'. The difficulty here,
again stemmed from a lack of information in the original incident reports.
It was difficult to ensure that all of the relevant error producing
conditions were identified and that there was no overlap between the error
producing conditions and original task descriptors. For example,
'unfamiliarity' appears both in a task descriptor and as an error producing
condition. A further problem was one of interpretation, for example, the
word 'feedback' could be interpreted as visual or auditory indications from
display devices, verbal feedback from co-workers or training feedback. The
precise meanings could no doubt be extracted from the source material
referred to by HEART, but advantages resulted from this flexibility of
meaning in that it was sometimes convenient to interpret some of the terms
as broadly as possible. A wider range of possible error producing
conditions were thus considered.
      Some of the solutions suggested by HEART were infeasible. For
example, in Incident 2, an error producing condition affecting the
decision to close the checkvalve was 'misperception of risk'. The solution
suggested was to check the operators' perception of risk against the actual
risk. In specific incidents such as this, it is not possible to predict in
advance what risks operators might consider taking. It may be possible to
assess operators in a more general sense and identify and retrain the
natural risk takers and the over-conservative, but in the context of the
present analysis, the remedy, as described is not applicable.The above
criticisms must be weighed against the relative immaturity of the HEART
technique, which has scope for future extension and elaboration and the
fact that the qualitative advice it provides is only a secondary facility
of the technique whose prime concern is to quantify error.
      Both techniques stimulated the author's thoughts during the analyses
and it was not difficult to generate ideas for additional CADA questions
and HEART remedies which suggests that some improvements to the techniques
could be implemented in the short term.

GENERAL CONCLUSIONS

Unstructured Analysis:
The informal analysis utilised a number of ideas proposed by previous
researchers as relevant to the study of cognitive behaviour in systems.
These ideas can be represented at three levels. At the individual level:
the notions of skill, rule and knowledge-based behaviour and the
accompanying decision making framework as described by Rasmussen (Reference

1). These ideas underpin the analysis of cognitive behaviour in all of the
incidents analysed. At the group level: the effects of group dynamics,
recently revived in the analysis of the Chernobyl accident (reference 3);
the notions of 'imperfect rationality' in groups (Incident 4); the
sociological implications of the 'trade defensive ideology' of the informal
organisation (Incident 2). At the organisation level: Reason's notion of
'resident pathogens' (reference 9) in systems (Incident 1).

Formal Analysis
CADA and HEART were applied in a rather unorthodox manner i.e. to incident
reports yet, despite this, still proved useful. CADA was fairly adaptable
and comprehensive in identifying credible error opportunities. HEART
proved more difficult to apply to the incidents, however, it is a recently
developed technique and the author was not experienced in its
interpretation and application. Despite this, HEART was able to suggest
mostly credible remedies to potential problems.
    Both techniques highlighted equipment design, procedures and training
as the key problems/focus of solutions. These compare with the
recommendations made in the incident reports. These recommendations
include: improvements to procedures (for monitoring displays, handing over
shift and recording information), additional training, changes to display
and control devices, interlocks, improved maintenance and reduction of
unofficial duties assumed by operators.

RECOMMENDATIONS

The incident reports referred to have already suggested changes within the
plant to eliminate specific errors. The present author cannot add to
these, but proposes some general recommendations targeted at the perhaps
more deep-rooted causes of cognitive errors.
    A general strategy for reducing error at the individual level may be
to attempt to reduce all potential cognitive tasks to proceduralised tasks.
The reasoning behind this is that, typically, when individuals are allowed
to decide how much and what type of information is required to make a
decision, they use inappropriate information. Thus, a general remedy is
suggested to reduce the task to one in which the operator has only to
consult a written document containing all the information needed to perform
the task and provides a checklist to keep track of the steps performed.
This recommendation has, however, inherent problems. Firstly, although,
the potential for cognitive error is reduced within the operator it is
simply transferred to the designer of procedures. Secondly, by making the
task purely procedural it is likely that the operator will become bored and
this may trigger further errors. The greatest problem with this approach to
reducing cognitive error concerns novel situations. Novel situations are
most likely to produce cognitive errors as they are often unexpected events
which require operators to reason in order to solve a particular problem.
Unexpected events by definition cannot be proceduralised other solutions,
therefore, are required in order to reduce cognitive error.
    Training provides a further solution for error reduction. It may be
useful to provide 'awareness training' in addition to operational training.
That is, make personnel aware of the types of error they are likely to make
in particular circumstances. For example, that humans in general are
likely, when under pressure, to fixate upon the first credible solution to
a problem, and to implement this without considering either its side
effects or the possibility that a better solution exists. The drawback

here is that a novice, when confronted with a problem, may fail to reach a
decision through fear of making a mistake. Training, therefore, must
include some advice on strategies. For example, an operator who is unsure
of a situation or of his solution to a problem, must seek the opinions of
colleagues, who, in turn, must examine the situation objectively and
uncritically (being aware of the error tendencies of groups), in order to
arrive at a best solution.

Other solutions involve improvements to interfaces and equipment which
should be aimed at reducing the cognitive load on operators, for example,
by analysing and prioritising information presented to the operator, or
by checking that the plant state does not violate procedures. Problem
arise as it is always possible for the operator to ignore priorities or
accept a violation in order to achieve an alternative goal. The
responsibility lies with the organisation to ensure that its personnel are
not set impossible goals and that the information provided is accurate.
Organisations, generally, should investigate how objectives are actually
achieved and not assume that they are being achieved legitimately. This
calls for improved consultation with those responsible for carrying out the
tasks to achieve those goals, and probably requires a more responsive, less
punitive image.

REFERENCES

1. Rasmussen, J. Information Processing and Human-Machine Interaction:
    An Approach to Cognitive Engineering. New York: North-Holland.
    1986.

2. Reason, J. ' A Framework for Classifying Errors'. In New Technology
    and Human Error. Rasmussen, J., Duncan, K. and Leplat, J. Chichester:
    John Wiley & Sons. 1987.

3. Reason, J. The Chernobyl Errors. Bulletin of the British
    Psychological Society., 1986, **40**, 201-206, .

4. Janis, I.L. Victims of Groupthink. Boston, Houghton Mifflin, 1972.

5. Doerner, D. 'On the Difficulties People Have In Dealing with
    Complexity'. In New Technology and Human Error. Rasmussen, J.,
    Duncan, K. and Leplat, J. Chichester: John Wiley & Sons. 1987.

6. Gall, W. The Assessment of Human Cognitive Error in Probabilistic Risk
    Assessment Studies: an Appraisal of Human Reliability Techniques. RTS
    88/120. Safety and Reliability Directorate, United Kingdom Atomic
    Energy Authority, Wigshaw Lane, Culcheth, Cheshire. September 1988
    (Draft).

7. National Centre of Systems Reliability. The Critical Action and
    Decision Approach (Longer Version). Human Reliability Assessment
    course notes. November 1987.

8

8.   Williams, J. C.   HEART – A Proposed Method for Achieving High
     Reliability in Process Operation by Means of Human Factors Engineering
     Technology.   Proceedings of a Symposium on Achievement of Reliability
     in Operating Plant.   Safety and Reliability Society.   September 1985.

9.   Reason, J.   Summaries of Five Case Studies Illustrating the Part
     Played by Latent Failure in Accident Aetiology.   From a lecture given
     to the Human Factors in Reliability Group at the National Centre of
     Systems Reliability, 1988.

Appendix 1 Summary of CADA Analysis

Incident	Errors	Decision Making Stages	Failure Modes	Possible Reasons	CADA Questions	CADA Conditions	Remarks
1	Fail to act on alarms.	Activation	Fail to notice or delayed detection.	Fail to follow procedure.	1,2,3,4,5	1	Alarms/indicators Warned that systems were unavailble.
				Alarms intermittent	6	2	
				Alarms obscured	7	3	
2	Leave checkvalve open	Observation	Acquire insufficient data	Acquire irrelevant data	10,11	1,6	
				Acquire too little data	6,7,10	4	
	Attempt to blow c/v shut	Task definition	Select an inappropriate strategy	Data gives conflicting results	6	4	
				Fail to know all possible plant states	12,12,20, 21	1,6	
				Fail to evaluate known plant states completely	12,13,18, 19	1,4	
3	Maintainer opens wrong valve	(Uncertain)		(See CADA technique)			Insufficient information
	Operators plan to restart RR pumps	Interpretation to Execution					A cognitive success rather than a cognitive error
4	Perform tests with scram system inoperable/untested	Observation	See above under Incident 2	-	-	-	A result of an error in planning rather than an operational error

Appendix 2 Summary of HEART Analysis

Incident	Task Type	Error Reduction (Task)	Error Producing Conditions	Error Reduction (Conditions)
1	Restoration of system state	Provide supervision and job aids	Unfamiliarity / Feature override allowed Misperception of risk Poor feedback / Impoverished information / Inadequate checking / Objectives conflict	Training / Interlocks with timeouts / Selection/Training / Task analysis/ergonomic study of feedback requirements. Eliminate procedures requiring accurate oral transmission / Independent checking regime / Reconcile Incompatible objectives
2	Leave checkvalve open Connect pressurised RHRS (Restoration of system state) / Attempt to blow c/v shut (Task type irrelevant)	Interlocks, supervision, job aids / Not applicble	Time shortage / Poor feedback / Misperception of risk	Training in strategies against time shortage / As 1 / Check operators' perception of risk (increase awareness of energy stores in the system)
3	Maintainer opens wrong valve (Highly practised?) / Operators plan to restart RR PUMPS (Unfamiliar task)	Difficult to improve - supervision? aids? training? / Reduce possibility of such tasks arising through training/work organisation	Irreversibility / Poor feedback / Time shortage Impoverished information	Ensure system is tolerant to error correction entailing reversal of error actions / Improve control identification / See 2 / Improve procedures, ensure they anticipate every eventuality
4	Performs tests with scram system inoperable/untested (Complex task)	Avoid such tasks where possible Training, ensure job aids used properly (checklist)	Feature over-ride allowed Inexperience / Impoverished information (operating order)	As 1 / Ensure experienced staff do not produce operating orders or supervise novices

COGNITIVE TASK ANALYSIS TECHNIQUES
IN THE DESIGN AND EVALUATION OF COMPLEX TECHNOLOGICAL SYSTEMS

Michael Carey and Susan Whalley
R M Consultants Ltd
Genesis Centre, Birchwood Science Park, Risley, Warrington WA3 7BH

ABSTRACT

Task analysis can provide an important contribution to ensuring the
reliability of human operators in complex technological systems. As with
advances in automation, the nature of operator activity has gradually changed
to include more decision making, supervisory and diagnostic activities.
Hence, there has been a need to develop task analysis techniques that analyse
cognitive behaviour. This paper reviews a number of embryonic task analysis
techniques specifically concerned with the assessment of the mental demands
of human interaction with computer systems. Discussion centres around the
practicality of such techniques, compared with existing task analysis
techniques, when applied in the design of complex systems. Suggestions are
given for the future development of effective cognitive task analysis
techniques.

INTRODUCTION

The nature of human tasks in the operation of large or complex systems is
considered in some form by most system designers. Operator task requirements
emerge from the process of defining the physical elements of a system and its
control philosophy. Indeed, explicit consideration may be given during
design to the role of the human operator and the types of demands system
functions will impose. However, such decisions are rarely formally recorded
or subjected to analysis to ensure that assumptions regarding operator
capabilities and reliability are well founded. Furthermore, it is important
to consider how tasks will be allocated amongst personnel and how operational
teams will be organised to maximize system performance and reliability. It
is easy to underestimate the combinational effect of; complex system events,

limited time in which to make a response and the concurrent requirement to communicate between multiple team members. Task analysis techniques address this problem by examining in detail what it is that personnel within a system are doing or will be required to do. This can be seen as a complementary perspective to that normally adopted in the design process, when attention is focused on how the technical process or overall system will function.

Typically, the process of performing a task analysis proceeds through three stages of activity; data collection, task description and task analysis. Whilst these three stages represent the logical sequence of activities, most practical applications of task analysis will involve some degree of iteration, where the need for further data collection is revealed during either the task description or analysis phases. Furthermore, an outline task description may be developed and utilised early in the design process and then subsequently refined and expanded as system design becomes more detailed.

The first activity involves gathering the necessary background information about the system context and task requirements. In some circumstances, it may also be possible to collect objective behaviourial data from existing related task contexts. The collected task information then forms the basis for the generation of a detailed task description. A variety of methods of task description are available, which, depending on the chosen technique, can be used to represent task information such as:

(a) The sequence of actions and decisions taken by an individual

(b) The immediate goals and overall objectives underlying task actions

(c) Information received, communicated and recalled at each stage of task performance

(d) The parallel and interrelated activities of a team

The task description techniques incorporate a variety of textual, tabular and graphical representational formats. The selection of a particular task description technique will depend upon the type of task being represented and the analytic objectives underlying the task analysis process.

In the final stage of the task analysis process, the described task is subjected to systematic analysis to generate the requisite design information. For example, an outline task description may be utilised early in the design process to assist in decisions regarding the allocation of control and supervisory activities between an operator and automatic control systems. As the design process progresses, decisions regarding the manning requirements for a new system can be made on the basis of task description documents. Also, it is possible to develop operator system interfaces utilising the information provided by task descriptions and assess both the potential adequacy and reliability of future operator task performance. In the commissioning phase, training requirements and the design of operational procedures can be developed on the basis of task descriptions. Eventually, once the system is in operation, the task description documents can act as a benchmark against which working practices can be assessed.

The formality of the analysis process will vary according to the type of design information that is required. An assessment of operator task loading, for example, may be developed in a systematic manner from a task description by comparing the sum of the estimated times required to perform each step in a task sequence against the total time available. In a similar manner, an assessment of information adequacy may be carried out by comparing operator task information requirements against the actual information that will be available. Other forms of analysis can be more intuitive, involving experts critically evaluating a task based on a knowledge of the system context, performance requirements and general ergonomic principles.

The overall efficiency of the task analysis process depends upon a number of factors. It is important, for example, to ensure that at each stage of the design process, the task description contains all the information necessary for all its intended uses, avoiding the need for further significant data collection activities. At the same time, it is inefficient to develop the task description to a level of detail or scope greater than that required. In this respect, efficiency depends upon achieving a balance between the information needed for analytic purposes and the collection and representation of further task information for contextual purposes. Simple factors such as the readability of the task description document and its ease of generation/revision also contribute to the effectiveness of the task analysis process. As is discussed in the remainder of the paper, such issues are of particular importance in the development of practical task analysis techniques which address the cognitive aspects of human behaviour.

THE COGNITIVE DIMENSION OF TASKS

The increase in the level of automation and scale of modern industrial and transport systems has fundamentally changed the nature of operator tasks. Much of the routine control and monitoring functions previously carried out by human operators are now performed by computer-based systems. Instead, the operator is called upon to make supervisory decisions, based on complex criteria, to diagnose and compensate for system failures and to reconfigure the system as required. In each case, the nature of the work involved is primarily cognitive. It depends upon considerable background knowledge of the controlled system, plus diagnostic and decision making skills which are combined together in both routine and novel situations to meet required system goals. In terms of analyzing operator tasks, it is no longer sufficient to examine tasks at the global level of 'diagnose fault in x' or 'monitor progress of operation y'. The complexity of each task and its potential for failure can only be accurately assessed by considering the underlying cognitive processes involved. As a result, there has been a considerable growth of interest in the development of methods of task analysis which address the cognitive aspects of tasks.

It has been found that there are inherent difficulties involved in the process of defining and analyzing cognitive activity. The most obvious difficulty is the covert nature of cognitive processes. Even when the task activity under analysis is available for direct observation and investigation, there are few data collection techniques that can be effectively applied to determine the mental processes involved. In most design contexts, there is not even the option of examining an appropriate existing task. The structure of cognitive processes and demands has to be synthesised from a knowledge of the task context, the requirements placed upon the human in the system and a general knowledge of the normative structure of human cognition in other similar task contexts. As this implies, the success of the endeavour relies heavily on the accuracy of the model employed of human cognition. Fortunately, whilst there are still large areas of uncertainty and debate regarding the structure of human cognition, a number of useful conceptual frameworks are emerging which are sufficiently well defined to be employed practically.

COGNITIVE TASK ANALYSIS TECHNIQUES IN HUMAN-COMPUTER INTERACTION

The need to consider both the context and the structure of cognition makes large demands upon the analyst. The accuracy of the process therefore demands techniques which are comprehensive and well-defined. This implies that the method of task analysis should embody a mechanism for defining the overall demands of the task context, in conjunction with theoretically-based structures for defining task cognition. In order to assess the described task, the technique must also define procedures for establishing the cognitive workload or complexity inherent in the described task and provide a method of assessing the potential for human error. Considering the maturity of current knowledge on human cognition in complex task contexts, it is not surprising that the majority of cognitive task analysis techniques are still at an early stage of development.

The most promising area for the development of effective cognitive task analysis in the short term is in the area of Human-Computer Interaction (HCI). A wide range of techniques broadly included under the title of cognitive task analysis have been developed for use in the design and evaluation of human interaction with computer-based systems. Some of these focus purely upon the assessment of the complexity of operating the computer system. So, for example, in a process control context, they might be used to evaluate the ease with which displays could be changed or control actions carried out, rather than for evaluating the overall complexity of the task of controlling plant operations. Other techniques recognise that the wider task context (i.e. what the computer system is being used for) must be considered in order to construct a form of human-computer dialogue which is both quick to learn and to use. In general, these techniques have a strong theoretical basis in current models of human cognition and are generally highly formalised in their notations and structures. It is worthwhile, therefore, considering whether similar techniques or approaches could be applied in the analysis of operator cognitive tasks.

A total of five HCI task analysis techniques are examined in the following discussion. For the purpose of examining the body of HCI cognitive task analysis techniques, a comparison will be made with the most widely used task analysis technique in process control applications; Hierarchical Task Analysis (HTA). Details of the main characteristics and key references for each technique are given in table 1. As can be seen there are wide differences in their intended purposes and forms of representation. A key underlying theme is that they attempt to represent various forms of operator knowledge related to performing actions on a computer based system.

Hierarchical Task Analysis (HTA)

Developed in the early 1970s, HTA was initially developed to assist with devising training programmes in the process control area [1]. The basic element in HTA is the 'operation' which was defined in the original paper on the technique as "any unit of behaviour, no matter how long or short its duration and no matter how simple or complex its structure, which can be defined in terms of its objective". Operations are actual units of behaviour which are determined by the goals the operator has to achieve and the resources, training, etc. the operator has received. The HTA technique recognises that goals can be hierarchically organised, such that the goal of transferring some fluid between two vessels might include sub-goals concerned with opening valves and starting pumps. The order in which goals are achieved are described by plans, which state the conditions under which sub-goals are attempted and their ordering. HTA, therefore, explicitly describes the units of activity an individual performs during a task, their sequencing and any decisions that need to be taken during task performance. The main form of representation in HTA is a graphical tree for the hierarchy of goals. This frequently includes the description of plans indicating the ordering and selection of sub-goals [2]. An additional tabular format may be included giving textual descriptions of goals, plans and associated information. In HTA, the details of the graphical and tabular format are not of overriding importance and are frequently changed by practitioners according to their specific needs (e.g. [3]). Methods of analyzing task information for the purpose of design or assessment are not generally well specified though there are specific examples of methods of human reliability analysis [4] and information requirements analysis [3] based upon HTA.

Technique	Purpose	Representations	Key References
Hierarchical Task Analysis (HTA)	Provides general purpose task description which has uses in interface design, reliability assessment and training analysis	Hierarchical diagram Tabular	[1], [2], [3], [5], [18]
Task Analysis for Knowledge Descriptions (TAKD)	Generates generic task description to underpin interface design.	Grammar and structured text representations	[8], [9]
Task-Related Knowledge Structures (TKS)	Generates generic task description from specific task descriptions and specifies mapping onto interface design	Graphic format and structured text representations	[10], [11]
Task-Action Grammar (TAG)	Analyses complexity of human-computer interface for performing a range of computer-based tasks	Grammar notation	[6], [7]
Goals, Operations, Methods and Selection Rules (GOMS)	Describes structure of routine operations on a computer-based system as a basis for predicting performance times	Structured text representation	[12]
Cognitive Complexity Theory (CCT)	Describes both task and interface structure for purpose of predicting performance times and complexity of operations.	Structured text representation and state transition networks	[13],[14]

Table 1 : Summary of Task Analysis Techniques

Task Analysis for Knowledge Descriptions (TAKD)

The starting point for a TAKD analysis is provided by observational data on existing tasks that bear some relationship to the tasks that will be performed using the system under design. The data on the existing tasks are initially recorded into standard task sentences. The objects involved in the task and the actions performed on these are then described separately in the form of a pair of taxonomic hierarchies. The tasks themselves are described in the form of a 'grammar' which relates to the object and action hierarchies, which enables both very specific task instances to be described and highly generic task descriptions to be generated. This then forms the basic structure of an interface dialogue which could be used for performing each individual task.

Task-related Knowledge Structures (TKS)

TKS, in contrast to TAKD, explicitly represents actions, objects, specific procedures, goals and for situations where more than one person will be operating a system, it considers individual roles. The first step is to construct a task-knowledge structure (TKS) model based on data collected from actual task instances. The TKS model includes various diagrammatic structures and a taxonomic structure of objects and actions performed. In essence, the TKS provides a 'database' for recording all the information about task performance in the correct system. This is then transformed into a second representation which is a 'generic' description of the task model. The generic task model describes each generic task goal identified in the data collection stage. In a final stage, the generic task model is transformed once more into a specific interface model for a computer-based system. Explicit decisions are made regarding the allocation of processes between the operator of a system and the computer system itself. In the context of process control, this might include decisions whether to implement control sequences to perform certain tasks or whether to rely upon the operator utilising written operating procedures.

Task-Action Grammar (TAG)

In contrast to TAKD and TKS, TAG models a much smaller component of the knowledge relevant to operating a computer. It examines the ease with which the task goals of a system operator can be accomplished utilising specific interface dialogues. It is therefore an evaluative technique, which requires some preliminary effort to identify relevant task goals (e.g. a task description generated by another cognitive task analysis method) and the existence of one or more specified interactive dialogues. The type of task that can be analyzed using this method, for example, would be 'altering a control loop setpoint', or 'switching to a standby pump'. Each of these tasks are 'rewritten' in the form of the system actions required for their performance. A formal grammar is specified for describing the rewrite rules. The strength of TAG is that it takes account of the consistency both within the system command structure and between the system commands and operator tasks. A highly consistent command structure generates fewer rewrite rules than a highly inconsistent, arbitrary command structure. The underlying theory asserts that the complexity of learning to use the computer system is related to the task consistency of the command dialogue. Therefore, the number of rewrite rules is a direct predictor of ease of learning and of use. The predictive accuracy of TAG has been assessed successfully in a number of laboratory studies [15], though the authors assert it is still essentially a research tool.

Goals, Operations, Methods and Selection rules (GOMS)
The remaining two techniques shown on Table 1 are closely related. The GOMS technique was one of the earlier theoretical developments in the discipline of human-computer interaction and is the most similar in outline structure to HTA. The technique represents the steps involved in carrying out specific routine tasks on a computer system. As with HTA, a hierarchy of goals is utilised, starting at the most general 'unit task' level (e.g. "shutdown plant"), breaking the overall goal down in to successive levels of subgoals. The technique recognises that standard methods will be applied by operators to achieve certain goals and the notation allows these to be represented. Also, where alternative methods can be applied to achieve the same goal, the technique allows a selection rule to be specified that indicates under what conditions each of the alternative methods would be applied. Operations in GOMS terminology represent the most basic unit of action at the bottom of the goal hierarchy. If performance times are available for each type of operation represented, it is possible to use GOMS to predict overall task performance times. For example, the technique might be used to estimate the times required to access items of information via system displays [16]. The bottleneck in this form of time estimation is obtaining accurate time estimates for the various fundamental 'operations'. The most useful aspect of GOMS is its representational format which is isomorphic with the hierarchical diagram format in HTA, but is much more concise for

representational purposes.

Cognitive Complexity Theory (CCT)

CCT brings together the task modelling concepts of GOMS with a state transition network format to describe the functioning of a device interface. The user of the technique is first required to describe a set of sample tasks that would be performed using a system. Then, for each task, the method for performing the task is specified using a task modelling language based upon the psychological structures employed in GOMS. Finally, the functioning of the device interface is described in a third representational format. The three forms of description can then be input to a simulation program that runs the task representation against the device representation and generates estimated learning and execution times. The algorithms for calculating the results of the analysis are admitted by the authors to be fairly crude, but have been validated for comparative accuracy in a number of empirical studies.

As with the GOMS technique, the task modelling language within CCT is designed to describe tasks within the context of the computing device, primarily consisting of a description of the sets of inputs required to achieve specific overall task goals. Therefore, taking as an example the assessment of a task concerned with transferring a fluid between two vessels, the task would first be described in terms of the major steps involved in the transfer task. For each step, the system commands required to achieve that step are then described using the task modelling language. A general description of the interface commands are then generated for the system commands involved in the task concerned. The simulation program would then provide an assessment of the ease with which the operation concerned could be performed and an estimate of the time required. The most useful application of the technique would be in assessing alternative control room interfaces to obtain a comparative assessment of their demands in terms of learning time and performance efficiency. It should be noted, however, that this technique is still under development and it is not clear whether the software involved in the analysis is openly available.

Comparison between HTA and the other techniques

Comparing the facilities offered by HTA with the cognitive task analysis techniques described above highlights a number of important practical and theoretical differences. These are outlined below in Table 2.

ATTRIBUTE	REQUIREMENT	HTA	COGNITIVE TASK ANALYSIS TECHNIQUES
Conceptual Complexity	-Easy to learn to use by non-specialists	-Simple and familiar representational formats -Terminology and concepts not overtly psychological	-Often utilise complex or special purpose representational formats -Knowledge of terminology and concepts of cognitive psychology often required
Readability	-Readable by non-specialists for validation and quality assurance	-Tree diagrams easily interpreted -Focus on structure and sequence of actions aids comprehension	-Representational formats can be cryptic -Task information dispersed amongst cognitive structures
Adaptability	-Adaptable to fit needs of application	-Lack of standard definition encourages adaption -Tabular formats can be easily modified -Text can be included for comments and to state assumptions	-Rigid structures not easily adaptable
Formality	-Formally and comprehensively defined to ensure consistency of use and interpretation	-No standard, formal definition -Concepts frequently loosely and inconsistently applied	-Employ formal set of structures, which in most cases are explicitly defined
Practicality	-Easy to generate and revise	-Both tree and tabular formats can be difficult to produce and revise	-Most methods utilise text based representations which are comparatively easy to produce and revise

Table 2: Comparison of HTA and Cognitive Task Analysis Techniques

The primary advantages of HTA when compared with the previously described cognitive task analysis techniques are its conceptual simplicity, the readability of its task representations and its adaptability. Its simplicity and readability offer important benefits during the initial data collection and task description phases of a project. Data on task structure is frequently elicited from task experts and the readability of the representational formats can prove important in the validation of the generated descriptions. Furthermore, the same representation needs to be effective in communicating a task overview to other specialists and managerial personnel. The conceptual simplicity of the technique also assists in the regard. Where non-specialists are required to use the task description technique, the conceptual simplicity of HTA ensures that it can be rapidly learnt and is therefore extremely useful in both the generation and validation of task-descriptions in multi-disciplinary projects. However, as with any analytic technique, significant skills are required before it can be applied in a uniform and efficient manner and some degree of training will still be required.

The adaptability of HTA arises in part from its lack of formal definition. As a result, most practitioners do not feel constrained in adapting it in a manner which suits their own style of working and project needs. If the accompanying tabular representation is employed, then these can be easily extended or customised by adding or removing columns. The ability to incorporate explanatory and descriptive text in tabular formats is also important, allowing complex contingencies and other vital contextual information to be stated.

In comparison to HTA, the cognitive task analysis techniques examined earlier are complex to learn, generate task descriptions which are generally more difficult to read and cannot be so easily adapted to new task contexts. This primarily is due to the formal and precise nature of the representations employed. The purpose of this is to reduce the ambiguity within a task description such that the resulting task information can be subjected to formal, analytic processes. This does have the benefit that inconsistencies and ambiguities must be resolved at the task description phase. For example, at one point during a task, an operator may be faced with a complex decision to make in selecting between a number of methods for achieving a goal. The lack of formal specification in HTA would allow an analyst to 'side-step' the difficult task of determining the exact decision process by stating a vague plan such as 'select one of n to m according to criterion x'. The representational methods in cognitive task analysis, however, require far more precision and force the appropriate information to be gathered or an assumption to be stated. At the same time, There is a danger that if the structures provided by the analysis technique are not appropriate for the type of task information that needs to be represented, the information will be made to 'force-fit' the structures provided. This can only be avoided if precise definitions of the intended contact of each structure are given, alternative structures can be grafted on to the technique as required and if there is provision to record any assumptions made.

A final practical aspect of task description techniques which needs to be considered is the ease with which the task documentation can be generated and edited. For example, the tree chart formats utilised in HTA can be notoriously difficult to generate and revise without appropriate tools. Some of the cognitive task analysis techniques have advantages in this regard as their task representation techniques are primarily text based. However, carefully designed diagrammatic formats can be easier to interpret. In this aspect, in particular, there is a role for computer-based tools to assist in the task analysis process.

FUNCTIONAL REQUIREMENTS FOR FUTURE DEVELOPMENTS OF COGNITIVE TASK ANALYSIS TECHNIQUES

Bearing in mind the time constraints inherent in any system design process, an effective cognitive task analysis technique would need to combine the following attributes:

(a) A method of task representation which is simple to understand for data collection and validation purposes.

(b) Comprehensive and rigidly defined task information structures to meet the requirements of a wide range of task types and contexts.

(c) Tailorability of task information structures to individual analyst requirements.

(d) Full support for the technique in form of documentation, production tools
     and a structured analysis approach.

These requirements combine the practical advantages of the HTA technique
as a flexible outline task representation tool with the precision and
comprehensiveness of the types of cognitive task analysis techniques reviewed
earlier. Preferably, it would allow the processes of data collection,
encoding, generation of a task model and analysis to be decoupled, such that
non-specialists could be utilised during the task description phase, reserving
the task modelling and analysis activities for specialists.

The complexity of such a combined and comprehensive analysis technique
suggests the need for appropriate computer-based support tools. This is
similar to the situation in systems analysis where there is considerable
effort being put into the development of Computer-Aided Software Engineering
(CASE) environments. The computer-based tool would primarily consist of a
task database combined with software to generate a variety of 'views' of the
task data for data entry, validation, analysis and exploration of the
computer-based model. The suggested contents for a comprehensive, fully
functional tool of this type are as follows:

(a) Modules allowing the specification of task activities in terms of
     sequences and hierarchies.

(b) Forms of representation for goal structures, plans, etc.

(c) A data dictionary for actions, objects, goals, etc.

(d) An integrated module to describe the logic of the system under control
     (e.g. signal flow diagrams, event trees, etc.).

(e) The ability to record notes and comments on the analysis for quality
     assurance purposes.

(f) Packages for various forms of analysis of task information.

The objective would be to provide a variety of interlinked
representational formats based on a common task database. The analyst would

be free to select whichever representational formats suited the particular application to begin building a task description. Cognitive elements such as goals, plans, operations and objects would only need to be specified once, at whichever point was appropriate during the task description process. To ensure efficiency, each analysis package would specify its minimum task data requirements. The task description software could then provide the analyst with a report on required task data items still missing from the description.

To conclude, the analysis presented in this paper has indicated the need to balance the practical aspects of well-established task analysis techniques such as HTA with the formalism and rigour of recently developed cognitive task analysis methods. The suggestion given in this paper is that this synthesis would most effectively be achieved in the development of an integrated computer-based tool.

REFERENCES

1. Annett, J., Duncan, K.D., Stammers, R.B. and Gray, M. J., Task Analysis, HMSO, London, 1971.

2. Shepherd, A., An improved tabular format for task analysis. Occupational Psychology, 1976, 49, 93-104.

3. Astley, J. A. and Stammers, R. B., Adapting hierarchical task analysis for user-system interface design. In New Methods in Applied Ergonomics, ed., J. R. Wilson, E. N. Corlett and I. Manerica, Taylor and Francis, London, 1987, 175-184.

4. Whalley, S. P., Factors Affecting Human Reliability in the Chemical Process Industry, PhD Thesis, University of Aston, Birmingham, 1987.

5. Piso, E. Task Analysis for 'process control tasks'. The method of Annett et. al. applied. Journal of Occupational Psychology, 1981, 54, 4, 247-254.

6. Green, T. R. G., Schiele, F. and Payne, S. J., Formalisable models of user knowledge in human-computer interaction. In Working With Computers : Theory Versus Outcome, G. C. Van der Veer, T. R. G. Green, J-M Hoc and D. M. Murray, eds. Academic Press, London, 1988, 3-46.

7. Payne, S. J. and Green, T. R. G., Task-action grammar : The model and its developments. In Task Analysis for Human-Computer Interaction, ed., D Diaper, Ellis Horwood, London, 1989.

8. Johnson, P., Diaper, D. and Long, J., Tasks, Skills and Knowledge : Task Analysis for Knowledge Based Descriptions. In Interact '84, IFIP, London, September 4-7, 1984, 1.23-1.27.

9. Diaper, D. Task analysis for knowledge descriptions (TAKD). In <u>Task Analysis for Human-Computer Interaction</u>, ed., D. Diaper, Ellis Horwood, London, 1989.

10. Johnson, P., Johnson, H., Waddington, R. and Shouls, A., Task-related knowledge structures : Analysis, Modelling and Application. In <u>People and Computers IV</u>, eds., D. M. Jones and R. Winder, Proceedings of the Fourth Conference of the British Computer Society Human-Computer Interaction Specialist Group, University of Manchester, 5-9 September 1988, Cambridge University Press, Cambridge, 1988, 35-62.

11. Johnson, P., Task knowledge structures. In <u>Task Analysis for Human-Computer Interaction</u>, ed., D Diaper, Ellis Horwood, London, 1989.

12. Card, S. K., Moran, T. P. and Newell, A. <u>The Psychology of Human-Computer Interaction</u>, Lawrence Erlbaum Associates, Hillside, N. J., 1983.

13. Kieras, D. E. and Polson, P. G. An approach to the formal analysis of user complexity, <u>International Journal of Man-Machine Studies</u>, 22, 365-394.

14. Bennett, J. L., Lorch, D. J., Kieras, P. E. and Polson, P. G., Developing a user interface technology for use in industry. In <u>Interact '87</u>, ed., H-J Bullinger and B. Shackel, Elsevier Science Publishers B. J., Amsterdam, 1987, 21-26.

15. Payne, S. J., Task action grammars : The mental representation of task languages in human computer interaction, PhD Thesis, University of Sheffield, 1985.

16. Carey, M. S. Design principles for hierarchical man-machine interfaces in process control. Report No. CR 2641, Warren Spring Laboratory, DTI, Stevenage, 1984.

17. Carey, M. S., Stammers, R. B. and Astley, J. A., Adapting hierarchical task analysis for determining user information needs. In, <u>Task Analysis for Human-Computer Interaction</u>, ed., D. Diaper, Ellis Horwood, London, 1989.

18. Shepherd, A. Analysis and training of information technology tasks. In <u>Task Analysis for Human-Computer Interaction</u>, ed., D. Diaper, Ellis Horwood, London, 1989.

EVALUATING HUMAN KNOWLEDGE AND THE KNOWLEDGE ASSESSMENT METHOD(KAM)
- REDUCING RISK ARISING FROM DEFECTIVE KNOWLEDGE

J.A. LOCKETT
Safety and Engineering Science Division,
AEE Winfrith, Dorset, UK

ABSTRACT

Under normal circumstances, plant operator decisions are made in
accordance with the operators understanding of written
procedures and of the plant which they are controlling. To
ensure plant safety, an operator must have sufficient knowledge
to respond correctly when both routine and unusual events place
demands on his knowledge and understanding of the plant and
procedures. In order that plant operating procedures or the
methods used in qualifying operators may be improved a
quantitative measure of understanding is required. This paper
proposes a methodology which could be developed to yield a
quantitative assessment of an individual's understanding. An
extension is outlined where the knowledge relevant at each stage
of a progressing task is asessed. This is intended to allow the
identification of those parts of a task which are only weakly
supported by an operator's background knowledge and where for
example additional training and/or improved documentation could
be introduced to greatest effect.

BACKGROUND

In addition to striving for the highest reliability, safety and
efficiency of the plant itself the task of an operator in
process control has long been an area for concern to the nuclear
power industry. Safety assessments must consider the system in
its entirety and it has been recognised that an understanding of
human factors is necessary and allows the operational task to be
improved, so increasing efficiency and safety.

Methods developed to improve the function of an operator must
both identify where improvements are necessary and have some
associated assessment which allows weaknesses to be quantified
in some way. This also allows subsequent improvements to be
monitored. The methods may either address behaviour directly
through the direct analysis of actions (observation, perhaps
using simulators) to improve the reliability of controlled

systems or may attempt to assess the likelihood of an operator making errorsby some indirect means.

The 'indirect' methods involve specific experiments which allow a researcher to 'target' specific areas associated with the function of an operator through carefully designed experiments. For instance, these may target the operator's knowledge of plant function or control desk operation. 'Direct' methods may simply be data collection exercises which measure the frequency of operator errors under various conditions. These have traditionally been used for risk assessment purposes. The data obtained can be useful, but it is only valid when collected over a long period of time and variety of operational conditions and this data gives no indication of particular conditions when the probability of error is particularly high. It is therefore difficult to assess the reasons for these errors using a direct approach since the estimated error rates are in fact a composite of error derived through a variety of sources which are likely to include environmental conditions (Eg. excessive heat or cold), interface confusion (control panel design, communication difficulties), errors or ommisions in written operating procedures or a misunderstanding of plant function.

In other words, the reliability data obtained result from the combination of an unknown number of parameters – each of which have some error associated with their estimate. It is therefore suggested that most benefit may be gained by the development and application of assessment techniques capable of addressing a single identified type of error source. The success of the application of such techniques resulting in changes to the organisation, operating conditions, training or written procedures may be assessed by conventional failure rate approaches such as that described above.

INTRODUCTION

This paper describes a methodology which may be applied directly to gain an assessment of the extent of understanding of an operator in a particular area of plant of plant function and thus provide operator input to a risk assessment which inludes all factors involved in a 'closed-loop' model of a power plant control system.

The role of the operator may be conventionally broken into several parts (Figure 1). This simplistically represents the flow and transformation of data in a closed single-operator system. The second and third boxes represented inside the 'Operator' box are related to the operator's understanding of plant and procedures and therefore must contribute to an assessment of the reliability of the operator and consequently to the system reliability.

The knowledge of an individual can be divided into many levels of abstraction. It may broadly be described as consisting of 'trees' where each level logically supports the one above. A single fact may be used by many trees and so it may be simplistically argued that knowledge does not exist as a

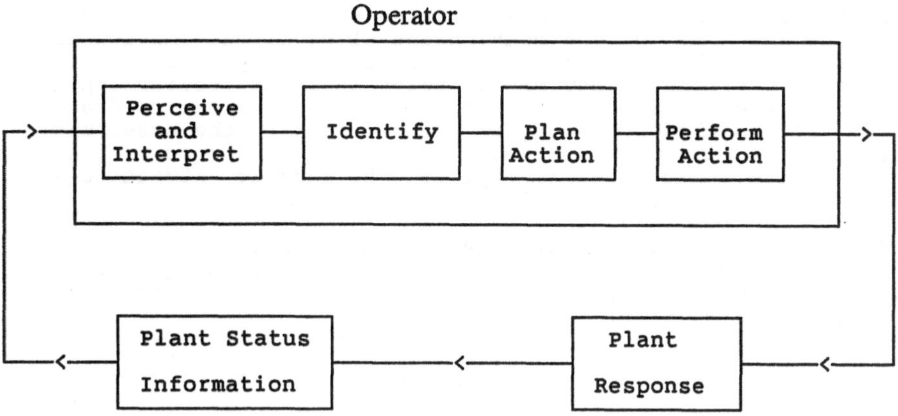

Operator

Figure 1 The operator as a component of a closed system

collection of disjoint facts. Knowledge about other knowledge
may be supporting evidence for a belief, rules which say when
other rules are or are not valid or any other knowledge which is
concerned with other knowledge. The requirement to quantify and
validate an isolated fact in a multi-level knowledge base is
therefore not a straightforward one.

Knowledge in different forms may contribute to a different
extent in various judgments but an operator cannot be asked to
say the extent to which each belief was used in making a
decision. This is partly because the method of knowledge
assessment would no longer be objective but also because he is
unlikely to be consciously aware of all the information he has
used. Some of the knowledge used in support of a reasoned
argument may have disappeared or become corrupted in time
leaving skills, or heuristic rules (rules of thumb) which have
little logical justification. For these reasons it is not
practical to obtain estimates of the 'strength of belief' of
individual facts which have been used by an operator to support
a particular course of action and to reproduce these in a
knowledge base which would be used in a simulation of his
reasoning process. An analogous argument may be used to explain
why the rules used in reasoning, several of which may be
considered by an operator before a balanced judgment is made,
may not be reliably defined along with their associated error.

In order that a quantitative measure of understanding may be
generated an alternative approach is required which does not
involve the combination of a large number of uncertain facts.
The Knowledge Assessment Method (KAM) is such an approach. It is
built around a novel application of Receiver Operating
Characteristic (ROC) curve generation and so the validation
process necessary for the corresponding part of the methodology
will be simplified.

The methodology is essentially one of comparison with an
'ideal', in this case the knowledge of an expert. An advantage

of this approach is that quantitative results are obtained
without the need for the combination of other information using
rules and so these do not need to be explicitly obtained. In
addition the methodology may be targeted at particular types or
areas of knowledge so that specific conclusions can be drawn.
Also, all knowledge used by an operator during the assessment
contributes to the quantitative assessment of the correctness of
his knowledge and so no attempt need be made to extract lower
levels of knowledge directly.

KAM provides a quantitative measure of the similarity of an
operators understanding to a 'correct' understanding. The
following section introduces and describes the methodology in
its unvalidated form.

A METHODOLOGY

The KAM (method) is described with reference to the ROC method
but sufficient detail has been included here to allow the ROC
method to be easily understood. KAM involves the generation of
graphs, or 'KAM Curves', which may be easily analysed.
The attraction of this type of approach is that meaningful
deductions may be made objectively from curves which have been
previously generated using the judgment of an individual on a
particular topic. Adaptation of the ROC method as described in
(1) has taken place in the following respects:

i)   The ROC method is applied in the evaluation of imaging
     systems whereas KAM assesses the knowledge of an individual.

ii) The ROC method uses judgments and confidence values given
     during image recognition tests. For instance, image
     recognition 'confidence values' (see later) can be effected
     by variations in contrast. In KAM, confidence values are
     obtained through questioning an individual, the 'subject' of
     the method.

iii) An additional stage shows how quantitative measurements may
     be made on the KAM curve and how these may be written using
     the support logic language FRIL (2). The clauses generated
     in this way could then be used by a knowledge-based system
     to provide answers to questions such as "How likely is it
     that this operator has the best all-round understanding of
     the plant ?".

It is believed that this method may be widely and beneficially
applied . It is restricted in its applicability however since it
is essentially based on comparison; KAM is an experimental
method which is based on a prior knowledge of whether
propositions are true or false. This information needs to be
provided by an 'expert' if the word 'assessment' rather than
'comparison' is justified. The knowledge of an individual, or
'the subject', acquired through KAM cannot form the knowledge
base of an expert system where reasoning, equivalent to that of
an operator during a task, may be performed. Instead, KAM
provides a balanced assessment of the subject's knowledge. KAM
will be used as a high level comparison method which provides
quantitative information as to the extent of an individuals

understanding in a certain area through comparing it with a 'correct' understanding. Common misconceptions may be identified through the analysis of results provided by several subjects.

KAM Curve Generation:

The procedure for the generation of curves is as follows:

Stage 1:

Generation of specimen questions:- A number of questions with corresponding answers are generated by a knowledgeable individual (The 'expert' or 'questioner'). These are in a particular area, for instance they may all be connected with the feedwater system of a reactor, and they should be chosen such that confident 'yes' or 'no' answers may also be provided by the expert. At this stage it is envisaged that approximately thirty questions would provide sufficient data for analysis. More than this would lead to slightly greater accuracy in results but is likely to be impractical due to the large quantity of data involved. Further work is being undertaken to assist the selection of appropriate questions and corresponding answers by the questioner. Guidelines are being developed to aid the generation of questions such that they both 'target' the required knowledge area effectively and lead to answers which may be checked for consistency. The questioner's data is subsequently used as a reference in the analysis process.

Stage 2:

A second individual, presumed to be less knowledgeable than the first is then asked the questions which were earlier generated by an expert. These are asked in a random order and every question must be answered by this person with a 'yes' or 'no' response in his own time. In addition to this, he is asked to give a value for the confidence he has in each answer. This is related to a statement in natural language and is chosen as a number from one to ten in increasing order of confidence. The subject would first be asked to divide the range of confidence values to reflect his own interpretation of five natural language categories. The following table might be typical.

Amount of Confidence:	Confidence Value Chosen:
Certain	10
	9
Very Confident	8
	7
Fairly Sure	6
	5
Unsure	4
	3
Very Unsure	2
	1

Stage 3:

It is possible to analyse both of the data sets obtained such that the data may be arranged in a form which will allow a KAM curve to be plotted in analogy with ROC analysis. This is done in the following manner:

Step 1:

The data obtained is split into categories according to confidence level. Five categories are defined. These are related to the answers in accordance with the degree of confidence the subject has assigned to them and each category corresponds to a range of Confidence Values.

Answers may therefore be placed in a category in accordance with a table such as:

Confidence Value:	Threshold Category:
9-10	Very Strict
7-10	Strict
5-10	Moderate
3-10	Lax
1-10	Very Lax

These are described as 'Threshold Categories' because each category takes in values above a certain Confidence Value. Each category is then considered in turn. For the 'Very Strict' threshold category, all answers assigned Confidence Values 9 and 10 are considered and the 'true positive' rate $P(S|s)$ and the 'false positive' rate $P(S|n)$ are worked out.

$P(S|s)$ is the probability that a question will be answered 'yes' given that the answer should be 'yes'.

$P(S|n)$ is the probability that a question will be answered 'yes' given that the answer should be 'no'.

These values are obtained by comparing the subjects answers with those provided by the 'expert'. $P(S|s)$ and $P(S|n)$ are calculated for each threshold range in turn. The answers included in each of these threshold ranges have associated confidence values either in the category being considered or higher.

Confidence Value:	Threshold Category:	$P(S|s)$:	$P(S|n)$:
9-10	Very Strict	5/17	1/13
7-10	Strict	9/17	2/13
5-10	Moderate	12/17	3/13
3-10	Lax	14/17	4/13
1-10	Very Lax	15/17	5/13

Stage 4:

The table shown above may be plotted using $P(S|n)$ as the abscissa and $P(S|s)$ as the ordinate (see Figure 2). A constant slope in this graph of 45 degrees would imply random replies to the questions asked, whereas a plot above the diagonal implies a degree of similarity between the knowledge of the expert and subject; ie. there is some evidence that they are in general agreement.

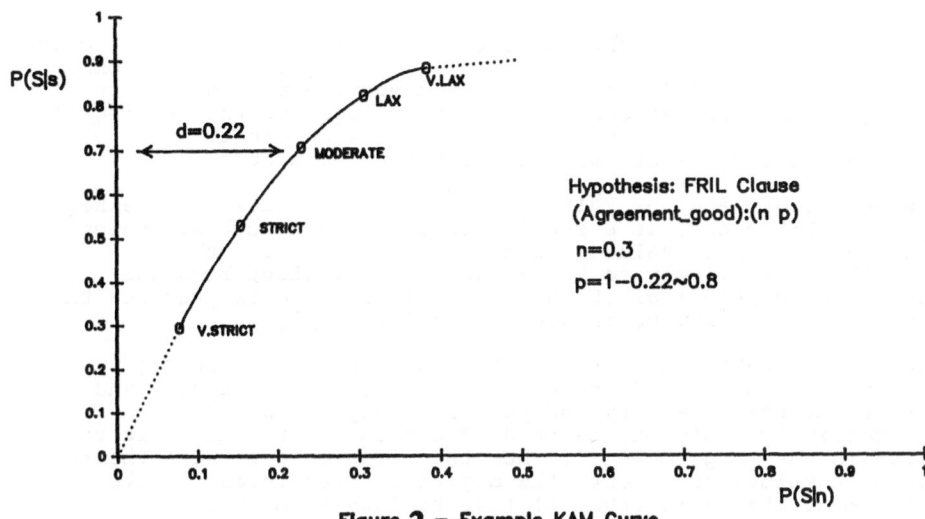

Figure 2 – Example KAM Curve

Analysis of Graphs:

A resulting graph may be qualitatively inspected and some
information gained.
a) If the whole of the KAM curve lies above the diagonal line
   then there is general agreement in the subject area.

b) The closer the curve is to the y-axis, the better the
   knowledge of the subject in this area.

c) It is apparent that $P(S|s)+P(N|s)=1$. In other words it is
   obvious that the number of questions to which the answer
   should have been 'yes' is the total of correct and incorrect
   answers when the answer should have been 'yes'. $P(S|s)$ is
   represented as the ordinate in the KAM curve. It may be seen
   that

   $$P(N|s)min = 1 - P(S|s)max$$

   and so the space above a drawn curve represents a deficiency
   in the knowledge of the subject.

By Quantitative Analysis:

a) Construction of a FRIL Clause:

The Fuzzy Relational Inference Language FRIL is similar to
Prolog in structure and can be used to construct knowledge based
systems. Simple statements are represented as 'clauses' and FRIL
allows the programmer to assign 'support values' to clauses
which represent both the strength of evidence for and against
the truth of a proposition.

Since the information contained in a KAM curve takes both
positive and negative results into account, a FRIL clause may be

constructed which allows the information implicit in a KAM curve
to be incorporated into a support logic programming environment
with little or no distortion in meaning. The method used in
bridging the gap between elicited knowledge and the meaningful
generation of support values is well defined and is based on
methods already established through imaging science.

Support logic and the FRIL environment are discussed elsewhere
(2). A FRIL clause is a representation of a hypothesis which
includes 'support values' which indicate how true that
hypothesis is likely to be. These values reflect both the
evidence for and against this proposition, and in practice the
support values may be interpreted quite simply.

Consider a FRIL clause ((True X)):(N P) where ((True X))
represents the simple statement 'X is True' and where N and P
represent the necessary and possible supports of FRIL
respectively. This may be read 'The probability that X is true
lies in the range N to P'. Thus the difference between the two
support values represents the degree of 'fuzziness' in the
statement that X is true. It also follows from this
interpretation that N<=P under all circumstances.

b)   Generation of Supports using a KAM Curve:

We may estimate the missing support values in the FRIL clause

((Knowledge_good   Subject)):(N P)

from the example KAM curve for the case in which an expert has
provided the initial reference questions and answers. This does
not conflict with the earlier statement that KAM is a method of
comparison since it is safe to assume that the experts knowledge
is good and so the subjects knowledge may also be said to be
good if the comparison is favourable.

N - Range 0 to 1 - Necessary support value. Its value reflects
    the minimum degree of support which may be attached to the
    proposition. If the necessary support is 1 then the
    proposition is certainly true.

P - Range 0 to 1 - Possible support value. (1-P) is a measure of
    the firm evidence against a proposition and so P may be
    interpreted as an upper boundary on the probability of the
    truth of a predicate. If P is 1 then no evidence is
    included which contradicts the truth of the predicate.

The way in which a support pair is generated is described using
the example KAM curve (Figure 2). It is claimed that the
techniques used are reasonable and may be applied provided that
the shape of the KAM curve is similar to that shown in the
diagram and that the curve is drawn above the diagonal (See
previous section). This forms a good visual check that the test
has been applied correctly. Techniques might be developed which
allow a valid re-scaling of confidence values for cases where
the operator has been consistently over or under-confident. The
assessment might also be improved if the subject were asked to

define the range of confidence values (and hence threshold
values at the analysis stage) which match his own interpretation
of the natural language categories such as 'very confident'
before he provides his answers.

Necessary Support (N):

This must represent a minimum value for the probability that the
knowledge of the subject is good. Following the previous
example, it was possible for the subject to make several correct
judgments with the highest degrees of confidence, this
information being contained in the first point of the KAM curve
and corresponding to a 'very strict' threshold. The better the
subjects knowledge is, the higher the value of $P(S|s)$ which
corresponds to a 'very strict' confidence value threshold. If no
decisions were made with a high degree of confidence then this
would be 0 and it is not possible to define a minimum level for
the necessary support that the subject has a good knowledge.

On the other hand if all of the questions were answered
correctly and with total conviction then every point on the
graph would appear at (0 1). In other words it is certain that
the subjects knowledge is good (If this occurs, the questions
asked were 'too easy' to highlight areas of weakness, and the
process must be repeated). The adoption of methods to aid the
selection of suitable question sets will minimise instances
where the KAM assessment is pitched at the wrong level.

The necessary support is therefore estimated as the value of
$P(S|s)$ which corresponds to the 'Very Strict' confidence value
threshold.

Possible Support (P):

An estimate of the maximum probability that the clause is true
can also be made from the KAM curve. The second statement in the
earlier 'Analysis of Graphs by Inspection' section is that a
curve moves closer to the y-axis if the knowledge of a subject
is greater. In other words the possible support for
(Knowledge_subject Good) increases. A measurement of this
characteristic distance of the curve from the y-axis is
therefore made close to the point corresponding to the Moderate
threshold data point.

Procedure:
A position is located on the y-axis of the KAM diagram 2/3 of
the way between the minimum and the maximum threshold points.
The distance of the curve from the y-axis is measured at this
point as d in Figure 2. (1-d) is then taken as the possible
support value.

The possible support changes from 1 (when there is total
agreement) to >0.5 since if random answers were provided the
graph would be a straight line, gradient 1. In this case the
support pair (N P)=(0 0.5) would be obtained and is interpreted
as: 'There is no evidence in support of the proposition but
there is some against it'. If the test answers agree with the

reference answers then the support pair (1 1) is obtained from the graph and the clause is true.

Combining Results:

If several curves are generated by an operator, their corresponding clauses may be combined into a more general overall statement using FRIL support logic following the construction of a suitable FRIL 'predicate' (rule). The support for the accuracy of the overall statement is also calculated through a mathematical combination of supporting evidence (2).

SUMMARY

In unusual circumstances where operators may be required to act outside laid-down procedures it is essential that operators have a clear and correct understanding of how the plant functions. The KAM methodology is being developed to assist in identifying operator misconceptions or any lack of understanding in a particular area of plant function. These may then be rectified by improvements in training, written operational procedures and operator aids.

The approach described may provide evidence of common misconceptions likely to lead to specific errors which, together with an assessment of the effect of erroneous operator actions on plant behaviour and consequently safety, would provide an input to plant risk assessments. If a representative sample of operators from individual plant types were available it would be possible to collect plant-specific data for use in plant reliability studies.

REFERENCES:

[1] Todd-Pokropek, A. ROC Analysis. In Physics Aspects of Medical Imaging. Eds.B Moores, R Parker, B Pullan. Wiley, 1981.

[2] Baldwin, Martin & Pilsworth. FRIL Manual, FRIL Systems Ltd. 184 Hotwell Rd. Bristol, 1987.

[3] Hunns, D.M. Discussions around a Human Factors Data-Base. An Interim Solution: The Method of Paired Comparisons. In: High Risk Safety Technology. Ed. A. Green, J.Wiley & Son 1982.

[4] Bersini, U.,Cacciabue, P.C., and Mancini, G. A Cognitive Model for Representing Knowledge, Intentions and Actions of Process Plant Operators. Commission of the European Communities, Joint Research Centre Ispra, Italy. Unpublished, January 1988.

[5] Colby, K.M. Simulations of Belief Systems. In 'Computer Models of Thought and Language, R.C. Schank (Ed.), Freeman & Co.

[6] Rasmussen, J. Strategies for State Identification and Diagnosis in Supervisory Control Tasks, and Design of Computer-Based Support Systems. In: Advances in Man-Machine Systems Research, Volume 1, pp 139-193, 1984.

[7] Morton, S.K. and Popham, S.J. Algorithm design specification for interpreting segmented image data using schemas and support logic. Image and vision computing, 1987.

[8] Baldwin, J.F. Support logic programming. Proc. NATO Advanced Study Institute on Fuzzy Sets (July 1985).

[9] Bersini U., Cacciabue P.C. and Mancini G.(1987): Cognitive Modelling: A Basic Complement of Human Reliability Analysis. Post-SMiRT 9 Seminar. Munich, August 24-25 1987.

[10] Knowledge elicitation.State of the art report on AI.

[11] Borys B. et al. 'Task and Knowledge Analysis in Coal-Fired Power Plants'. IEEE Control Systems, Vol.7 No.3, June 1987.

[12] Duboir, D and Prade, H. Processing of Imprecision and Uncertainty in Expert System Reasoning Models. Management Expert Systems, Ernst. C.J. Ed. Addison-Wesley Publishing Company 1988.

TIME-DEPENDENT STOCHASTIC MODELLING OF FIRE SPREAD IN SEGREGATED STRUCTURES

A. VEEVERS and P. MANASSE
Department of Statistics and Computational Mathemtics,
University of Liverpool
P.O. Box 147, Liverpool, L69 3BX.

ABSTRACT

A segregated structure is a unit comprising smaller, distinct 'compartments' or 'volumes' which are separated by various 'barriers'. If a fire starts in one of these volumes it may spread to other volumes in the structure by breaching the interposed barriers. A general model to assess the risk from fire spreading through a structure in which one or more volumes is in some sense critical is introduced. The model is time-dependent, based upon failure-time probability distributions assigned to each barrier and embracing extreme value theory in the determination of the probability distribution of the time taken for a fire to reach a particular critical volume.

INTRODUCTION

It has long been accepted that serious fires, although relatively infrequent, can cause tremendous human suffering as well as material damage and financial loss. It is no longer acceptable to consider fire and its consequences to be inevitable, instead we should try to familiarise ourselves with its nature so that we may attempt to predict how a fire will behave under certain conditions and accordingly modify those factors under our control. This attitude is reflected in the recent tightening up of the fire safety specifications of many common potential hazards ranging from the foam content of settees to the design of aircraft.

The work described in this paper is a development of the work of Veevers,

Boffey and Yates [1], and is concerned with the class of problems whose features are described as follows :

1)   A piece of critical equipment located in one or more 'target' volumes of a given segregated structure is at risk from damage by flame (but not from smoke alone).

2)   Initial ignition occurs in only one volume of the structure.

3)   Following ignition the fire can spread to other volumes in the structure by breaching common barriers.

4)   Each volume contains sufficient combustible material for flashover intensity to be reached.

5)   Any barrier will be breached in finite time when exposed to a fire of flashover intensity.

6)   No significant fire-fighting action takes place, e.g. the sprinkler system is inadequate or in a failed state and the time-period involved is prior to any external fire-fighting action being brought to bear.

Since it is not known in advance which volume will be the one in which the fire starts, all possibilities must be included.

Illustration

Throughout the paper reference will be made to the illustrative example displayed in Figure 1. The target volume is taken to be the room numbered 9 and particular attention is paid to the consequences of a fire starting in volume number 1.

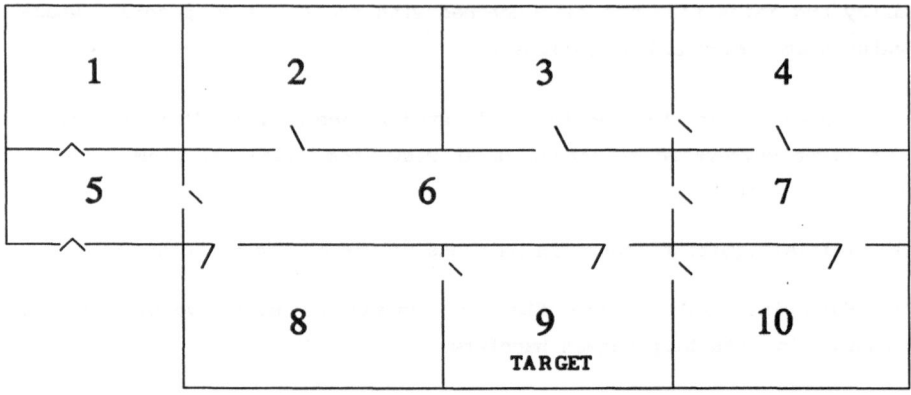

Figure 1.   Plan of a single-storey ten volume building.

Construction of the Model

The first step in the construction of the model is the allocation to each
volume, i, of an ignition probability, p_i, representing the likelihood of
fire starting in that volume. These values are generally derived from
existing fire frequency data by professional assessors. Next, it is
necessary to assign breach time probability distributions to each
barrier. These specify the probabilities of a fire breaching the barrier
within any time interval given that it exists in one of the adjacent
volumes. There will be two such probability distributions for each
barrier, one for each direction of burn-through. The ability of a fire to
breach a barrier is clearly dependent on the severity of the fire in the
volume concerned, and some work has been done to model accurately the
growth of fire in generally specified volumes [2]. The models of spread
presented here may be used in tandem with existing models of growth or be
adapted so that the time taken for a fire in a volume to become
sufficiently severe to threaten a barrier is subsumed into the barrier
failure-time probability distribution. For the purposes of the present
development, no distinction is made between the two cases.

The next step is the identification of all distinct paths from each possible ignition volume to the target volume. In the example there are 99 different paths to the target room 9; there being 1 for which fire starts in the target itself, 15 for which fire starts in room 1, 14 for which fire starts in room 2, and so on. The paths from each potential ignition volume are classified according to 'path length' - the number of barriers in a path, and then for each path a probability distribution of the time for fire to travel from source to target is derived.

Models of barrier failure

The fire resistance of a barrier in a volume in which there is a fire obviously has an extremely important influence on the time taken for a fire to spread from the burning volume to an adjacent one, and thus from the volume in which the fire starts to any target volume.

Whilst many barriers are British Standard graded to be, for example, 10 minute or 20 minute barriers, it is clearly only reasonable as a first approximation to take these graded values as nominal breach times. It is preferable to model the uncertainty using probabilistic methods. For example, experts might say that a door rated at 10 minutes will not fail before 8 minutes' but will certainly fail before 13 minutes' exposure to a fire of flashover intensity. Using such expert knowledge we are able to produce an acceptable probability distribution of failure time for each barrier. Reasonable distributions for a barrier rated at α minutes include those in Table 1.

42

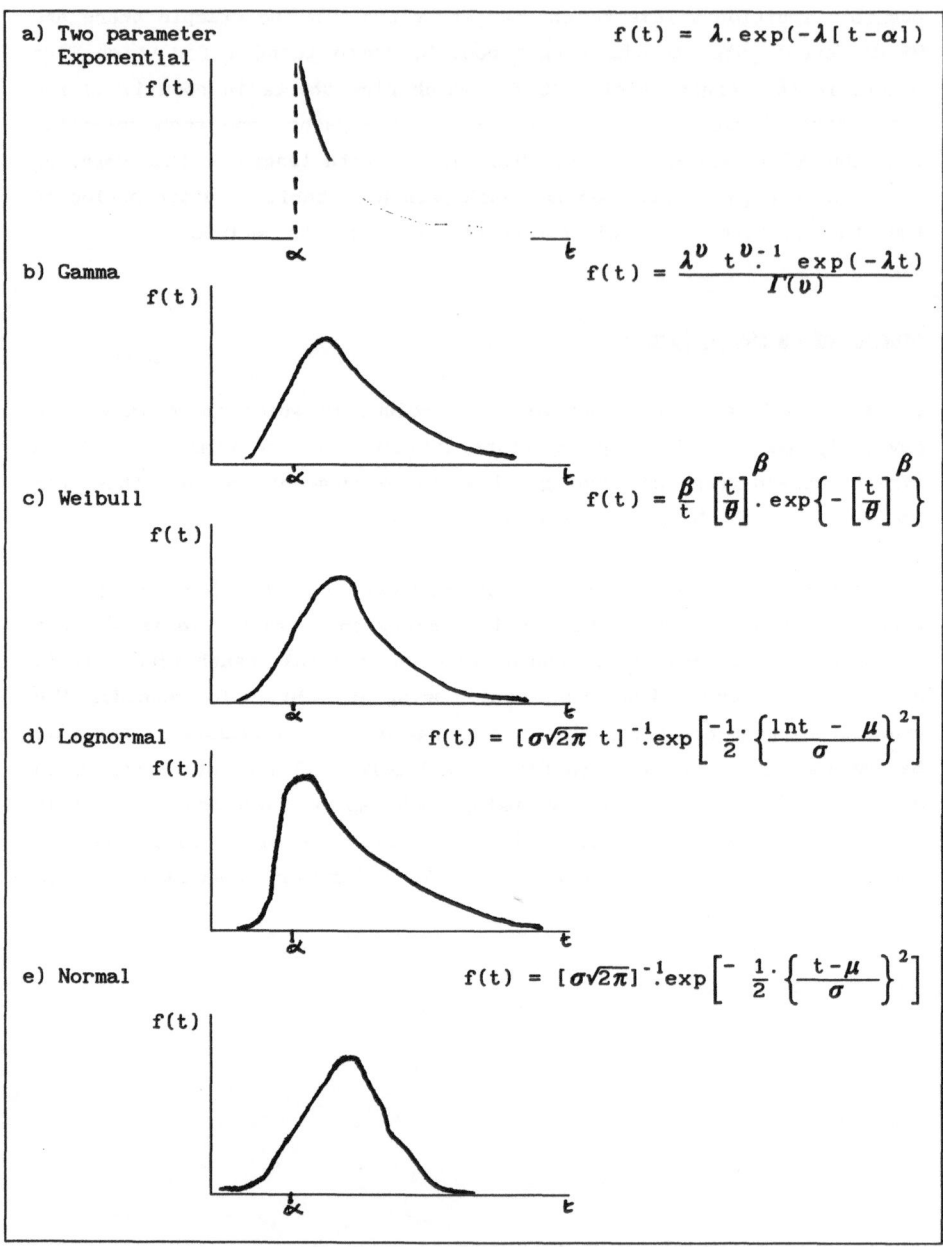

Table 1   Some candidate Breach-time probability distributions
for a barrier rated at α minutes.

Model (a) represents the situation in which the barrier is secure up to the rating value and then burn-through may take place soon afterwards.

Models (b),(c),(d), & (e) can all represent the situation in which the barrier may fail before the rating value but is nonetheless more likely to fail after it. Note that in (c) there can be a rapid drop in probability density to the right of the mode. There is some empirical support for the use of the Weibull distribution from data given by Ling and Williamson [3] and the Lognormal distribution provides a good fit to data quoted in a confidential report from the Fire Research Station. In (e) the variation in breach-time is symmetrically distributed about a nominal value which may, in some cases, be the rating value. Its applicability is supported in AEA funded work currently underway at Edinburgh University [4]. Further support for these models is provided by some unpublished data obtained from tests done in Australia.

Time to reach the Target Volume

Once a reasonable breach-time distribution has been assigned to each barrier, the properties of the paths from each potential ignition volume to the target can be explored. The complexity of path identification rapidly increases as the number of volumes increases, and enumerating the paths is certainly not trivial except in a very simple structure having only a few volumes. We are much aided, however, by the existence of the ARSSUN computer program [1] and it is straightforward to classify the paths listed in its output according first to starting volume and subsequently to length.

For each route, the probabability distribution of the time taken to reach the target is that of the sum of the breach-times along its path.

In the example, suppose that all the barriers are identical and that each has, independently, a two-parameter exponential (0.1, 10) breach-time distribution. Consider one of the two shortest routes from volume 1 to volume 9, [1-2-6-9]; we see that the distribution of time for fire to travel along that path is the sum of three independent identical

exponential times. This we know to be distributed as Gamma (0.1, 3) +30.
The collected results of repeating this for all paths from volume 1 to
volume 9 are shown in table 2.

Path number 1,k	Route	Number of Barriers, $c_{1,k}$	Distribution of Travel Time, $F_{1,k}(t)$
1,01	1-2-6-9	3	30 + Gamma(0.1, 3)
1,02	1-5-6-9	3	30 + Gamma(0.1, 3)
1,03	1-2-6-8-9	4	40 + Gamma(0.1, 4)
1,04	1-2-3-6-9	4	40 + Gamma(0.1, 4)
1,05	1-5-6-8-9	4	40 + Gamma(0.1, 4)
1,06	1-2-3-6-8-9	5	50 + Gamma(0.1, 5)
1,07	1-2-6-7-10-9	5	50 + Gamma(0.1, 5)
1,08	1-5-6-7-10-9	5	50 + Gamma(0.1, 5)
1,09	1-2-3-4-7-6-9	6	60 + Gamma(0.1, 6)
1,10	1-2-3-4-7-10-9	6	60 + Gamma(0.1, 6)
1,11	1-2-3-6-7-10-9	6	60 + Gamma(0.1, 6)
1,12	1-2-3-4-7-6-8-9	7	70 + Gamma(0.1, 7)
1,13	1-2-6-3-4-7-10-9	7	70 + Gamma(0.1, 7)
1,14	1-5-6-3-4-7-10-9	7	70 + Gamma(0.1, 7)
1,15	1-5-6-2-3-4-7-10-9	8	80 + Gamma(0.1, 8)

Table 2:   Distributions of travel times along routes from vol.1 - vol.9

Similar tables may of course be produced for all the other possible
ignition volumes.

To summarise, what we now have is a collection of tables, one for each
volume, which contain information about the travel times for all the

45

possible routes which a fire may take to the target volume. Taking the probability of simultaneous initiation of fire in two or more volumes to be negligible, this comprises all the necessary information.

Fire may spread simultaneously along many of the identified routes from a given ignition volume. However, we are exclusively interested in the time of **first** arrival of fire at the target. Since the coincidental breaching of two or more barriers admitting fire to the same volume is an event with probability zero in continuous time, the path by which the fire first arrives at the target is uniquely identified.

If we suppose that a fire breaks out in volume 1 of our example structure and ultimately spreads to the target volume 9, then exactly one of the 15 paths in Table 2 will be the one associated with fire first entering the target.

Extreme Value Theory

The study of extremes is concerned with the analysis of the behaviour of the largest or smallest value among a number of observations. In our case, as stated above, we are interested in the time to first arrival of the fire at the target.

In statistical terms, the time sought in the example is the smallest in a sample of one each from the 15 random variables having the distributions shown in Table 2. Since these are neither identical nor independent distributions the standard results of extreme value theory do not immediately apply. However, it has been shown, [5], that the standard results approximately apply under much weaker assumptions about the contributing random variables. This robustness is appealed to in the present application to produce the results which follow.

In the general case suppose the barrier breach times are independently, but not necessarily identically, distributed continuous random variables $T_{m,n}$ with distribution functions $F_{m,n}(t)$, where m and n are typical adjacent volumes. The time $T_{i,k}$ for fire to travel from ignition volume i to the target along the path labelled k is the sum of the $T_{m,n}$ values corresponding to the barriers defining the path.

Once we have identified all the paths from the ignition volume i to the target volume we have K_i, say, random variables $T_{i,k}$ with distribution functions $F_{i,k}(t)$, $k=1,2,\ldots K_i$, and we require the distribution of the associated smallest order statistic, T_i. Using a standard result, e.g. David [6], we know that this is:-

$$\Pr\{T_i \leq t\} = 1 - \prod_{k=1}^{K}[1 - F_{i,k}(t)] \qquad (1)$$

When this distribution has been derived for each ignition volume the results can be aggregated for the whole structure, yielding the distribution of breach-time, T, conditional on fire starting somewhere in the structure as :-

$$\Pr\{T \leq t\} = \sum_{i=1}^{V} \left\{ \left[1 - \prod_{k=1}^{K_i}\left(1 - F_{i,k}(t)\right)\right] \times q_i \right\} \qquad (2)$$

where $q_i = \dfrac{P_i}{\sum\limits_{i=1}^{V} P_i}$

and V is the number of volumes in the structure.

Example Application

In the example, with the $F_{1,k}(t)$ as specified in Table 2, the distribution of time to first arrival from volume 1 is, from (1) above,

$$\Pr\{T \leq t\} = 1 - \prod_{k=1}^{15} \left[\exp[-\lambda(t-c_{1k}\alpha)] \sum_{j=0}^{c_{1,k}-1} \frac{\lambda^j}{j!}(t-c_{1k}\alpha)^j \right] \qquad (3)$$

where $\lambda = 0.1$, $\alpha = 10$ and
$c_{1,k}$ = number of barriers in path k from volume 1 to the target.

47

Also (2) becomes :-

$$Pr\{T \le t\} = \sum_{i=1}^{10} \left\{ \left[1 - \prod_{k=1}^{K_i} \left[\exp[-\lambda(t-c_{ik}\alpha)] \sum_{j=0}^{c_{ik}-1} \frac{\lambda^j}{j!}(t-c_{ik}\alpha)^j \right] \right] \times p_i \right\} \quad (4)$$

where $\lambda=0.1$, $\alpha=10$ and

$c_{i,k}$ = no. of barriers in path k from volume i to target.

So, from (3), if a fire starts in volume 1 of the building we calculate the distribution of time to first arrival at the target to be as shown in Figure 2.

Whilst from (4), and for illustrative purposes taking the scaled ignition probabilities to be 0.1 for each volume, we calculate the probability distribution shown in Figure 3. This is the distribution of time to reach the target when it is known that a fire has started but it is not known in which volume ignition occurred.

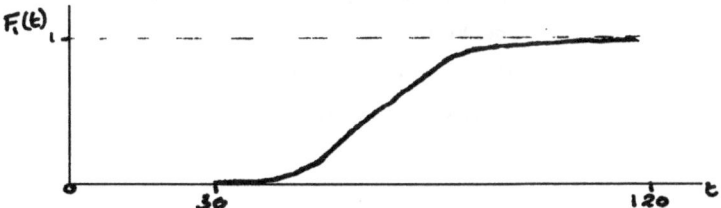

Figure 2.   Distribution of time to first arrival at target of a fire starting in volume 1.

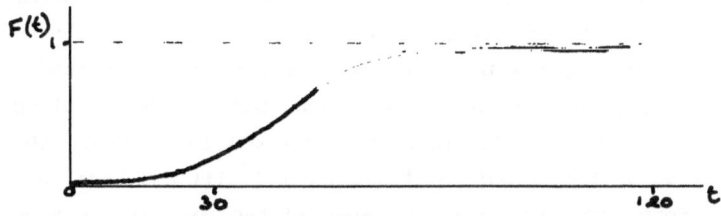

Figure 3.   Distribution of time to first arrival at target of a fire, given that a fire starts somewhere in the structure.

Results of this sort can be used in probabilistic risk assessments at the building design stage, or for existing structures. Furthermore, because of the time dependency, the assessment can include a provision for the time for fire-fighting action to be initiated. This can be influential in the decision about how much to invest in local appliances and how much to rely on public ones. In safety critical situations in which the target volumes need ultra-high protection against fire (amongst other things), use of the above quantitative methodology will increase the confidence in the assessment of this aspect of safety.

Further Aspects of the Theory

Returning to the general case, it seems reasonable to categorise the possible relationships between the individual independent barrier breach distributions according to the following:-

Category ...	breach-time distributions are ...
1	identical
2	same family, different parameters
3	different families

In category 1 the path travel times are all distributed as the sums of independent, identical (iid) random variables. Thus, if the barrier breach-times follow the Gamma distribution (includes the Exponential distribution) or the Normal distribution, (1) will have a simple closed form, the distribution of the sum being of the same family as the breach distribution itself. If, on the other hand, the barrier breach-times follow either of the other two distributions suggested in Table 1, the Lognormal distribution or the Weibull distibution, (1) will not have such a straightforward form. In that case one must either evaluate a very complicated expression or appeal to the Central Limit Theorem, as applied to iid random variables, which enables one to approximate the sum of the random variables by a Normal distribution.

In category 2 explicit forms of (1) exist for the Normal family of distributions, and for the Gamma family with a common value of λ. In the other cases, use may be made of the Central Limit Theorem for the sum of independent, non-identical random variables if one wishes to avoid complex mathematics.

In category 3, straightforward forms of (1) do not exist for the distributions given here, but it may often be the case that in a particular building the number of different types of barrier is small, so that some of the breach-time distributions may be dealt with as in category one, finally yielding perhaps only two or three different distributions in category 3. Thus we may have, for example, several barriers in a path whose breach-times are identically Exponentially distributed whilst the others are Normally distributed. We then have two sets of random variables, each of which falls into category 1 and whose sums are straightforward to evaluate. What is finally required in this case is then the sum of one Normal random variable (the sum of the Normal breach-times) and one Gamma random variable (the sum of the Exponential breach times) which is the $F_{i,k}(t)$ in (1). The mathematics is once again complex but not intractable!

Acknowledgements

Grateful thanks are expressed to the SERC and the UKAEA(SRD) for providing the funding for the PhD research project whose pursuit has enabled this paper to be written.

REFERENCES

1   Veevers, Boffey and Yates. Probabilistic Risk Assessment in Segregated Structures, 10th ARTS (1988).

2. Mitler and Rockett. Users guide to FIRST, Report NDS IR86 (September 1986).

3. Ling and Williamson. Modelling of Fire Spread through Probabilistic Networks, Fire Safety Journal 9 (1985).

4. A. Beard, University of Edinburgh, Private Communication (1989).

5.   Leadbetter, Lindgren, and Rootzen. **Extremes and Related Properties of Random Sequences and Processes**, Springer, New York (1982).

6.   David. **Order Statistics**, Wiley, New York [2nd edition] (1982).

51

RELIABILITY OF THE SPENT FUEL IDENTIFICATION FOR FLASK LOADING PROCEDURE USED BY COGEMA FOR FUEL TRANSPORT TO LA HAGUE

EID Mohamed
Commissariat à l'Energie Atomique, Saclay,   91191 Gif sur Yvette Cedex, France

ZACHAR Michel
Nuclear Transport Limited , Risley, England

PRETESACQUE Patrice
Compagnie Générale des Matieres Nucléaires, COGEMA, France

ABSTRACT

The Spent Fuel Identification for Flask Loading, SFIFL, procedure designed by COGEMA is analysed and its reliability is calculated. The reliability of the procedure is defined as the probability of transporting only approved fuel elements for a given number of shipments. The procedure describes a non-coherent system. A non-coherent system is the one in which two successive failures could result in a success, from the system mission point of view. A technique that describes the system with the help of its maximal cuts (states), is used for calculations. A maximal cut contains more than one failure can split into two cuts, (sub-states). Cuts splitting will enable us to analyse, in a systematic way, non-coherent systems with independent basic components.

INTRODUCTION

The SFIFL procedure has been analysed. This procedure is a very good example of a non-coherent system, whose basic events are independent. In non-coherent systems, two successive failures might result in a success. A systematic analysis for that type of systems is not always possible. We propose to analyse it with the help of the maximal cuts representation technique. A maximal cuts representation is a truth table like representation. We, some times, alter between the two terminologies : maximal cuts or modified truth table representation. The main point of the technique is to be exposed in section 2.

Firstly, the system and its basic events will be defined. Secondly, a maximal cut card will be deduced and cuts will be grouped according to their consequences. Thirdly, the probability of each group of sequences and conditional probabilities will be determined. Finally, the reliability of the whole procedure is to be calculated and conclusions will be drawn.

This work has served to elaborate a document presented to the French safety authorities to describe the Spent Fuel Identification for Flask Loading procedure used by COGEMA and NTL.

I. PROCEDURE DEFINITION

The Spent Fuel Identification for Flask Loading procedure, {2}, will be defined by its basic events (actions). These basic events are :

M1 : Read the coordinates (X,Y) on the list of approved fuel elements, and point the reading device right on the fuel element.
M2 : The present fuel element corresponds to the position.
M3 : Read the actual identification number on the fuel element and notify it on the list.
M4 : Compare the notified number with that given in the hidden part on the list .
D1 : Stop if comparison is negative, otherwise continue.
M5 : Take the fuel element out of its pond.
M6 : Measure the γ-emission level of the fuel element.
D2 : Stop if test is negative, otherwise continue.
D3 : Go to **M1** if total number of fuel elements to be loaded in the flask has not yet been attained, otherwise continue.
M7 : Having completed the flask loading, Proceed to a second identification checking.
M8 : Read identification numbers and compare, while fuel elements are in the shipping flask.
D4 : Stop the whole process if test is negative, otherwise seal the flask and proceed to transport.

The SFIFL procedure contains 12 events. These events are either missions to be executed, (M1, M2, ...)., or decisions to be taken, (D1, D2, ...). Each mission will be given a failure probability, table (4). Only missions can fail. Decisions may be either **STOP** or **CONTINUE**. In our study, only a

CONTINUE decision can be inadequate. This is simply a conventional matter. From a safety point of view, **STOP** decisions are always adequate as far as our present study is concerned. **CONTINUE** decision is adequate if it is associated to an approved fuel element, otherwise it is inadequate.

We suppose, also, that the operator will proceed in a chronological manner and will never repeat a mission twice. However, it is not the case in practice. For example, if decision **D1** is stop, the operator will not go back to mission **M3** or to mission **M4** to make sure of his readings or of the comparison before deciding definitively to stop. From a safety point of view, this hypothesis is acceptable.

II. MAXIMAL CUTS

A maximal cut card, table (1), is a card contains all possible combinations (sequences) between events (actions). Each sequence of events will result in a specific consequence. Sequences are, then, to be grouped according to their consequences. In this card, 1 may assign a failure of a mission and 0 may assign a success. A sequence contains more than one failure may split into two subsequences. Splitting assigned by (-1) is said to be in the negative sense, otherwise, it will be a positive sense splitting, (+1). Negative and positive senses are so by definition. A positive sense splitting factor is to be associated to each mission failure. It is the probability that a definit splitting will have a positive sense consequence. In this study, three types of consequences have been distinguished :

1. STOP,
2. CONTINUE with an approved fuel element, (adequate decision), or
3. CONTINUE with a non-approved fuel element, (inadequate decision).

Thanks to the notion of splitting, non-coherent systems with independant components could be examined.

III. DEFINITION OF PROBABILITIES

The quantification will be carried out in four steps, {4}:

Step(1) : Establishing a maximal cut card for the missions M1, M2, M3, M4 and M5, table(1). Two probabilities are to be calculated, P1 and P2. P1 is the probability to grip a fuel element from the pool. P2 is the probability that the gripped fuel element will be non-approved, (inadequate D1).

	M5	M4	M3	M2	M1	conseq.		M5	M4	M3	M2	M1	conseq.
C1	0	0	0	0	0	T1 = CAE	C21	1	0	1	0	0	T32= STO
C2	0	0	0	0	1	T2 = STO	C22	1	0	-1	0	1	T33= STO
C3	0	0	0	1	0	T3 = STO	C22	-1	0	+1	0	1	T34= CNE
C4	0	0	0	-1	1	T4 = STO	C22	+1	0	+1	0	1	T35= CAE
C4	0	0	0	+1	1	T5 = CAE	C23	1	0	-1	1	0	T36= STO
C5	0	0	1	0	0	T6 = STO	C23	-1	0	+1	1	0	T37= CNE
C6	0	0	-1	0	1	T7 = STO	C23	+1	0	+1	1	0	T38= CAE
C6	0	0	+1	0	1	T8 = CNE	C24	1	0	-1	-1	1	T39= STO
C7	0	0	-1	1	0	T9 = STO	C24	-1	0	+1	-1	1	T40= CNE
C7	0	0	+1	1	0	T10= CNE	C24	+1	0	+1	-1	1	T41= CAE
C8	0	0	-1	-1	1	T11= STO	C24	1	0	1	+1	1	T42= STO
C8	0	0	+1	-1	1	T12= CNE	C25	1	1	0	0	0	T43= STO
C8	0	0	1	+1	1	T13= STO	C26	-1	1	0	0	1	T44= CNE
C9	0	1	0	0	0	T14= STO	C26	+1	1	0	0	1	T45= CAE
C10	0	1	0	0	1	T15= CNE	C27	-1	1	0	1	0	T46= CNE
C11	0	1	0	1	0	T16= CNE	C27	+1	1	0	1	0	T47= CAE
C12	0	1	0	-1	1	T17= CNE	C28	-1	1	0	-1	1	T48= CNE
C12	0	1	0	+1	1	T18= STO	C28	+1	1	0	-1	1	T49= CAE
C13	0	1	1	0	0	T19=CAE	C28	1	1	0	+1	1	T50= STO
C14	0	1	-1	0	1	T20=CNE	C29	1	1	1	0	0	T51= CNE
C14	0	1	+1	0	1	T21=STO	C30	-1	1	-1	0	1	T52= CNE
C15	0	1	-1	1	0	T22=CNE	C30	+1	1	-1	0	1	T53= CAE
C15	0	1	+1	1	0	T23=STO	C30	1	1	+1	0	1	T54= STO
C16	0	1	-1	-1	1	T24=CNE	C31	1	1	+1	1	0	T55= STO
C16	0	1	+1	-1	1	T25=STO	C31	-1	1	-1	1	0	T56= CNE
C16	0	1	1	+1	1	T26=CAE	C31	+1	1	-1	1	0	T57= CAE
C17	1	0	0	0	0	T27=CNE	C32	-1	1	-1	-1	1	T58= CNE
C18	1	0	0	0	1	T28=STO	C32	+1	1	-1	-1	1	T59= CAE
C19	1	0	0	1	0	T29=STO	C32	1	1	+1	-1	1	T60= STO
C20	1	0	0	-1	1	T30=STO	C32	1	1	1	+1	1	T61= CNE
C20	1	0	0	+1	1	T31=CNE							

TABLE (1) : The Modified Truth Table of the first 5 missions

CNE : Chargement of Non-approved Element, CAE : Chargement of Approved Element, STO : Stop

	St1	M6	conseq.		St2	M7	M8	conseq.
C1	0	0	T1 = AEC	C1	0	0	0	T1 = AET
C2	0	1	T2 = AEC	C2	0	0	1	T2 = STO
C3	1	0	T3 = STO	C3	0	1	0	T3 = AET
C4	1	1	T4 = NEC	C4	0	1	1	T4 = AET
				C5	1	0	0	T5 = STO
				C6	1	0	1	T6 = NET
				C7	1	1	0	T7 = NET
				C8	1	1	1	T8 = NET

TABLE (2) : The Truth Table for Step1 and M6   **TABLE (3) : The Truth Table for Step2, M7 and M8**

A/NEC : Approved/ Non-approved Element to be Charged in the flask,

A/NET : Approved/ Non-approved Element to be Transported.

Mission	Lower Limit	Higher Limit	Splitting Factor
M1	1.00E-3	5.00E-3	0.00
M2	1.00E-3	5.00E-3	1.00E-2
M3	1.00E-3	5.00E-3	1.00E-2
M4	1.00E-4	5.00E-4	0.00
M5	1.00E-4	3.00E-4	1.00E-2
M6	1.00E-2	1.00E-2	0.00
M7	3.00E-3	1.00E-2	0.00
M8	5.00E-3	1.00E-2	0.00

TABLE (4) : Mission Failure Probabilities and Splitting Factors

Step(2) : Establishing a maximal cut card for the event "gripping an approved fuel element, step(1)" and the mission M6, table(2). Two probabilities will be calculated, P3 and P4. P3 is the probability that a gripped fuel element will be loaded in the shipping flask. P4 is the probability that this fuel element is non-approved, (inadequate D2).

Step(3) : Establishing the maximal cut card for the event "a fuel element loaded in the shipping flask is approved, step(2)" and the missions M7 and M8, table (3). Then, we calculate P5 and P6. P5 is the probability that a fuel element loaded in the flask will be transported. P6 is the probability that a transported fuel element will be non-approved, (inadequate D4).

Step(4) : Calculate P7, the probability to transport at least one non-approved fuel element in the shipping flask, (the unreliability).

To demonstrate the basic idea, we are going to make use of table (3). Firstly, we are interested in calculating the probability P5 of the event (E) : "a fuel element that has already been loaded in the shipping flask will be transported":

$$E \quad = \quad T1 \cup T3 \cup T4 \cup T6 \cup T7 \cup T8,$$

Secondly, we will calculate the probability of the event E*: "a non-approved fuel element that has already been charged in the shipping flask will be transported":

$$E^* \quad = \quad T6 \cup T7 \cup T8.$$

Finally, the probabilty P6 , that a transported fuel element will be non-approved, is:

$$P6 \quad = P(E^*) / P(E) \quad = P(E^*) / P5$$

The application of the same scheme to calculate P2 and P4 is direct.We notice then that P2, P4 and P6 are all conditional probabilities.

IV. NUMERICAL APPLICATION

Calculations are carried out with data provided mostly by NTL and COGEMA, {3}. In case of data insufficiency, expert judgement was asked for. The previously defined probabilities were calculated :

- P1 : The probability that D1 will be a continue decision :

$$9.97E\text{-}01 > P1 > 9.84E\text{-}01$$

- P2 : The probability to be inadequate :

$$1.00E\text{-}04 < P2 < 3.06E\text{-}04$$

- P3 : The probability that D2 will be a CONTINUE decision,

$$P3 \approx 1.0$$

- P4 : The probability to be inadequate,

$$9.95E\text{-}05 < P4 < 3.06E\text{-}04$$

- P5 : the probability that D4 will be a CONTINUE decision,

$$9.95E\text{-}01 > P5 > 9.90E\text{-}01$$

- P6 : The probability to be inadequate,

$$8.02E\text{-}07 < P6 < 6.15E\text{-}06$$

- P7 : The probability to transport at least one non-approved fuel element in the flask :

$$P7 = 1. - (1.\text{-}P6)^n$$

This probability , with its higher and lower limits, has been calculated for shipping flasks with the capacity of 11 fuel elements by flask.

$$8.82E\text{-}06 < P7 < 6.77E\text{-}05$$

V. CONCLUSION

The probabilities preceedingly calculated could be presented in different forms. As the objective is to control the quality of the SFIFL procedure, some other values seem to be more proper for quality control use:

1° The unreliability of the procedure, within 20 years, may be more significant for safety use. The unreliability will be calculated as follows :

58

$$\underline{R}(20) = 1 - (1.-P6)^n$$

where, n = Number of fuel elements transported in 20 years, (n = 11 fuel elements x 20 flasks x 20 years = 4 400). We find that :

$$3.52E-03 \quad < \quad \underline{R}(20) \quad < \quad 2.67E-02$$

2° Or, <u>the mean number of shipments between failures</u>, where a shipment will be considered failed if it contains at least one non-approved fuel element, (out of 11 fuel elements/ shipment) :

$$1.134E5 > N > 1.477E4 \qquad \text{shipments}$$

The unreliability or the mean number of shipments between failures may be a good criteria to control the quality of the procedure.

Secondly, the maximal cuts analysis technique enables us to quantify non-coherent systems with independant basic components. The non-coherency will be described by sequence splitting. Only sequences with more than one failure can split. Splitted sequences have identical binary structure, only consequences are different. This technique seems powerfull when it is a matter of systems having few missions. It is actually the case of many practical applications. However, with systems having more missions, analysis may be possible thanks to expert systems.

REFERENCES

1. "Elements de Méthodologie pour l'Etude de Défaillances des Grands Systèmes, Calculs Logiques et Algebriques", VERGEZ Pierre, Note CEA-N° 2262, Janvier 1982, CEA/ DEIN/ SIAI, CEN-SACLAY, France.
2. "Fuel Identification at Loading, General Instruction", 5 PG 01 / COGEMA and NTL.
3. "Arbres de Défaillance Particuliers", 5 RAE 103 NTL.
4. "Analyse Probabiliste de Procedure de Chargement d'Assemblages Irradiés", M. EID, DEMT 89/005, DEMT/ CEN-Saclay, France.

MANAGEMENT AT RISK

ANGELA M JENKINS
Human Factors Unit
AEA Technology, SRD
Wigshaw Lane, Culcheth, Warrington, WA3 4NE

ABSTRACT

It is evident, from the concern of the public, expressed through the media, and from the recent moves to prosecute organisations for corporate manslaughter, that management is AT risk as a result of the potential consequences of a major disaster. The purpose of this paper is to describe the work being carried out by SRD towards producing a guide to corporate management on the management of risk. The guide, based around a systems model and drawing support from the analysis of major disaster case studies, will help corporate management to identify not only the problems which lead to risk but also some of the tools which could help to alleviate those problems. This paper describes the approach to modelling and case study analysis.

INTRODUCTION

Management is AT risk owing to heightened public awareness and concern about disasters arising from the activities of organisations. The politicians and the legislature have responded to popular demand and public opinion that organisations should be made to account for the disastrous consequences of their activities. Obviously, in order to reduce the risk to themselves, corporate management must succeed in the management OF potential risks to people and the environment as a result of the pursual of their intentions or policies.

SRD is producing a guide to corporate management which identifies problem areas in safety or risk management and identifies the tools and aids available to resolve those problems. The document will call on evidence from disaster case studies.

Both the guide and the analysis of the case studies will be based on systems models and the following sections describe two separate approaches to devising the appropriate model or framework. In order to define the direction of the study, however, two issues first need to be given some consideration:

1.   What do we mean by a 'disaster'?
2.   How do we deal with many different management styles and structures?

WHAT IS A DISASTER?

There have been a number of attempts to define disasters according to the number killed, the number injured and so on; but a disaster is not just a statistic since it may have emotional, social and political dimensions. It is quite evident, from media coverage, that the scale or importance of a disaster is as much about who is affected and where is affected as about the number of casualties. There is a very high degree of subjectivity.

One could even suggest that if an incident were classified as 'insignificant' because only one person was injured, this could be a disincentive to investigate and resolve some very fundamental organisational weaknesses. Those same weaknesses could result in the death of many people days, weeks or years later. Essentially one should not mask the significance of weaknesses in an organisation or system by simply observing the apparent insignificance of the consequences of a failure of that system. Are definitions of 'disaster' desirable or relevant, therefore?

For this project, the relevance of a definition relates more to a definition of our committment as to how we intend to analyse disaster case studies. From the point of view of emergency planning, death and injury statistics might be useful, but such statistics do not reveal causality. From the point of view of corporate management a disaster can be viewed financially, politically, legally, socially etc and our definition must be global enough to capture this.

Corporate Management, from an organisational point of view could be regarded, like light, as having a dual nature. Early organisation theorists defined an organisation as (1):

"the planned coordination of the activities of a number of people for the achievement of some common explicit purpose or goal, through division of labour and function and through a hierarchy of authority and responsibility".

In this static context, corporate management is the spearhead of the organisation. Modern theorists, however, regard organisations as open, complex systems. Within this context, corporate management could be seen as being at the hub of the organisation. In either case, corporate management sets out, within a set of dynamic environments, to achieve its multiple intentions.

In 1978, Barry Turner (2) suggested that a disaster always involves a failure of intention. Consider the following, for example:

Mr X killed in a road accident, may represent a failure of intention on the part of

Mr X?

the driver of the other car?

the road planners?

N people killed at a sports stadium, may represent a failure of intention on the part of

the stadium designers?

the controlling sports body?

the fire and security controllers?

M people killed in a chemical plant explosion, may represent a failure of intention on the part of

the operator to carry out a task?

supervisory management to carry out checking?

the design engineer?

CORPORATE MANAGEMENT

      to meet health and safety standards

      to make a product at maximum profit?

Corporate Management uses a hard or soft system in order to carry out its intentions or policies. These policies are formulated by observation and analysis of the environment. In this context, any failure of the system can be regarded as a failure of intention and there lies the potential disaster. From the point of view of the public, failures to ensure minimum risk and maximum safety to people and the environment are the most noteworthy types of disaster.

The environment is dynamic and so intentions, policies and the complex systems designed to achieve them, together with their internal environments, are highly susceptible to change. The concept of change is another important element in disasters. A failure to respond to changing environments and needs or a failure to control change is a consistent feature of major disasters.

MANAGEMENT STRUCTURES

Just as there are many kinds of hard and soft system, for example, football matches, trains, boats, planes and stock markets, there are also many kinds of organisational and management structures. Within these structures there are also formal and informal groups.

Since there are so many management styles and structures, it is necessary to ask whether the same risk reduction or safety management rules or problems apply. To become enmeshed in organisation structure issues raises some very fundamental questions about, for example, how safety might be viewed. Should safety be an objective just as producing widgets at a profit is an objective? Alternatively, should safety be regarded as an essential attibute permeating the whole philosophy of the organisation; that is, should there be a safety culture? To answer this is beyond our remit but one suspects that the concept of safety being an objective suggests that it could always be

said that safety is someone elses responsibility; this is not
desirable.

The study of organisations as open, dynamic, complex systems is
made difficult because there are not yet good taxonomies of all the
factors that relate to the organisation and its environments; concepts
such as organisation structure, environment, organisational process,
for example. An organisation has constant interaction with its
multiple dynamic environments, has multiple purposes or functions, and
consists of many dynamically interacting subsystems. The problem of
risk reduction and safety represents just one more factor into a
complex decision-making environment. The problems for organisational
leaders are essentially how to weight safety in decision-making
processes and how to ensure its presence in the complexities of an
organisation with its internal environment.

We begin with the premise that there are some general concepts
relating to the management of risk and safety which must prevail
irrespective of structure or style and irrespective of the type of
system. In order to draw together these concepts, it is necessary to
find a suitable framework or model to use in the analysis of those
disaster case studies which will be used to illustrate the issues
raised.

Although disaster analyses have been carried out before, the
analyses have been for different purposes such as emergency planning
or human factors issues. We are interested in issues affecting
corporate management and thus there is a need for a different
perspective.

SAFETY MANAGEMENT - PLAYING WORD GAMES

The vehicle for disaster case study analysis was originally envisaged
as a structured questionnaire.

There are already safety management models and questionnaires
used for safety audits such as, for example, the International Safety
Rating System (3). Various safety management documents were read and

the main concepts and key words used in those documents were identified. An abbreviated list of keywords is given in Table 1. As a result of some word games it was seen that most of the concepts could be grouped under six major headings: structure, resources, environment, communication, job, control.

Retrospectively, a comparison with management theory and organisational psychology showed that these key issues are central to all organisational management, operations and decision-making.

For example, Galbraith's (4) model considers the primary problem of an organisation as its relationship with the environment and the acquiring and utilization of information. The design of the organisation structure concerns decisions about division of labour, coordination of units and integration of individuals within the organisation (job concept). Galbraith postulates that organisations have limited capacities to process information and adopt different organisational modes or structures to deal with task uncertainty.

Kotter (5) also pulled together basic elements for analysing short-run organisational dynamics which included organisational processes such as information gathering, communication, decision making; formal organisational arrangements such as structure and operating systems; external environments such as task environment and wider environment; the social system incorporating social structure and culture; employees and other tangible assets, which we refer to as resources.

Having identified some fundamental concepts it was then necessary to consider how to link the concepts and frame a coherent set of questions relevant to safety management. The key linking elements are control and communications since these relate to decision-making and change.

The questions need to reflect those issues which corporate management would need to address during their decision-making process and ensuring the implementation of decisions.

TABLE 1
Key Words

STRUCTURE	Hierarchy Management Level
RESOURCES	Men, money, material, time
ENVIRONMENT	Politics, society, culture, law, commerce, geography, economics etc.
COMMUNICATION	Report, publicity, information, answer, question, interaction

JOB	Position
	    role
	    authority
	    status
	Requirement
	    accountability
	    responsibility
	Attribute
	    capability
	    skill
	    expertise
	    expectation
	    qualification
	    awareness etc.

CONTROL	Plan
	    Target, boundary, phase, procedure, activity
	Decision-making
	    Definition, assignment, priority
	Observation
	    Monitor, detection
	Change
	    Feedback, addition, diminution, maintain
	Implementation
	    Objection, rejection, acceptance, influence, assumption, imposition, interference, penalty, groupthink, constraint, freedom, protection
	(Technical System)
	    Assurance, maintenance, design, test

Since there is no corporate management decision model, we can only talk of decision-making as a very simple model for change, as shown in Figure 1. Corporate decision-makers monitor the changing environments and the system and resources they have available. They can manipulate the resources including people and the jobs they carry out, and the structure of the organisation to achieve their objectives.

The decision process has controls imposed on it in a variety of ways: both from external and internal sources. Similar controls also apply to the successful implementation of that change. Change implies a new system and potentially new environments and the monitoring process continues.

The structure of the questionnaire is based around this simple model. For example, possible internal controls on decision-making are illustrated in Table 2.

This list can then be converted into a general set of questions which could then be made specific to safety management as illustrated in Table 3.

For example, at Flixborough, a change in the technical system resulted in a disaster. One of the major causes, from a management point of view was severe undermanning and the absence of a properly qualified engineer. The question that needs to be asked is, what was corporate policy on resourcing personnel and was any thought given to the safety implications of not having qualified personnel? Had key posts (jobs) been identified as having major inputs to plant safety and were those posts properly resourced?

The questionnaire we have developed, therefore, is based around the change/decision-making algorithm and pulls together the key concepts discussed in relation to current safety management models. We are currently using the questionnaire to examine case studies.

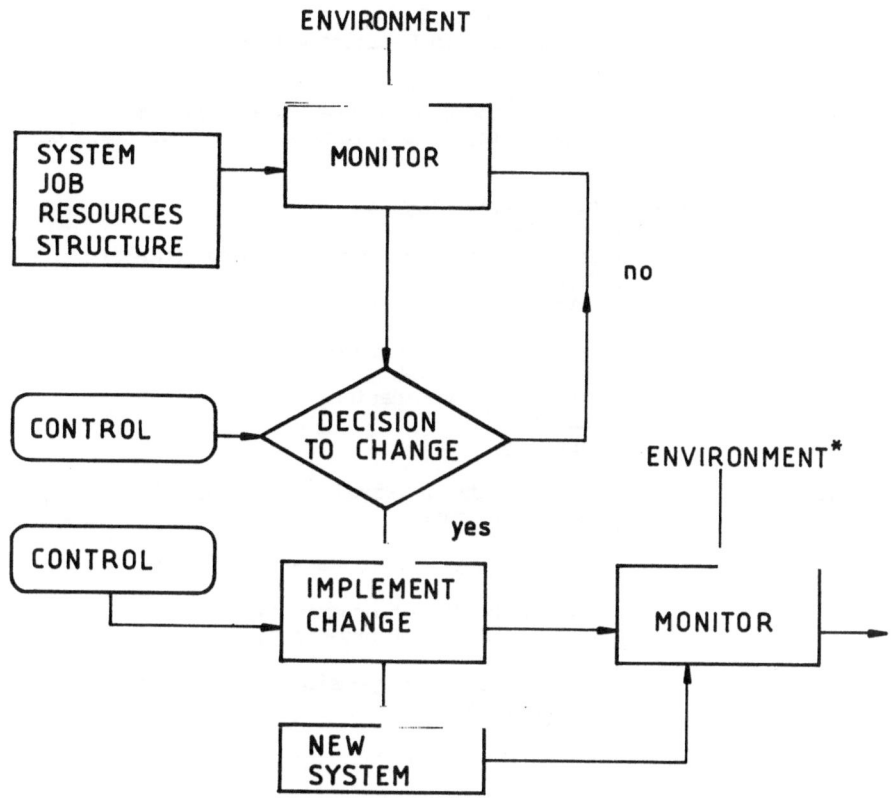

*NOTE THAT THIS ENVIRONMENT MAY BE THE SAME AS BEFORE OR
A NEW ENVIRONMENT

FIGURE 1   A SIMPLE MODEL FOR CHANGE

TABLE 2
Internal Controls on Decision Making

STRUCTURE	Information Flow – Bureaucracy/Freedom Flexibility of Organisation to Change Image of Structure to those within it
PEER GROUP PRESSURE	Personality clashes Politics Expectations (leading to say groupthinks) Dominance Freedom (of speech)
COMMUNICATION	By virtue of structure Resources available Speed of communications Quality Quantity
RESOURCES	Availability of experts/knowledge Managerial time Money Availability of men, materials, time
JOB	Accountability of organisation Responsibility of organisation
PERSONAL	Accountability/responsibility of individual Status/position Self interest Personal Attributes
PERCEIVED OUTCOME OF DECISION	Group Individual Available alternatives

TABLE 3
Questions Relating to Internal Controls on Decision Making

IN RELATION TO SAFETY ISSUES ...

1. How flexible is the organisation?
    Does the hierarchy interfere with the flexibility?
    (constrain, freedom, protect against)

2. How adaptable is the organisation to change? Does, for
    example the bureaucracy mean that any change takes a long
    time. This may deter suggestions from within the system as
    the response (the system's feedback) is not immediately
    forthcoming.

3. Is the Chairperson's expectations expressed and known to the
    decision makers? (Groupthink)

4. Is there a dominating "force" within the decision-making
    process?

5. Do management have sufficient time to reach a proper decision?
    (bad time-planning/need for quick decisions etc).

6. How does the availability of money affect the decision?
    (cutting of corners to meet budget, etc)
    (similarly; men, material, time)

7. Is the organisation accountable for its decisions? How does
    this affect the decision-making process?
    (responsible)

8. Does individual responsibility/authority affect the decision-
    making process?
    (status, position)

9. Does self-interest affect the decision-making?

10. Are there personal differences that affect the decision-making
    process? (political differences) (penalise, constrain,
    freedom, influence)

11. How do the personal attributes of the decision-making body,
    and the members of that body, affect the decision? (capability,
    skill, expertise, qualification, etc..?

12. How does the perceived outcome of the group/individual affect
    the decision-making? (of business, image, environment)

13. How does the available information affect the decision-makers?

14. What, for instance, would the decision-makers do if they
    thought that they did not have enough information, or the
    information was not of suitable quality, or it contradicted
    their initial thoughts and ideas?

A SYSTEMS APPROACH

Although we regarded the development of the questionnaire as successful, we decided to pursue another route which had come to our attention. This second route lead to very similar conclusions but provides a more flexible and formal model.

As mentioned above, organisational psychologists now regard organisations as complex, dynamic, open systems. It is, therefore, sensible to treat the disaster case studies in a similar way. Using a systems approach to analysing the case studies allows us to explore and test for relationships and processes at work in the development of the disaster.

Having collected data about the disaster and defining our committment to the analysis as being 'to identify and understand the role played by management instituting the conditions necessary for the disaster', then we try to identify systems present in the scenario. It is possible to set trial system boundaries thus determining what environmental influences are at work.

The components in the system and the environment are listed and inputs and outputs can be defined. The systems' state variables can be listed and structural relationships between components and between state variables can be given. Finally, having defined all of these issues the significant system failures can be identified.

At this point we can then use the formal systems, control and communications paradigms suggested by Bignell and Fortune (6). The systems paradigm is extremely powerful. It can be made specific to Safety Management Issues as shown in Figure 2. The control and communications paradigms illustrated in Figures 3 and 4 can then be regarded as relating to specific legs of the systems paradigm. For example, the 'provision of resources' by the 'Management of Safety System', which for our purposes refers to Corporate level management, has to be communicated to the 'Working Management System'. The control paradigm would relate closely to the 'performance monitoring' sub-system.

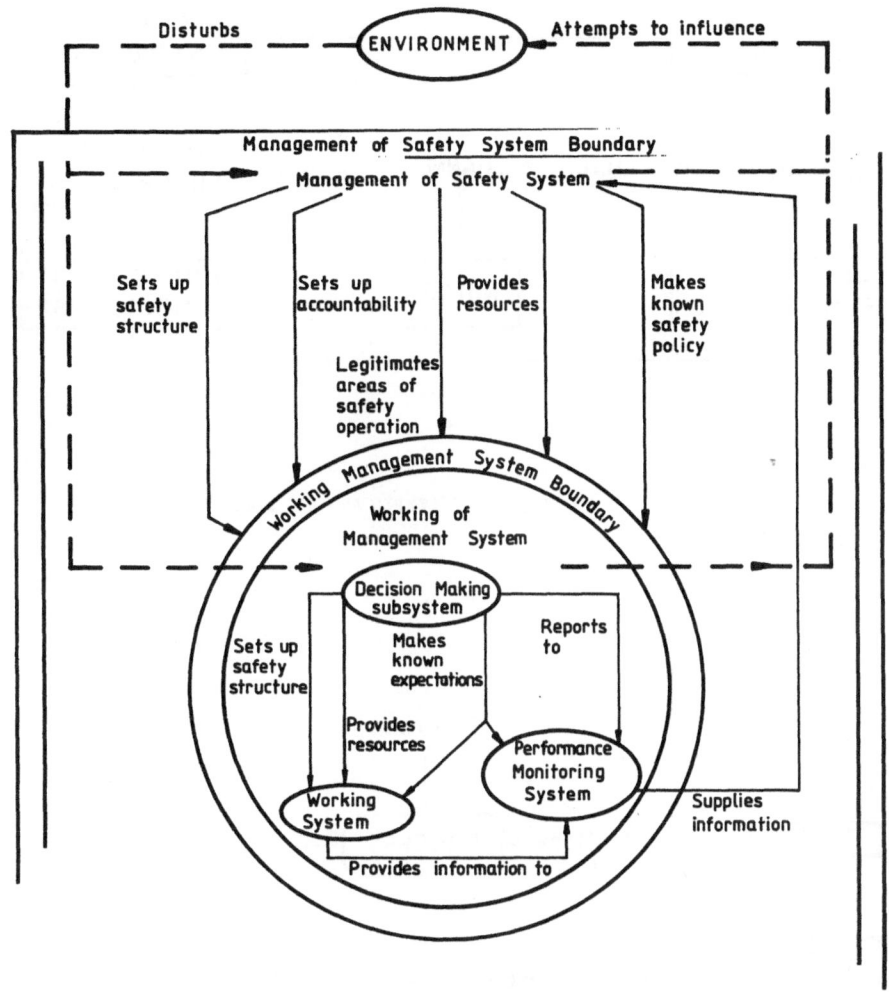

FIGURE 2   FORMAL SYSTEM PARADIGM MADE SPECIFIC TO SAFETY
MANAGEMENT ISSUES

72

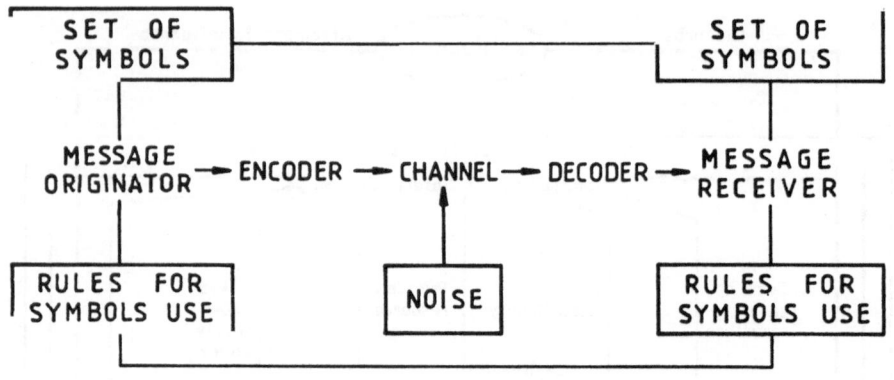

FIGURE 3   COMMUNICATIONS PARADIGM

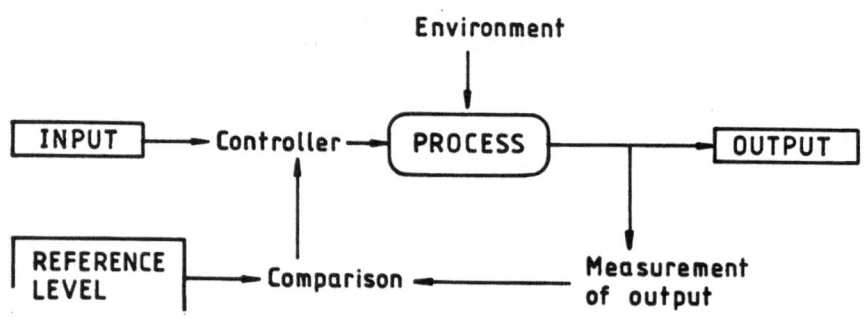

FIGURE 4   CONTROL PARADIGM

The systems failures can be mapped onto the paradigms. For example, in the Flixborough disaster, one of the 'failures' was that the new working system (ie the modified plant) was not tested properly. There were several issues here:

1. British Standards information available in the environment did not affect the system.
2. Management failed to supply the system with correct specifications and methods for safe testing. (Legitimation of area of safe operation).
3. There was no communication between the system and environment to acquire testing instructions.
4. The (British Standards) message had no channel to flow through because of the absence of a qualified mechanical engineer. (Failure to provide adequate resources).
5. A lack of a mechanical engineer meant that (even if the information had got through) there was no 'decoder'.
6. The controller did not have, or did not use personnel with the right qualifications.
7. No information from the environment affected the testing.
8. In the testing no comparison was made to British Standards or any other reference level.

The paradigms, in particular the systems paradigm are thus a vehicle for us to focus on the essential key concepts of structure, control, resources, environment, communication and job in relation to safety management and the concepts can be systematically illustrated by pulling together the information from the disaster-case studies.

As well as discussing issues specific to corporate management, and illustrating them using the case studies, the management guide will draw together an overview for corporate management of the organisational aids and techniques for risk and safety assessment available to executive management.

FUTURE WORK

In order to aid decision making with regard to risk and safety, corporate management need some measurements, notably financial measurements, in order to judge the value and merits of taking a particular course of action. What constitutes a good company in terms of risk and safety and how much does it cost to be a good company? What are the penalties (financial) for failing to be safe or failing to keep risk as low as reasonably practicable in the eyes of the law? The ability to answer to these questions would help in the rational aspects of corporate decision-making. There are still the apparent irrationalities to contend with but that is for the decision theorists to tackle.

ACKNOWLEDGEMENTS

This work is being conducted by SRD on behalf of the Human Factors in Reliability Group which recognised the need for the Management at Risk Guide.

The work of Mr R Overton, SRD, and Mr P Stephens, a student from The University of Wales, is gratefully acknowledged.

REFERENCES

1. Schein E H, Organisational Psychology, Foundations of Modern Psychology Series, Prentice-Hall, Englewood Cliffs, 1980.

2. Turner, B A, Man-made Disasters, Wykeham Publications, London, 1978.

3. International Safety Rating System, International Loss Control Institute, 1986.

4. Galbraith, J, Organisation Design, Reading, Mass., Addison-Wesley, 1978.

5. Kotter, J P, Organisational Dynamics' Diagnosis and Intervention, Reading, Mass., Addison-Wesley, 1978.

6. Bignell, Fortune, et al., Systems Paradigms, Studying Systems Failures, Course T301, The Open University, 1984.

A developed Transient-behaviour Method

Isa S. QAMBER   &   A. Z. KELLER
EE & CS Dept.   Ind. Tech. School
College of Eng., University of BAHRAIN   University of BRADFORD

ABSTRACT

Transient solutions are important when it is required to study behaviour immediately after start-up or repair of a system or the behaviour is required over a fixed period during which the system is required to perform a given mission. There are several methods available to solve the first order differential equations. Some of these are:

  1. Unmodified Euler method
  2. Modified Euler method
  3. Fourth-order Runge-Kutta method
  4. Milne's method
  5. Adams method

The second method advantage over the first one is the error term reduction. But,when looking to the third method it can be found that it is more accurate than the previous two methods.

The Milne's method is a multistep predictor-corrector method. The disadvantage of this method is that it is not a self-starting and also it produce instabilities in numerical evaluation.

The Adams method is also not a self-starting method as the Milne's method, but because of its superior accuracy and stability it was chosen as a comparison method for further study with a developed method to obtain both the steady-state and transient solutions of different state models.One advantage of the developed method is its simplicity to programmed. The solutions are investigated and discussed.The particular investigated methods are applied to a range of calculations and their results sre obtained. A comparison between the different methods with regard to their stability and accuracy of their solutions are performed.

SYSTEMS OF DIFFERENTIAL EQUATIONS

Many engineering problems analysis produces not one but several differentail equations. These equations needs to be solved simultaneously[1]. Most conveniently,this set of first order differential equations can be written in the following form[2]:

$$\frac{dP_1}{dt} = f_1 \ (t, P_1, P_2, \ldots, P_n)$$

$$\frac{dP_2}{dt} = f_2 \ (t, P_1, P_2, \ldots, P_n)$$

$$\vdots$$

$$\frac{dP_n}{dt} = f_n \ (t, P_1, P_2, \ldots, P_n)$$

In many applications such as Markov modelling these equat ons reduce to a linear form:

$$\begin{aligned}
\frac{dP_1(t)}{dt} &= -a_{11}P_1(t) + a_{12}P_2(t) + \ldots\ldots + a_{1n}P_n(t) \\
\frac{dP_2(t)}{dt} &= a_{21}P_1(t) - a_{22}P_2(t) + \ldots\ldots + a_{2n}P_n(t) \\
&\ \ \vdots \qquad\qquad \vdots \qquad\qquad \vdots \qquad\qquad\quad \vdots \\
\frac{dP_n(t)}{dt} &= a_{n1}P_1(t) + a_{n2}P_2(t) + \ldots\ldots - a_{nn}P_n(t)
\end{aligned}$$

Equations (1) can be re-written in the simple matrix form a

$$d\underline{P}/dt = A\ \underline{P}(t)$$

where A is an n by n matrix with all elements independent time and components Pi.

MATRIX MULTIPLICATION METHOD (MM)

In many practical problems [3],a natural unit of time oft suggests itself, where it could be the year.This natural dis etisation of time also often has advantages in numerical eva uations. Transformation from the continuous time to the dis rete time domain can be accomplished as follows:
Considering

$$d\underline{P}(t)/dt = A\ \underline{P}(t) \tag{3}$$

where A is the transition rate matrix, one can approximate

$$d\underline{P}(t)/dt \quad \text{by} \quad (\underline{P}(t+h)-\underline{P}(t))/h \tag{4}$$

where h is a sufficiently small interval of time. This give

$$\begin{aligned}
\underline{P}(t+h) - \underline{P}(t) &= h\ A\ \underline{P}(t) \\
\text{i.e}\quad \underline{P}(t+h) &= \underline{P}(t) + h\ A\ \underline{P}(t) \\
&= (I + h\ A)\ \underline{P}(t) \\
&= T\ \underline{P}(t) \tag{5}
\end{aligned}$$

where T is a transition matrix defined now by:

$$T = I + h\ A \tag{6}$$

and I is the unit matrix.

The algorithm for the present method can be written as follows:

$$\underline{P}_{i+1} = h\ A\ \underline{P}_i'$$

$$P_i' = P_i\ /\ \sum_{i=1}^{N} P_i \tag{7}$$

The second formula ensures that all state probabilities at each step sum to unity and as a result as will be shown subsequently significantly stabilises the solution, but this equation was not included in the program built using the University of Bradford CYBER. This equation is added later on using the University of Bahrain VAX. To stabilise the results the following steps are added to the computer program (Univ. of BAH):

$$n > \left[\frac{|v_1 - v_2|}{\epsilon |1 - m^{(1-p)}|} \right]^{1/(p-1)} \tag{8}$$

$$h_3 = h_1/n$$

where: h_1 is the 1st step-size used to obtain estimate value v_1
h_2 is the 2nd step-size used to obtain estimate value v_2
h_3 is the 3rd step-size used to obtain the true value to be estimated
m is an integer number used to obtain h2, h2=h1/m
p the order of accuracy which is 2 for the developed method
n is a number calculated using equation(8) to obtain h3

Because the arithmetic performed with any computer involves numbers with only a finite number of digits, a rounding error is always associated with any arithmetical calculations. For this reason, the step-size h and the accuracy E cannot be chosen arbitrary small. They must be chosen so that corresponding changes in the values of the components of the state vector Pi are significantely greater than the rounding error[6].

The MM method is programed using the University of Bradford CYBER in FORTRAN 77. Also, the same computer is used for Adams method.

RESULTS & DISCUSSIONS

In the present section, some applications of the transient solution methods, i.e. the stabilisation and Adams methods, are given. The main objective of this section is to compare solution for a certain models using both methods.

The results obtained using both methods will be discussed. The criteria used to compare them is the accuracy, the CPU time and the numerical stability of the solution.

(a) <u>Three-state</u> Models:
Four three-state models studied by BOGGERSTAFF and JACKSON[5] for a system comprising of a number of generators will be considered. The common transition rate matrix for each model can be written in the following form:

$$
\left.
\begin{aligned}
\frac{dP_1(t)}{dt} &= -(a+b)P_1(t) + cP_2(t) + dP_3(t) \\[2mm]
\frac{dP_2(t)}{dt} &= aP_1(t) - (c+e)P_2(t) + fP_3(t) \\[2mm]
\frac{dP_3(t)}{dt} &= bP_1(t) + eP_2(t) - (d+f)P_3(t)
\end{aligned}
\right\}
\qquad (9)
$$

The models are differ by nature of different numerical values for the transition rates(these are given in Table 1). The data for these models are based on data [4] relevant to plant operated by Tennessee Valley Authority (TVA). The system considered is in one of the following states (Fig.1):

i - S1: System operates at full capacity
ii- S2: System operates at less than full capacity due to generator outages
iii- S3: because of forced outages no power at all is generated.

Both methods were used to calculate the state probabilities variation with time for each model and compared with the results of previous study obtained by Biggerstaff and Jackson [5] for time t= 6 hours and t=24 hours which are reproduced in Table (2). The results for the two methods are summarised in Tables (3 and 4), where for convenience the % error in values obtained by the methods and the original TVA results of [5] are given by P_{up} , P_{down} and $P_{derated}$ respectively.
It can be seen from Tables 3 and 4 that the Adams method has a very small difference than the TVA results, while the difference for the same models using the MM method is increased as the time interval h increases.

(b) <u>Seven-State</u> Model:
The following example is taken from a paper written by Barlow and Hunter[7]. The example considered is a system consisting of two transmitters A and B and power supply C providing a high voltage with a modulator and modulation programmer. The two transmitters are run from the same power supply (C).
If one of the two transmitters fails, the other one can perform the system task satisfactorily but with a decrease in exp-

ected time.
The system is considered as being in one of seven possible states; these are as follows:
S1: all three units are working properly
S2: B and C are working properly, but A fails
S3: A and B are working properly, but C fails
S4: A and C are working properly, but B fails
S5: B is working properly, but A and C are failed
S6: C is working properly, but A and B are failed
S7: A is working properly, but B and C are failed

The four states S3, S5, S6 and S7 represent the system failures when the system shuts down, requiring repair. In case of the power supply unit failure, the system will fail as a whole. Similarly, if both transmitters fail, the system serves no purpose and operation is discontinued until repaired. For above two reasons, the state represented by failures of all three units can be ignored (the state space diagram is given in Fig.2). The system equations are:

$$
\begin{aligned}
\frac{dP_1(t)}{dt} &= -0.0283P_1(t) + 0.5P_2(t) + 0.2P_3(t) + 0.5P_4(t) \\
\frac{dP_2(t)}{dt} &= 0.0133P_1(t) - 0.525P_2(t) + 0.2P_5(t) + 0.5P_6(t) \\
\frac{dP_3(t)}{dt} &= 0.005P_1(t) - 0.2P_3(t) \\
\frac{dP_4(t)}{dt} &= 0.01P_1(t) - 0.525P_4(t) + 0.5P_6(t) + 0.2P_7(t) \\
\frac{dP_5(t)}{dt} &= 0.005P_2(t) - 0.2P_5(t) \\
\frac{dP_6(t)}{dt} &= 0.02P_2(t) + 0.02P_4(t) - P_6(t) \\
\frac{dP_7(t)}{dt} &= 0.005P_4(t) - 0.2P_7(t)
\end{aligned}
\right\} \quad (10)
$$

Both methods were applied to the seven state system. Reference calculations in this case are taken from [7] and reproduced in Table 5(for the first four states only). The results obtained applying the MM method is given in Table 6. Using the Adams, it is found that the percentage errors are zeros.
It can be seen from the results of the 7-state model that the Adams method produced virtually identical results for the model, surprisingly there are small differences in the results of the four earlier three-state models.

CPU Time:
Despite the relative inaccuracy of the Matrix Mult. method it does have advantages of simplicity of use and a high stability unless the time interval exceeds a certain critical value. From the study, it is clear that the time interval should not exceeds 54.9, 18.9, 49, 19.4 and 2 hours for the 3-state models and 7-state model, respectively when the developed method is used without using the stabilisation step.

To study the effect of the time interval h on CPU time, the two methods were applied to four state and sixteen state models.

All units are in arbitrary units; a time interval t = 0 - 5 units was taken for the two models. A range of runs were made for h. The number of steps M is accordingly varying by M= 5/h. The results for the two models are given in Figs.3 and 4. From the two figures, it is seen that the relationship between the CPU time and M is an approximately linear for the two methods. The two methods were also used to investigate the relationship between the CPU time and the number of states in the model(the results are given in Figs. 5 and 6. It is seen that the curve obtained by the developed method for all evaluations appear to pass through or near the origin. Also, as expected the CPU time increases nonlinearly as the number of states increase.

CONCLUSION

It can be conclude out of the study that Adams method was stable for all values of h. The minor disadvantages of Adams method that it is relatively complex to program and it requires a subroutine to generate starter values. While the MM method has a number of advantages. One of its advantages is its simplicity of formulation. Also,it is easy to program and incorporate as a subroutine in other programs. The principal disadvantage is that : if a step length h is chosen significately greater than the reciprocal of the smallest element in the transition rate matrix, numerical instability of the solution can occur.

REFERENCES

1. Morris, W.D,"Differential equations for Engineers and applied Scientists", McGraw-Hill, UK, 1974.

2. Campbel, S.L,"An Introduction to differential equations and their Applications", Longman Inc., 1986.

3. Keller, A.Z,"Markov chains and Monte Carlo Simulation Methods", Lecture Note No. 4, School of Industrial Technology, University of Bradford.

4. Billinton, R & Jain, A.V,"Unit derating in spinning reserve studies",IEEE Trans. on PAS, Vol.90,No.4,pp. 1677-1687, 1971.

5. Biggerstaff, B.E & Jackson, T.M,"The Markov process as a Means of determining generating-unit state probabilities for use in spinning reserve applications", IEEE Trans. on PAS, Vol.88, No.4, pp. 423-430, 1969.

6. Gerald, C.F & Wheatley, P.O,"Applied Numerical Analysis",Addison Wesley, 1984.

7. Barlow & Hunter, "System Efficiency and Reliability", IRE National Convention Proceedings, Vol. 7, No. 6, 1959, pp. 104-109.

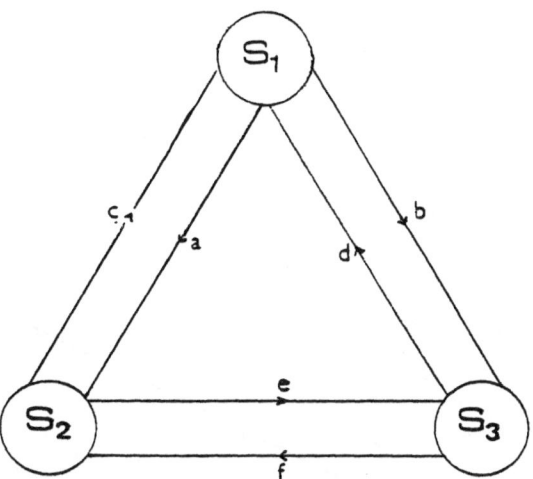

Fig.(1) Three-state models

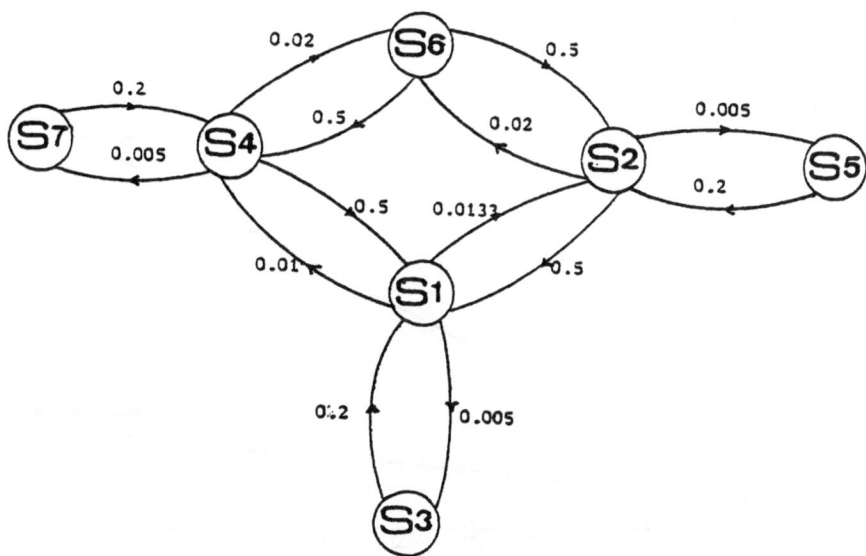

Fig. (2)   State-space diagram for three-unit system

82

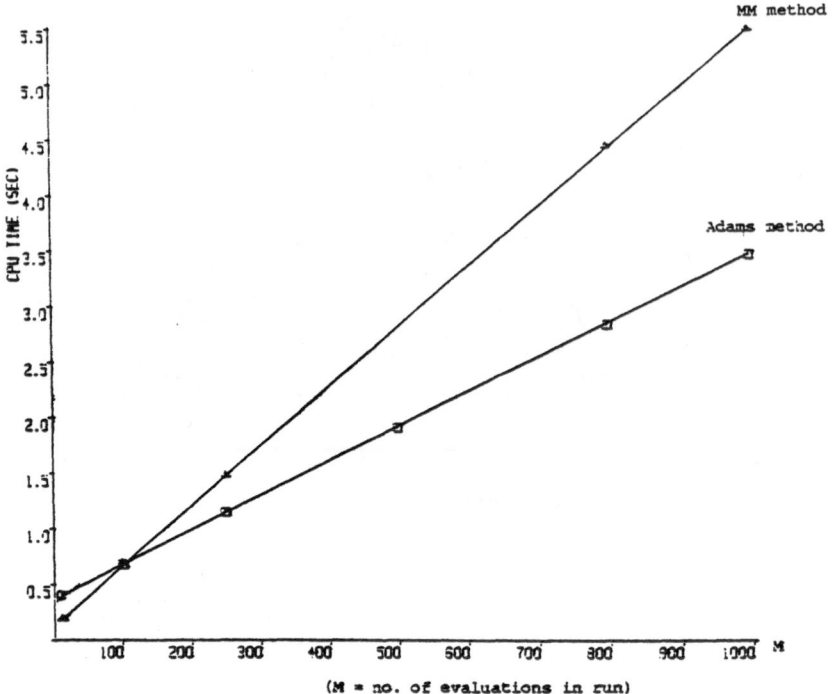

(M = no. of evaluations in run)

Fig. (3)   Variation of the CPU time with M for case (1)

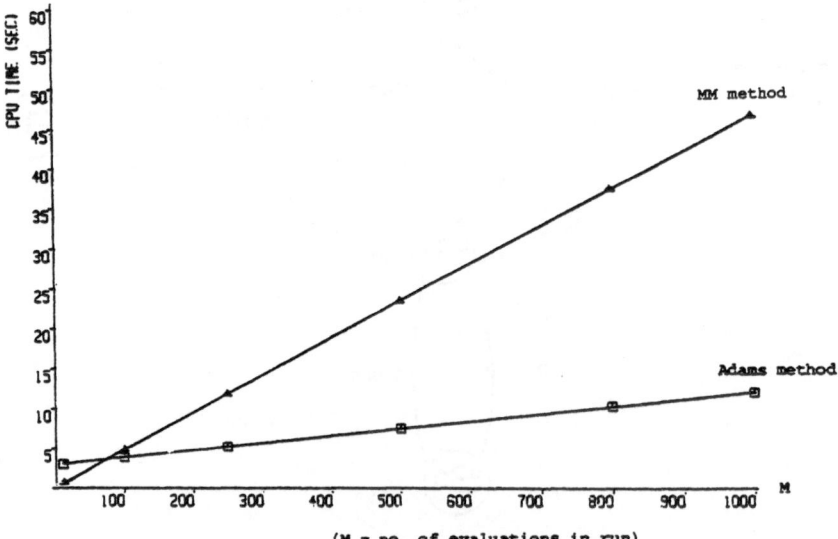

(M = no. of evaluations in run)

Fig. (4)   Variation of the CPU time with M for case (2)

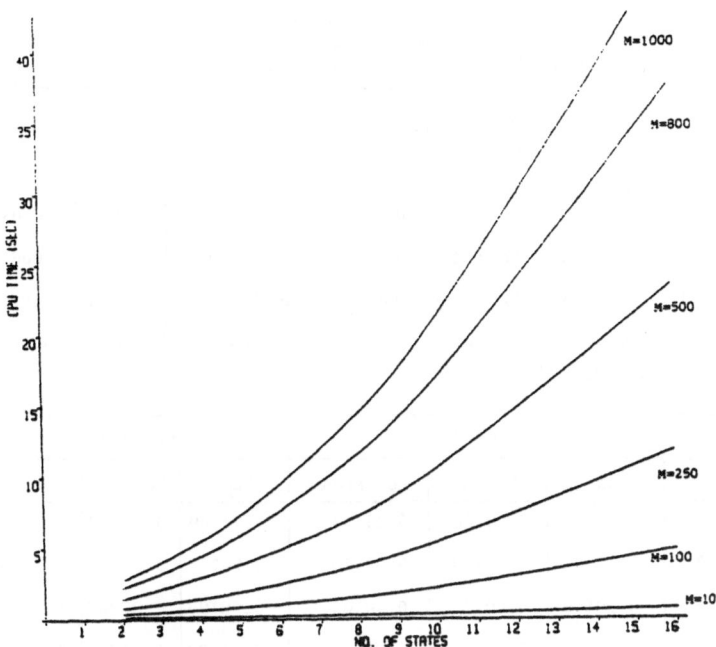

Fig. 5   Variation of the CPU time with N
(using MM method)

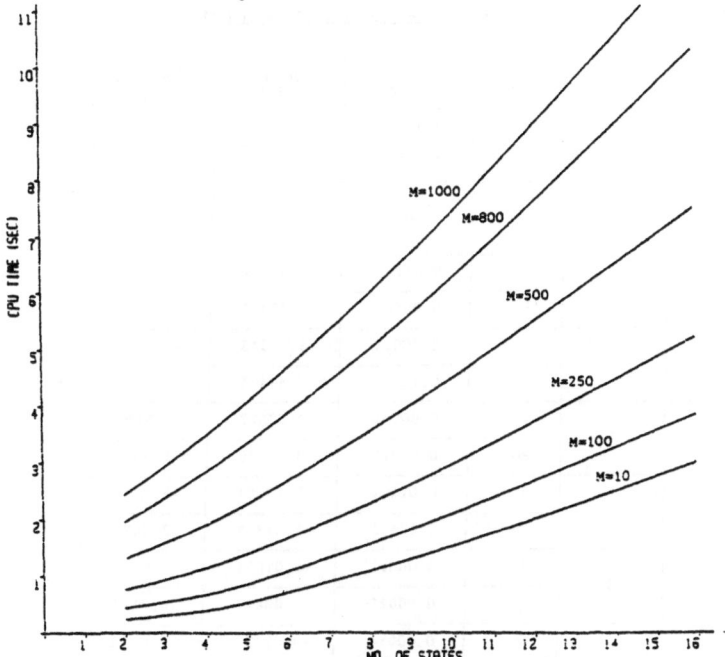

Fig. 6   Variation of the CPU time with N (using Adams method)

Table 1   TVA System Configuration

Model	Transition Rates (1/hr)					
	a	b	c	d	e	f
1	0.0003	0.0010	0.0225	0.0350	0.0008	0.0004
2	0.0006	0.0050	0.0400	0.1000	0.0004	0.0004
3	0.0005	0.0002	0.0240	0.0430	0.0001	0.0001
4	0.0010	0.0006	0.0200	0.1000	0.0002	0.0020

Table 2   TVA Results

Model	$t=6$ hours			$t=24$ hours		
	P_{up}	P_{down}	$P_{derated}$	P_{up}	P_{down}	$P_{derated}$
1	0.99293	0.00168	0.00539	0.97849	0.00552	0.01599
2	0.97462	0.00318	0.02220	0.94756	0.00906	0.04338
3	0.99616	0.00279	0.00106	0.98798	0.00905	0.00298
4	0.99168	0.00564	0.00268	0.97582	0.01888	0.00530

Table 3   MM method (Results for Models 1, 2, 3 and 4)

Model	h (hour)	t (hour)	P_{up} (% error)	P_{down} (% error)	$P_{derated}$ (% error)
1	1	6	0.01111	-1.19048	-1.66976
		24	0.02351	-0.72464	-1.18824
	2	6	0.02216	-2.38095	-3.33952
		24	0.04803	-1.63043	-2.37649
2	1	6	0.09747	-1.88679	-4.00901
		24	0.06543	-1.21413	-1.15261
	2	6	0.20623	-3.77358	-8.51351
		24	0.12981	-2.53863	-2.30521
3	1	6	0.00602	-1.07527	-1.88679
		24	0.01316	-0.88398	-1.00671
	2	6	0.01104	-2.15054	-3.77358
		24	0.02530	-1.87845	-2.34899
4	1	6	0.00017	-0.01064	-0.04104
		24	0.00023	-0.00847	-0.01132
	2	6	0.00034	-0.02128	-0.08582
		24	0.00045	-0.01695	-0.02453

Table 4   Adams Method Results for Models 1, 2, 3 and 4)

Model	t (hour)	P_{up} (% error)	P_{down} (% error)	$P_{derated}$ (% error)
1	6	-0.00002	0.00000	0.00371
	24	-0.00001	0.00181	0.00063
2	6	-0.00002	0.00000	0.00045
	24	-0.00001	0.00000	0.00023
3	6	-0.00002	0.00717	0.00943
	24	0.00000	0.00110	0.00000
4	6	-0.00001	0.00000	0.00000
	24	0.00000	0.00000	0.00000

(Note: Values for h = 1 hour and h = 2 hours are identical)

Table 5 State-Probabilities for the First 4 States (7-State Model)

t (hours)	$P_1(t)$	$P_2(t)$	$P_3(t)$	$P_4(t)$
0	1.000	0.000	0.000	0.000
1	0.977	0.010	0.004	0.008
2	0.963	0.016	0.008	0.012
4	0.948	0.022	0.013	0.016
8	0.937	0.024	0.019	0.018
16	0.932	0.025	0.022	0.019

(Note: Results taken from ref. [79])

Table 6   "S." Method (Results for 7-State Model)

t (hour)	P_1 (% error)		P_2 (% error)		P_3 (% error)		P_4 (% error)	
	h=0.5	h=1.0	h=0.5	h=1.0	h=0.5	h=1.0	h=0.5	h=1.0
1	0.002	0.005	-0.100	-0.300	-0.250	-0.250	-0.125	-0.250
2	0.003	0.006	-0.125	-0.188	0.000	-0.125	-0.033	-0.167
4	0.003	0.005	-0.145	-0.045	-0.077	-0.077	-0.063	-0.125
8	0.001	0.002	-0.042	-0.042	0.000	-0.053	-0.056	-0.056
16	0.000	0.000	0.000	0.000	-0.045	-0.045	0.000	0.000

RELIABILITY GROWTH MODELS AND APPLICATION TO A PROCESSOR

Dr U D PERERA, IBM (UK) LIMITED, HAVANT PLANT, P O BOX 6,
HAVANT, HAMPSHIRE, PO9 1SA, ENGLAND

SUMMARY

This paper describes reliability growth models and application
to a medium processor. The field reliability of the processor
was monitored by collating field failure and related data from
a sample (180) of products. The field reliability data of the
processor was monitored for approximately two years, in order
to assess its field performance. During this monitoring period
the reliability of the product improved significantly.

The objective of this paper is to review and assess the
validity of a number of reliability growth models for the
processor failure data. The applied growth models include
Duane, Weibull process, IBM, Aroef, Lloyd-Lipow and Arinc.
The reliability growth phenomena is reviewed and the validity
of different reliability growth models for field reliability
prediction is assessed.

INTRODUCTION

In the development of complex electro-mechanical systems, the
early machines manufactured are likely to contain various
design and engineering deficiencies. To evaluate performance,
reliability and latent early-life problems so that corrective
actions can be taken, these prototypes are typically
subjected to a development and manufacturing test or a 'Test
Analyse And Fix' program. During this program, product
deficiencies are identified and corrective actions are
introduced and as a result the reliability of the product
improves or the product is said to be subject to reliability
growth.

The reliability growth phenomenon is not limited to the reliability demonstration phase, but can be observed in the field, and can be due to similar reasons.

PRODUCT DESCRIPTION

The product considered in this study is one of a family of data processing systems designed to operate in a cooperative network. This processor offers the customer a variety of alternative functions, applications and costs through its systems approach to distributed data processing. These processors can be installed in a variety of ways and can be used for many business and industrial applications. The system's computer can be connected to host computers, others of the same family or run as stand-alone processors.

The processor provides controls, storage, diskette storage, communications and input/output capabilities for the family of systems. The processor storage can be up to 8.2 MBytes, and basic storage can be expanded by adding disk storage, magnetic tape units, communications and input/output capabilities. Communications and input/output adapters attach multiple printers, display terminals, and card readers, either directly or by data links.

The processor also incorporates hardware, programming and network error detection and monitoring facilities aimed to improve its availability and problem identification in the event of a failure.

PROCESSOR FIELD DATA

Field data provides the final performance measure for any product. Hence, it is normal practice to collate and monitor the field performance of products in order to compare with expectations and to take early actions against unexpected failures.

In order to assess the field performance, the field failure data of a sample (180) of processors was monitored and collated. The collated information included, date of installation, calendar time to failure, area and type of failure, repair action and repair time. The monitored period was up to 110 weeks. However, as the minimum monitored period of the sample was 90 weeks, in all models but the Weibull process model, the sample data was limited to 90 weeks.

88

During this period the reliability of the processor was
improving. In this paper the collated data is modelled to
evaluate the effectiveness of reliability growth models and
the processor repairability.

The distribution of the observed processor field failures
is shown in figure 1. This shows all the hard failures
observed during the monitored period. As would be expected,
the majority (59%) of field failures is related to logic.
This is followed by cables/connectors (22%). Failures
related to external sources, interconnected equipment are not
significant.

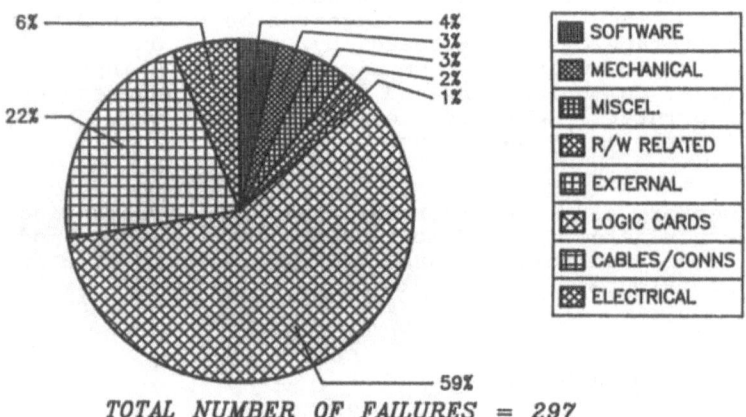

FIG 1 DISTRIBUTION OF PROCESSOR FIELD FAILURES

RELIABILITY GROWTH

Reliability growth or reliability improvement with time is a
common phenomenon observed for complex repairable electro-
mechanical systems during development and early-life in the
field. This phenomenon can be due to a single or a
combination of aspects such as:

Natural Screening

It is common for most system components, in particular, electronic components to have a small proportion of weak or early-life populations. These would fail early during the field usage and most likely be replaced by good components. This in turn introduces reliability growth.

Corrective Actions

The field performance of most systems is monitored and in order to prevent recurring or undesired failures often design modifications or preventative maintenance procedures are introduced through failure analysis. The effort put in to this process is a function of the deviations in the failure rate, repair costs and safety levels compared to expectations.

Maintenance Personnel and Operator Learning

Often the maintenance of complex systems demands some degree of familiarity with the system and system diagnostics. The acquisition of maintenance and operator skills follow the familiar learning curve. This is usually a function of familiarity with similar systems, training and the effectiveness of operating manuals and procedures.

The failure times of a system is stated to follow a homogeneous process (with a constant intensity f) when during a small interval of time (t), the probability of an event occurring (f t) is independent of how long the system has been operated prior to the beginning of the time interval. In this situation, the failure rate is constant and the time between failures is exponentially distributed. For a system with failure times which follow a non-homogeneous process, the intensity function is not constant, but varies with time. This process is applicable to a system showing reliability growth.

RELIABILITY GROWTH MODELS

The earliest and most widely used reliability growth model proposed by Duane (1). Since, a number of alternative models have been developed. These are based on different assumptions and it is likely that for a given data set a certain model will prove more suitable than the others.

In this paper the Duane, Weibull process, IBM, Lloyd-Lipow, Aroef and Arinc models are outlined and the application to the processor field data is presented.

CRITERIA FOR THE GOODNESS OF FIT OF THE MODELS

There are many mathematical models which can be used to describe reliability growth. One approach to fitting growth curves is to apply a large number of models and select the one that fits BEST. This approach is grossly unsatisfactory, as the criteria of 'best' may be somewhat arbitrary. This is due to the fact that, for any data set, an arbitrary good fit can be obtained by selecting a mathematical function with many parameters. Thus, for even a 'misbehaved' data set, it is possible to find a mathematical model which will fit it. In essence this amounts to passing all decisions solely in the data, ignoring any other physical-engineering information. Moreover, the problem of interpreting the physical meaning of more than three parameters is difficult.

The usual methods for judging the goodness-of-fit of a model are the application of the 'F and 't' tests and the evaluation of the coefficients of multiple determination. These are based on the assumption that the data is multivariate normally distributed, which is not always applicable. Thus two measures of goodness -of-fit were used. Details of these can be found in reference {2}.

However, in the paper t-test statistics for each of the model parameters are given. The overall goodness-of-fit of the models are considered in a later section.

DUANE MODEL

Duane modelled the reliability data of several systems in order to find some consistent pattern for reliability improvements during the development phase. He observed a linear relationship between cumulative failure rate and the cumulative test time, when plotted on a log-log scale. Applications of this model are discussed in references (3,4).

Expressed mathematically, the Duane model is:

$$f(0,t) = Kt^{-a} \quad . \quad . \quad . \quad . \quad (1)$$

where $\quad f(0,t)$ = Cumulative failure rate
$\qquad t$ = Total cumulative test time
$\qquad K$ = A constant
$\qquad a$ = A measure of reliability growth or growth rate

The Duane model parameters for the processor data are given
in table 1. Parameters were estimated using the least square
method, based on all the data (without grouping).

MODEL COEFFs	ESTIMATE	STD ERROR	T-TEST	SIG LEVEL
K	12.229	0.10548	115.94	0
a	-0.356	1110215	-165.52	0

TABLE 1 - DUANE MODEL PARAMETERS AND T TEST
STATISTICS

Plot of observed and predicted cumulative MTBF curves is
shown in figure 2.

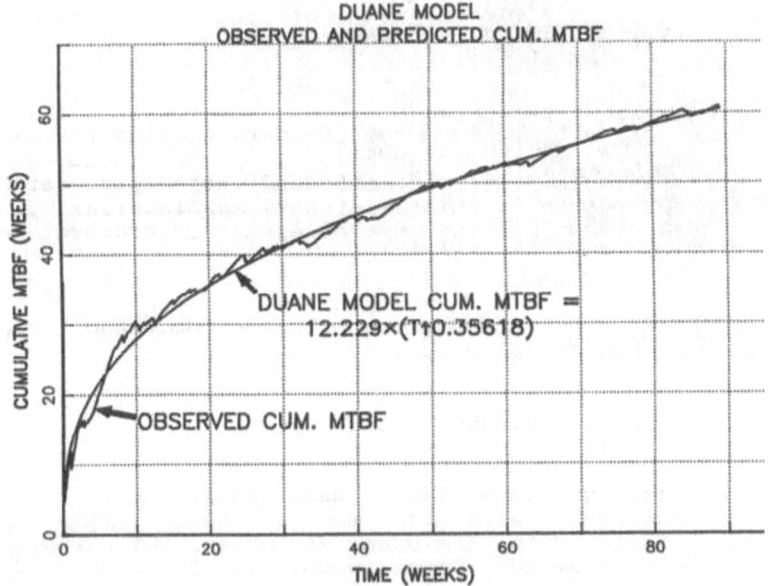

FIG 2- PLOT OF OBSERVED AND PREDICTED CUMULATIVE
MTBFs BASED ON THE DUANE MODEL

Observations are:
- The processor reliability improvement is
significant (Duane slope - 0.356).

- Based on t-test, a significantly good
correlation to the model can be observed.

- The cumulative MTBF at week 90 is in the
order of 61 weeks.

WEIBULL PROCESS MODEL

This model was proposed by Crow (5). He observed that reliability growth is equivalent to a non-homogeneous poisson process with a Weibull intensity function. He expressed the Duane model outlined above with a Weibull intensity function of the form:

$$f(t) = ab\,t^{b-1}\exp(-at^b)\quad .\quad .\quad (2)$$

where b = shape parameter
a = scale parameter

The instantaneous failure rate R(t) is given by:

$$R(t) = \frac{f(t)}{1 - F(t)} = ab\,t^{b-1}\quad .\quad .\quad .\quad .\quad (3)$$

The method of estimating model parameters using the maximum likelihood method, which was applied in this study can be found in reference {6}. The method of estimating confidence levels are also given in this reference. Applications of this model to a communications controller and a processor can be found in references {7} and (11).

The estimated model parameters based on the maximum likelihood method are given below:

Shape - 0.6656
Scale - 13.206

The Chi-square goodness-of-fit test statistics-1.89, a high level of agreement with the model. Also these model parameters agree with the Duane estimates, for example, the shape parameter from the Duane model is 1-0.356 = 0.644.

The plot of instantaneous failure rate against time in field is shown in figure 3. Also shown are the actual MTBFs for grouped data.

93

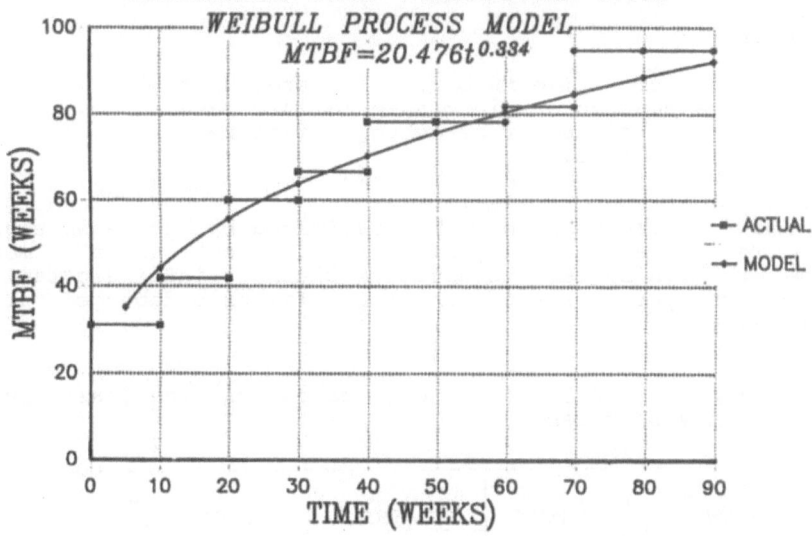

FIG 3 OBSERVED AND PREDICTED INSTANTANEOUS MTBFs BASED
ON THE WEIBULL PROCESS MODEL

IBM MODEL

This model {8} is based on the following
assumptions:

(a) Two types of failures are possible, inherent
failures occurring at constant failure rate f,
and non random failures due to design or
manufacturing defects.

(b) At the beginning of testing the number of non-
random failures is unknown but is a fixed amount.

(c) If N(t) is the number of non-random failures
remaining at a time t, the rate of change is
proportional to N(t) at t.

That is,

$$d\ N(t)/dt\ =\ -\ B\ N(t)\ .\ .\ .\ .\ (4)$$

$$\text{hence}\quad N(t)\ =\ A\ e^{-Bt}\ .\ .\ .\ .\ (5)$$

Defining V(t) as the cumulative number of failures
up to time t, then

$$V(t) = ft + A \{1 - \exp(-Bt)\} \ . \quad . \quad . \quad (6)$$

where,

V(t) = Cumulative number of failures up to t
   f = constant or intrinsic failure rate
   A = total number of debugged failures
   B = a constant

The model parameters can be estimated by the least square method.This model is of particular interest as the model parameters can be used to estimate ultimate intrinsic failure rate and the percentage of non-random failures debugged at a given time.

The estimated model parameters based on the least square method is given in table 2.

MODEL COEFFs	ESTIMATE	STD ERROR	T-TEST	SIG LEVEL
f	113.09	1.6643	67.49	0
A	73.91	2.1472	34.42	0
B	4.31	0.2117	20.35	0

TABLE 2 - IBM MODEL PARAMETERS AND T TEST
STATISTICS

From above, the following inference can be made:

- The processor intrinsic failure rate is approximately 0.62 failures per year.

- Total non-random failures in the sample is 74

- 30% of non-random failures occur in the first month.

Plot of observed and predicted cumulative number of failures is shown in figure 4. A good agreement between the observed and predicted can be seen. An application of this model can be found in reference (7).

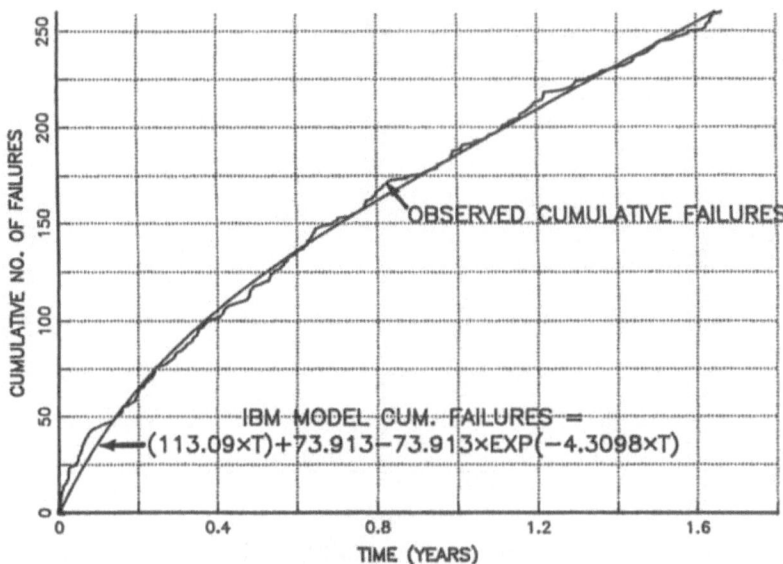

IBM MODEL — PROCESSOR
OBSERVED AND PREDICTED CUMULATIVE FAILURES

OBSERVED CUMULATIVE FAILURES

IBM MODEL CUM. FAILURES =
(113.09×T)+73.913−73.913×EXP(−4.3098×T)

FIG 4 OBSERVED AND PREDICTED CUMULATIVE FAILURES BASED ON
THE IBM MODEL

ARINC MODEL

Details of this model can be found in reference {10}. The estimated model parameters based on the least square method is given in table 3. Estimation is based on an initial MTBF of 5 weeks. Based on the t-test, the goodness-of-fit of the model is not significant.

MODEL COEFFs	ESTIMATE	STD ERROR	T-TEST	SIG LEVEL
A	4347.6	76873	0.0550	0.958
B	0.000446	0.0081	0.0550	0.957
C	0.39048	0.0111	35.171	3.52E-8

TABLE 3 - ARINC MODEL PARAMETERS AND T TEST
STATISTICS

The observed and the predicted cumulative MTBFs based on the Arinc model are shown in figure 5. In spite of the low goodness of fit, a good agreement between the observed and predicted MTBFs can be seen.

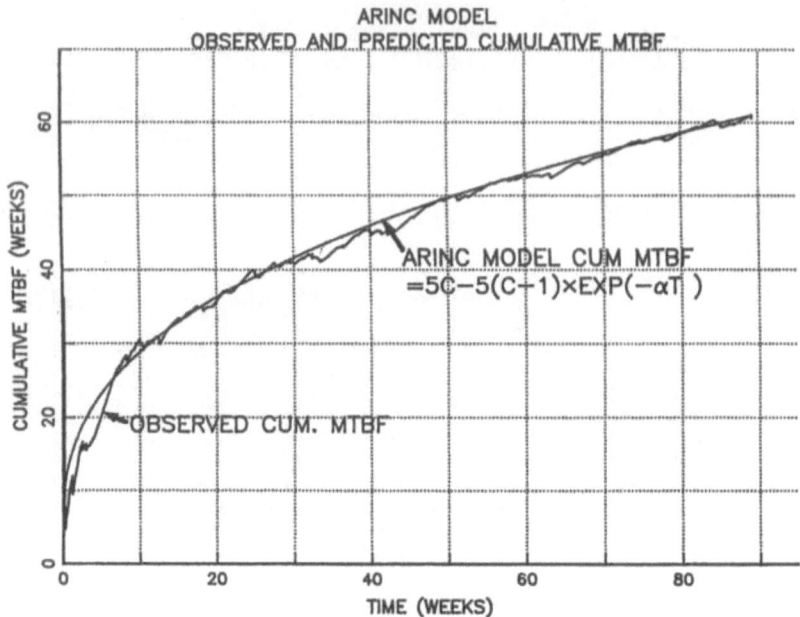

ARINC MODEL
OBSERVED AND PREDICTED CUMULATIVE MTBF

FIG 5 OBSERVED AND PREDICTED CUMULATIVE FAILURES BASED
ON THE ARINC MODEL

AROEF MODEL

Aroef {2} assumes that the growth rate is jointly
proportional to the growth achieved at t, i.e. Y(t), a
constant multiplier (growth rate parameter) B and inversely
proportional to t squared. That is;

$$dY(t)/dt = B\ Y(t)/t^2 \quad . \quad . \quad . \quad . \quad . \quad (7)$$

This differential equation has the solution

$$Y(t) = A \exp(-B/t) \quad . \quad . \quad . \quad . \quad (8)$$

Since $\lim_{t \to \infty} Y(t) = A$

Reliability growth limit in cumulative MTBF is A

Also $\lim_{t \to 0} Y(t) = 0$

Since $\ln Y(t) = \ln A - B/t \quad . \quad . \quad . \quad . \quad (9)$

The least square method can be used to estimate the model parameters A and B.

The model parameters based on the least square method are given in table 4. Estimation is based on data grouped in ten weekly periods. The ultimate achievable cumulative MTBF is approximately 64 weeks. Also the t-test appears to be satisfactory.

MODEL COEFFs	ESTIMATE	STD ERROR	T-TEST	SIG LEVEL
A	64.56	2.6248	24.59	4.67E-8
B	10.00	1.7936	5.57	8.35E-4

TABLE 4 - AROEF MODEL PARAMETERS AND T TEST STATISTICS

The observed and predicted cumulative MTBFs based on the Aroef model are shown in figure 6. The deviation between the observed and predicted MTBF curves can be seen.

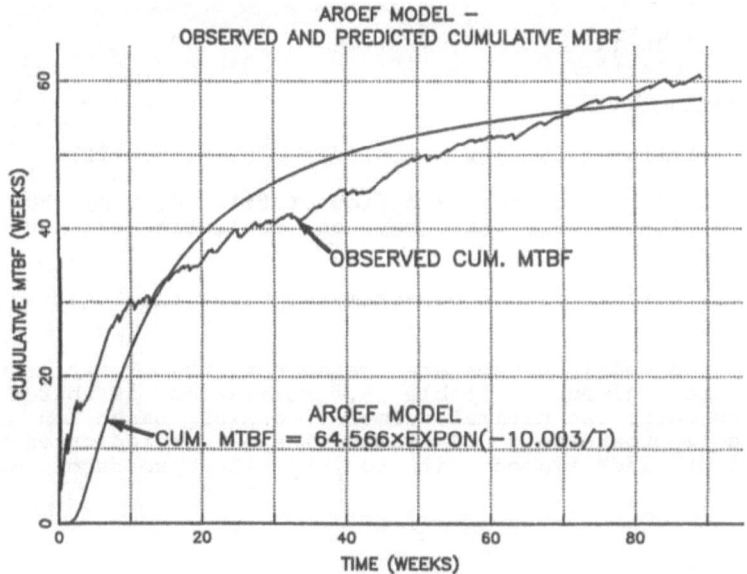

FIG 6 OBSERVED AND PREDICTED CUMULATIVE FAILURES BASED ON THE AROEF MODEL

LLOYD-LIPOW MODEL

This model was proposed by Lloyd and Lipow {9}, and supposes that the growth rate is inversely proportional to the square of cumulative time.
That is:

$$dY(t)/dt = B/t^2, \quad B > 0 \quad . \quad . \quad . \quad . \quad (10)$$

Then clearly,

$$Y(t) = A - B/t \quad . \quad . \quad . \quad . \quad . \quad (11)$$

where A is constant of integration, but it should be noted that

$$\lim_{t \to \infty} Y(t) = A$$

Thus A is the limiting value of cumulative MTBF.

The parameter B is a growth rate parameter which also affects the location of the curve. Since Y(t) cannot be negative and

$$\lim_{t \to 0} Y(t) \to -\infty$$

We must define

$$Y(t) = 0, \quad 0 < t < B/A \quad . \quad . \quad . \quad (12)$$

This definition provides a period of time (0, B/A) when the cumulative MTBF is 0. This may be realistic for certain systems.

The model parameters can be estimated using the least square method.

The estimated model parameters based on the least square method are given in table 5. Estimation is based on continuous data. The ultimate cumulative MTBF based on this model is approximately 64 weeks. The t-test indicates that the model provides a good fit to the the processor data.

MODEL COEFFs	ESTIMATE	STD ERROR	T-TEST	SIG LEVEL
A	61.18	0.840	72.80	0
B	366.41	5.655	64.78	0

TABLE 5- LLOYD-LIPOW MODEL PARAMETERS AND T TEST
STATISTICS

A plot of the observed and predicted cumulative MTBFs based
on the Lloyd-Lipow model is shown in figure 7. In spite of
the good agreement with the t-test, a marked deviation
between the observed and predicted cumulative MTBFs can be
seen. To a certain degree, this is likely due to the
assumption that the cumulative MTBF up to 5.98 weeks (B/A)
is zero.

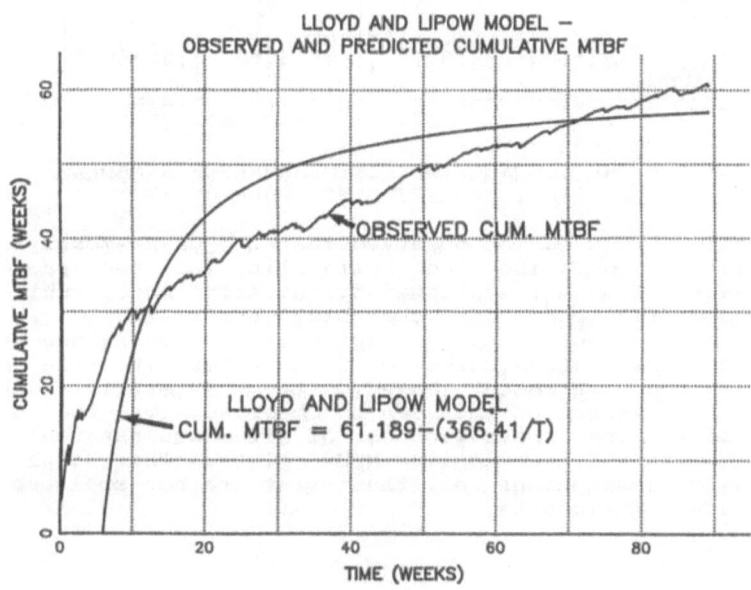

FIG 7 OBSERVED AND PREDICTED CUMULATIVE FAILURES BASED
ON THE LLOYD-LIPOW MODEL

COMPARISION OF THE GOODNESS FIT OF THE MODELS

In order to compare the goodness-of-fit of the growth models to the processor data, the method described in reference [2] and t-test statistics (standard error and residual sum of squares) (R) were applied. The calculated goodness-of-fit measures for each model, in the decreasing order of fit is given in table 6.

MODEL	R (%) Error	RELATIVE VAR	T-Test STD Error	R
DUANE	5.37	0.0229	2.774	0.998
ARINC	8.55	0.0278	2.114	0.999
IBM	20.90	0.0963	4.785	0.996
AROEF	23.15	0.1847	13.660	0.965
LLOYD-LIPOW	28.10	0.2923	33.306	0.945

TABLE 6 - THE GOODNESS OF FIT MEASURES FOR MODELS
IN THE DESCENDING ORDER OF FIT

From table 6, it can be observed that the Duane model is the best model to describe the reliability of the processor field data. This is followed by the Arinc model, while the least good fit model is the Lloyd-Lipow model. It is interesting to note that, in spite of the poor goodness-of-fit of model parameters based on the t-test to the Arinc model, a good agreement based on the R (% error) and RE is observed. A reversal of this can be observed for the Lloyd-Lipow model. The above goodness of fit measures are also in agreement with the cumulative MTBF plots. This highlights that basic assumptions of the t-test are not applicable to the processor field data.

CONCLUSIONS

1) There relioability of the processor improvevement is significant and a variety of models can be used to model the failure data.

2) The Duane and Arinc models are most suitable to
describe the reliability growth of the processor.
The Lloyd-Lipow model gives the poorest fit.

3) The goodness of fit based on absolute error and
relative variability appears to be superior to the
traditional methods.

4) The majority, 30% of non-random failures occur during
the first month.

REFERENCES

1) Duane, L H: Learning curve approach to reliability
monitoring. IEEE transactions on Aerospace, Vol 2, No 2,
(1964), P563-566

2) Rome Air Development Center, Reliability Growth Study,
RADC-TR-75-253, New York.

3) Clarke, J M: No growth growth curves, IEEE proceedings
1979 annual reliability and maintainability symposium.

4) Bezat, at al; Growth modelling improves reliability
prediction, IEEE proceedings 1975 annual reliability
and maintainability symposium.

5) Crow L H: Reliability analysis for complex repairable
systems in: F Proshan and R J Serfing, Eds, Reliability
and Biometry   (SIAM Philadelphia, 1974) P379-410.

6) Crow L H: Confidence interval procedures for the Weibull
process with applications to reliability growth,
Technometrics, Vol 24, No 1, February. 1982.

7) U D Perera: Reliability demonstration and field follow-up
analysis of a communications controller, Proceeding of
the 6th Eueidata conference, Siena, Italy, March 15-17,
1989.

8) Rosner N: Systems analysis-non linear estimating
techniques, Proceedings national symposium on reliability
and quality control, P203-207, New York, 1961.

9) Lloyd D K and Lipow M: Reliability management, methods
and mathematics, second edition, published by the authors.
1977.

10) Balaban H S: Reliability growth models, The journal of
environmental sciences, Jan 1978.

11) U D Perera: Reliability Growth and repairability
modelling of a processor, Fifth IASTED International
conference on Reliability and Quality Control, held
at Lugano, Switzerland, June 1989.

APPENDIX

GOODNESS OF FIT CRITERIA OF MODELS

(a)   R - Average Absolute Percent Error

$$R = \left\{ \sum_{i=1}^{n} \left| \frac{Y(t_i) - \overline{Y}(t_i)}{Y(t_i)} \right| / n \right\} 100$$

where $\underline{Y}(t_i)$ = Observed Cum MTBF at time t
$\overline{Y}(t_i)$ = Calculated Cum MTBF at time t
n = No. of failures in data set
i = 1, 2, ...n

(b)   R.E. - Relative Error (Variability)

$$R.E = \frac{S_e^2 / u-2}{S_{Y(t_i)} / n-1} = \frac{(\text{ Residual Error })^2}{\text{Variance of observed Cum. MTBF}}$$

where $S_e^2 = \sum_{i=1}^{n} \left[Y(t_i) - \overline{Y}(t_i) \right]^2$

and $S_{Y(t_i)} = \sum_{i=1}^{n} \left[Y(t_i) - \overline{Y(t_i)} \right]^2$

and $\overline{Y(t_i)} = \sum_{i=1}^{n} Y(t_i) / n$

ESTIMATION OF THE MEAN NUMBER OF FAILURES
FOR REPAIRABLE SYSTEMS

MICHAEL J. PHILLIPS

Department of Mathematics,
University of Leicester,
Leicester LE1 7RH, UK

ABSTRACT

The estimation of the mean number of failures for repairable systems can be
approached either by a parametric approach or by a non-parametric approach
recently suggested by Nelson. This paper compares these two approaches. Their
features are illustrated by considering the analysis of a data set collected from the
performance of an electronic system constructed of several replaceable modules.

INTRODUCTION

Many industrial systems are repairable. After failing to perform satisfactorily
the systems are restored to fully satisfactory performance by a method other
than complete replacement of the entire system. Such a system can fail more
than once and the pattern of times between successive failures is important.
The rate at which failures occur is fundamental in determining such matters as
preventive maintenance, replacement decisions and the assessment of reliability
growth. Let $N(t)$ be the cumulative number of failures in $(0,t)$. The distribution
of $N(t)$ is of interest and is usually summarized by the mean (expectation) which
is often denoted by $H(t)$. Frequently it is more useful to deal with the rate of
occurrence of failures $h(t)$ at time t. It is usual to assume instantaneous repair
which is realistic if the repair times are very small compared to times between

failures or if there is adequate time to make repairs to the system before it is next required for operation.

Standard theory exists to model such systems when times between failures have negative exponential distributions and in this case it is possible to obtain $H(t)$ or $h(t)$ in terms of the parameters of the negative exponential distributions. So it is possible to estimate $H(t)$ or $h(t)$ by estimating the parameters of the negative exponential distribution, using the observed and censored failure times. However for other distributions, like the widely used Weibull, the situation is not quite so straightforward, as it is necessary to resort to the use of asymptotic formulae. This has recently been demonstrated by Ansell and Phillips [1].

An alternative to the parametric approach is the non-parametric. Nelson [2] has suggested a simple non-parametric graphical method for estimating the cumulative number of failures. Recently he has proposed a method of calculating confidence intervals for this estimate.

This paper compares these two approaches. Their features are illustrated by considering the analysis of a data set collected from the performance of an electronic system constructed of several replaceable modules, which was considered by Ansell and Phillips [1].

AN EXAMPLE OF A REPAIRABLE ELECTRONIC SYSTEM

A manufacturer of electronic equipment decided to assess the performance of the N systems supplied to a customer. The supplier's data in this case consisted of the failure times. The customer's data consisted of the failure free times. This is because whilst the manufacturer is under an obligation to repair failed systems for his customer and will therefore have information about these failed systems the customer is under no obligation to supply information about systems which are performing satisfactorily. This information must be obtained from the customer in some direct or indirect way. The information from both sources was combined to produce the data. This was done for N = 100.

Another aspect of the data is that calendar time is not a useful metric since the system was not used throughout the period. The time taken to repair the system was not taken into consideration as these times were short compared with the average time between failures. The aim of the study was to decide whether the systems satisfied specification, needed a major modification or a complete redesign.

When a failure of the system occurs in service the electronic module (sub-system) which caused the failure is identified and is then replaced by a new module and the system is returned to service. So there is a record for the nth system which consists of the failure times $t_{n,j}$ for the jth ordered failure with the appropriate serial number of the replaced module. The record also contains an estimate of the last time τ_n at which the *nth* system was withdrawn from service before the end of the period of data collection.

There are two possible levels of analysis, the systems level and the module level. Since the failures are occasioned by the modules it seemed appropriate to start the analysis at that level.

Starting with simple plots a Pareto plot of the total number of failures of a module against module number ranked by frequency of failure seemed initially most helpful. This immediately highlighted for the management of the manufacturing company the modules which were performing badly and hence those where initially resources should be focused. Four modules contributed 50 % of the failures, whereas three modules did not fail at all during the study.

The next stage was to concentrate on the modules with most failures.

THE PARAMETRIC APPROACH

Ansell and Phillips [1] continued the analysis of the data from this study by using a parametric approach. The failure times of the module with the second highest number of failures were used to fit a Weibull distribution. Both Total Time on Test (TTT) plots and Weibull plots were obtained. These indicated that the distribution was Decreasing Failure Rate (DFR) rather than Increasing Failure Rate (IFR) and that a Weibull distribution with a scale parameter equal to 283.3 and shape parameter equal to 0.9 should be fitted to the data if Maximum Likelihood Estimation was used.

The failure data for another 17 of the modules were analysed in a similar way to that outlined above for the second module. As a result of this analysis of the 18 highest failing modules the Maximum Likelihood Estimates (MLEs) indicated that there was no significant difference between the failure time distributions of some of these modules. The 4 highest failing modules (1-4) were treated separately but the next 14 highest failing modules (5-18) were grouped into four groups. The previous analysis of failure times was repeated for these four groups and the results are given in Table 1.

The remaining 14 modules had 4 or fewer recorded failures per module. With such sparse data it did not seem reasonable to perform any elaborate model fitting. Instead these modules were assumed to all have failure times from a common negative exponential distribution. This result is also included in Table 1.

There are two groups (A and B) of fairly reliable module which can be fitted by negative exponential distributions with medians of about 800 and 1,000. Then there are two groups (C and D) of reliable modules, which exhibit a high initial hazard rate, which can be fitted by Weibull distributions of about 0.6 and 0.5 respectively and medians of about 2,000 and 8,500.

TABLE 1
MLEs of the parameters of the Weibull and negative exponential distributions
for the failure times of the grouped modules of the repairable electronic systems

Group	Serial no. of module	Number of failures	Shape parameter	(s.e.)	Scale Parameter	Median
	1	52	0.71	(0.09)	275.3	164.3
	2	41	0.90	(0.12)	283.3	188.5
	3	27	0.76	(0.13)	593.8	365.8
	4	19	0.93	(0.18)	623.4	419.8
A	7,9,14	26	1.22	(0.21)	792.3	586.8
B	8,10,13 15,16	36	0.91	(0.14)	1854.5	1240.4
C	5,6,12	31	0.64	(0.11)	3462.0	1957.4
D	11,17,18	20	0.53	(0.11)	17039.6	8490.1
	19-32	25	1.0		5845.3	4051.7

The analysis of these failure data illustrates some practical problems. Firstly any system of electronic modules will typically contain many modules and some of these will have few failures. In practice to do any parameter estimation it is necessary to group together similar modules. There are some dangers in doing this on the basis of the data rather than for prior physical/engineering reasons. Secondly there is the problem of highly censored data. Any plotting methods used for goodness-of-fit must take this into account. Thirdly it is possible to obtain estimates of the parameters of Weibull distributions easily but obtaining standard errors of the estimates is not straightforward.

If this model fitting is acceptable the final problem is obtaining the failure rate for each module and combining them to obtain the failure rate of the system assuming that the modules operate independently. The failure rate is far from easy to obtain for certain distributions. Baxter et al [3] have produced tables for the Weibull distribution but these unfortunately are incomplete and not easily accessible. Hence for many distributions using asymptotic formulae [4] which may not be sufficiently accurate is the only possibility for a parametric approach. This was done by Ansell and Phillips [1]. A Superposed Renewal Process model for the system was assumed and hence summing over the five negative exponential distributions gave a failure rate of 0.014, ie 14 failures per

1000 time units. For the four Weibull distributions the expected cumulative number of failures $H(t)$ using the asymptotic formula was estimated by

$$\hat{H}(t) \sim 0.005t + 8.4 \ .$$

So there was a failure rate of 0.005, ie 5 failures per 1000 time units, plus 8.4 failures due to early failures because of the form of the Weibull distribution. Combining the two failure rates the total expected cumulative number of failures for the electronic system was estimated by

$$\hat{H}(t) \sim 0.019t + 8.4 \ .$$

So the asymptotic failure rate for the system was estimated as 19 failures per 1000 time units.

THE NON-PARAMETRIC APPROACH

An alternative is to use a non-parametric approach. Nelson [2] suggested estimating $H(t)$ by plotting the mean cumulative number of failures $C(t)$ at time t. This is calculated by dividing the time axis into N intervals by using the last times at which the systems were withdrawn from service before the end of the period of data collection. These are used after they have been ordered in decreasing magnitude so that

$$0 \leq \tau_N \leq \tau_{N-1} \leq ... \leq \tau_2 \leq \tau_1.$$

Then interval i is defined by $(\tau_{i+1}, \tau_i]$, with $\tau_{N+1} = 0$. Let $C_{i,n}$ be the number of failures in interval i for system n. Also let t be contained in the Ith interval. Then let $C_{I,n}(t)$ be the number of failures in the interval $(\tau_{I+1}, t]$ for system n.

With these definitions $C(t)$ is given by

$$\begin{aligned}
C(t) \ = \ & [C_{N,N} + C_{N,N-1} + ... + C_{N,I} + ... + C_{N,1}]/N \\
& + [C_{N-1,N-1} + ... + C_{N-1,I} + ... + C_{N-1,1}]/(N-1) \\
& + [C_{N-2,N-2} + ... + C_{N-2,I} + ... + C_{N-2,1}]/(N-2) \\
& + ... \\
& + [C_{I,I}(t) + ... + C_{I,1}(t)]/I
\end{aligned}$$

Nelson also calculated the variance of this estimator $Var(C(t))$ as

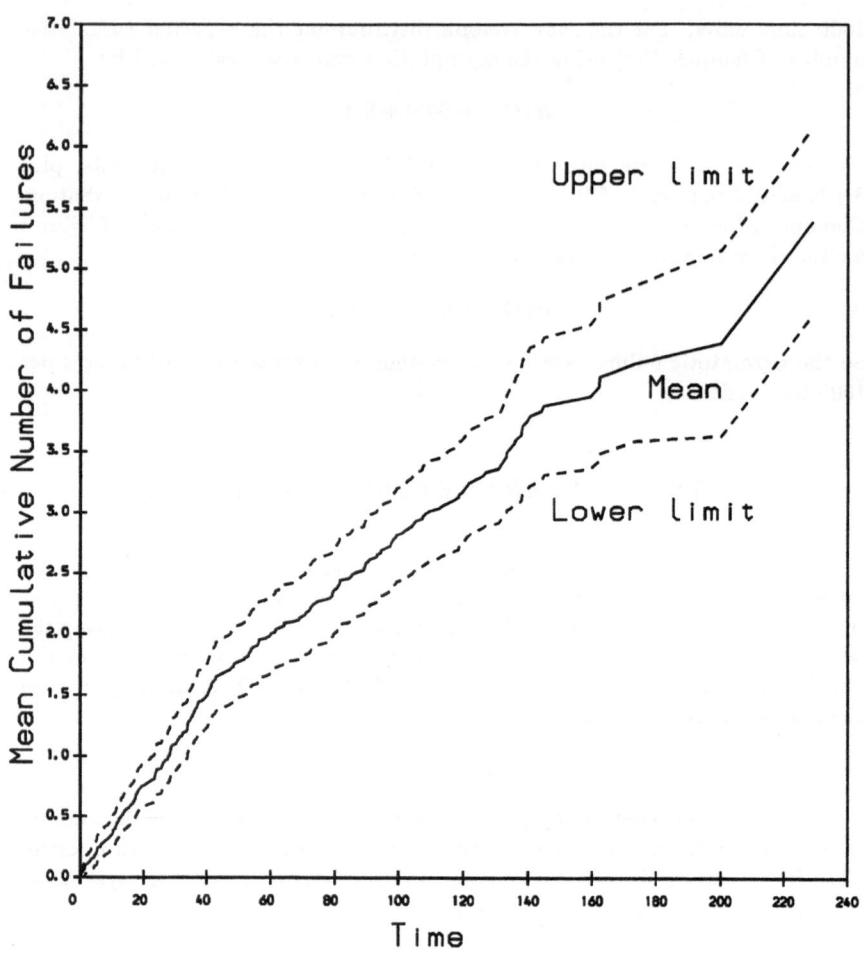

Figure 1. Mean Cumulative Number of
Failures with Upper and Lower Limits

$$
\begin{aligned}
Var(C(t)) \;=\; & Var(C_{N,.})/N + Var(C_{N-1,.})/(N-1) + ... + Var(C_{I,.}(t))/I \\
& + 2[Cov(C_{N,.}, C_{N-1,.}) + ... + Cov(C_{N,.}, C_{I,.}(t))]/N \\
& + 2[Cov(C_{N-1,.}, C_{N-2,.} + ... + Cov(C_{N-1,.}, C_{I,.}(t))]/(N-1) \\
& + ... \\
& + 2[Cov(C_{I+1,.}, C_{I,.}(t))]/(I+1)
\end{aligned}
$$

The variances $Var(C_{*,*})$ and covariances $Cov(C_{*,*}, C_{*,*})$ are estimated by the usual sample estimates. Nelson suggested that this is a non-parametric equivalent to Greenwood's variance for recurrence data. Limits can be calculated by adding and subtracting twice the standard error, the square root of the estimated value of the variance $Var(C(t))$.

The plot for the electronic data for the mean cumulative number of failures is given in Fig. 1 with upper and lower limits. This plot suggests that the failure rate of 19 per 1000 time units is about right but that the 8.4 failures due to early failures is much too large (pessimistic).

DISCUSSION

This analysis illustrates the problem of estimating the failure rate as it is difficult for the Weibull distribution. The usefulness of the asymptotic formulae applied by Ansell and Phillips [1] depends on whether they can be used for reasonably small t or not. As they have been obtained by considering different failure rates and the rate of convergence varies over the different functions it is difficult in any application to be sure. But until methods of easily evaluating the failure rate are readily available this seems to be the best that can be done using a parametric approach. The alternative is to use the non-parametric approach suggested by Nelson. This approach has been used for analysing the system but it can also be applied to the analysis of individual modules as was done with the parametric approach.

REFERENCES

1. Ansell,J.I. and Phillips,M.J. Practical problems in the statistical analysis of reliability data (with Discussion). *Appl. Statist.* , 1989, **38**, 205-247.

2. Nelson,W. Graphical analysis of system repair data. *J. Quality Tech.* , 1988, **20**,No.1,24-35.

3. Baxter,L.A., Scheuer,E.M., McConalogue,D.J. and Blischke,W.R. Renewal Tables : Tables of Functions arising in Renewal Theory. *Technical Report* , Univ. of S. Calif, 1981.

4. Cox,D.R. *Renewal Theory*, Methuen, London, 1962.

Dependency Modelling

JI Ansell
Department of Management Systems and Sciences
University of Hull

and

LA Walls
Department of Mathematics and Statistics
Paisley College of Technology

Abstract

Since dependency can influence the performance of a system, it is crucial to investigate and to understand the consequences of dependency when designing, operating and maintaining a system. To do this requires a clear understanding of various types of dependency. For example, it is important to distinguish between dependency in the times between failure and the practically important area of common-cause. The type of dependency will also effect the nature of any data analysis to be carried out.

By reviewing the literature in the area, this paper attempts to categorise the main types of dependency. Methods of identification of dependency are examined. This is done primarily through data analysis. Ways of incorporating dependency into model-building are also described.

INTRODUCTION

In recent years engineers working in various industries, such as nuclear, aerospace and chemical, have been increasingly concerned with dependency since it can degrade the performance of engineering systems. This is particularly the case for multiple redundant and diverse systems where these built-in defences can perform well below expectation or even be rendered void with potentially catastrophic consequences. Thus it is important to identify the extent to which dependency influences system behaviour and to assess its effects.

Ideally dependency should be recognised at the system design

stage so that appropriate defences can be built in to avoid or
reduce its effects. However in practice it is common for most
information about dependency to emerge during later stages of
the system's life cycle. Under these circumstances methods are
required to identify, assess and model dependency.

Before describing such methods it is necessary to clarify what
is meant by dependency. In the reliability literature
dependency is a term used in many different circumstances.
Often a particular definition of dependency is used within a
specific context.

The Shorter Oxford Dictionary defines dependency as 'the
condition of being dependent; contingent, logical or causal
connection'. The idea of causality is consistent with the most
common interpretation of dependency in reliability
engineering.

In statistics dependency is defined probabilistically. For
example, consider a simple redundant system with two identical
components A and B. This system will fail if A and B both
fail. If the failure of A is statistically dependent on the
failure of B then the probability of A failing is conditional
on the probability of B failing. This can be written as
$P(A|B)$. The probability of B failing is denoted by $P(B)$. Thus
the probability of system failure is the product of the
probabilities of A and B failing, $P(A|B).P(B)$. However if the
failure of A is statistically independent of B then the
probability of system failure is $P(A).P(B)$, where $P(A)$ denotes
the probability of A failing.

The concept of statistical dependency is consistent with the
common usage of the term in that knowledge about B affects the
knowledge about A. However statistical dependency, if it
exists, is not so powerful as to establish that there is a
causal link between the failures of A and B.

Thus the statistical definition of dependence is not entirely
satisfactory for general usage in reliability engineering.
Rather than attempt to produce a better definition it appears
more plausible to progress by examining the various types of
dependency found in reliability. With a clearer understanding
of the forms of dependency it is then possible to describe
appropriate methods for the identification, assessment and
modelling of dependency.

In this paper statistical methods for the identification of
dependency are considered. Engineering methods are discussed
in detail elsewhere. Reviews are given by, for example,
Amendola [1], Ballard [2], Crellin et al [3], Edwards and
Watson [4], Fleming et al [5], Games et al [6,7], Humphreys et
al [8], Watson [9]. Methods for assessment follow directly
from identification. The major problems with modelling
dependencies are discussed.

113

TYPES OF DEPENDENCY

In this section the types of dependency of prime concern in reliability studies are examined. In particular those aspects of dependency which may affect identification, assessment and modelling are considered.

Causal links between system components can result in dependency. For example, component behaviour can be affected by their common design, manufacture or maintenance. The more important causal links are summarised in Table 1. These are generic descriptions which may encompass a range of particular links. For example, environmental links includes external factors (eg. temperature, pressure, humidity, weather) and internal factors (eg. tool wear). A link can be labelled deterministic if that cause always results in failure. Otherwise a causal link can be considered stochastic.

The effect of such causal links between components may be immediate or delayed. For example, the failure of a component may result in the failure or unavailability of another functionally dependent component in what is often known as a cascade effect. This need not be immediate or even occur within a limited time period since the second component may only be damaged or more stressed by the failure of the first. The major effect of such causal links, whether immediate or delayed, is to impair either system or component performance. Instantaneous failures are obviously potentially most serious for safety-critical systems.

A system whose behaviour is changing through time due to, for example, repair, maintenance, demand, can be considered dynamic. While a system can be regarded as static if its behaviour is stable through time (ie. stochastically stationary) or the only characteristic of interest is the probability that the system works under specific circumstances.

Table 1.

Causal Link	Environmental, Structural, Component, Manufacture, Procedural, Design
Nature of Link	Deterministic, Stochastic
Effect on Item	Instantaneous Failure, Delayed Failure
Systems Process	Dynamic, Static
Status of Resulting Failure	Revealed, Unrevealed
Description	Common Cause (CCF), Common Mode (CMF), Time Between Failure (TBF)

In the past most interest has been centred on two generic

forms of dependency which can be described as 'common cause failures' (CCF) and 'times between failures' (TBF). Both of these terms require definition.

Several authors, including Amendola [1], Crellin et al [3], Edwards and Watson [4], Games et al [7], Humphreys et al [8], Virolainen [10], Watson [9], have discussed at length the various inconsistent, and often contradictory, definitions of CCF. Indeed some authors (eg. [4], [7] and [8]) have abandoned a formal definition altogether and instead have proposed some classification scheme that encompasses the range of failures due to common causes. Here the term CCF is simply used in its most intuitive sense and covers all cases of multiple failures for which it is possible to discover a common causal event - a root cause. Often CCF are assumed to be simultaneous or coincident failures. However here CCF are extended to include failures which need not occur at the same time provided that they can be linked back to the same root cause. This extends the definition to include cases where there has been a lack of accurate time measurement as well as delayed or unrevealed failures of items.

The term TBF broadly describes dependency through time, both within and between items. For example, the lifetime or life experience of one component can have an effect on other components. The system can then be described as stochastically nonstationary. Ascher and Feingold [11] have pointed out that in the past little attention has been given to such dependency. They suggest that this may be because engineers have not been made aware of the need to consider such dependency, as well as the complexity of the problem and the relatively large amounts of data needed to investigate it.

There is a large literature on both of these forms of dependency, although for the latter this is primarily in statistical journals.

DATA AVAILABILITY

Information about dependency is often not available until a system is operational. It is important to record and to use such information so that the effects of the identified dependency in the system of interest can be assessed. Such information can also be exploited to encourage good design practice to reduce the effect of dependency in new systems which are to perform similar functions under nominally identical conditions. Without relevant data a source of dependency can very easily remain hidden.

Unfortunately data specific to the operational history of the system of interest tends to be sparse. Therefore generic data based on the operational history of diverse plants is often used. However in the past dependency has not been well understood and consequently few databases have made allowance for it. This has resulted in a paucity of data. Games et al

[6] discuss some of the problems encountered in attempting to
use generic databases for CCF analysis. They point out that
the most useful information is often contained in some
unstructured remarks section. Similarly Walls and Bendell [12]
outline the shortcomings of generic databases containing
summary measures, such as failure rates, for identifying
dependencies in the TBFs.

Recently there has been growing awareness of the importance of
dependency and special databases and classification schemes to
support data collection have been constructed. These include
the NCSR DEFEND database described by Games [13] and
classification schemes such as those described by Crellin et
al [3], Humphreys et al [8] and Watson [9]. These aim
primarily to support CCF analysis.

In response to the need for information about dependency,
operators of generic databases have also attempted to
incorporate data concerning the event history of items and
causal links. See, for example, Cross and Stevens [14], Gibson
et al [15], McIntyre et al [16].

Deciding what information about dependency should be included
in a database is difficult. For example, if a particular type
of dependency is not perceived as important then it will not
normally be included in the design of the database. On the
other hand, collecting all relevant data on all possible types
of dependency will possibly present the analyst with too much
and too complex information. Ideally data concerning:

> 1) components manufacturing details, batch no etc
> 2) time as accurately as possible
> 3) environmental conditions
> 4) design features of components and systems
> 5) stresses
> 6) failure modes and causes

would be useful for dependency studies. Further discussions of
the sort of data required to support CCF analysis are given
in Games [13] and Humphreys et al [8].

In the not too distant future modern data capture methods and
'automatic' data analysis may permit the analyst to have the
full set of data in a manageable form.

The form of the data will depend on the system under study and
its failure process. Figure 1 illustrates two simplified cases
corresponding to nonrepairable and repairable systems.

The data for nonrepairable systems will consist of the failure
times of the components and the subsystems. Some of these
failures will be multiple failures. Failure of some of the
components may lead to failure of the system. The data will
therefore consist of the lifetimes of components and/or
systems, possibly truncated at some point in time or by the
failure of a certain component or system. The dependency will
be between the components and/or the systems.

116

Since repairable systems are repaired on failure these give
rise to repeated cycles of failure and repair. Typically
repair times are unknown or ignored because they are not
relevant to the system under study. Therefore attention is
given to the sequence of times between failures of the system
observed during its life or to the sequences of lifetimes of
the components which comprise the system. These times may also
be truncated by failure of a certain item or after a fixed
period of time. The dependency in this case may be within a
component or between components.

The form of data available will obviously have an effect on
the model which can be used to analyse the data.

(a) Non-repairable systems

components/subsystems

```
    1 | -------------------x
    2 | ------------x
    3 | --------------------x
    : |
    i | ------------------------|truncated
    : |
    n | ------------x

        ------------------------------------>time
```


(b) Repairable components/systems

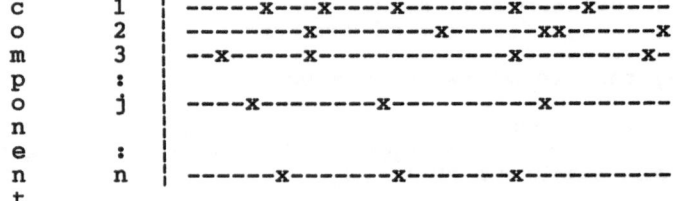

```
    c   1 | ------x---x----x-------x----x-----
    o   2 | --------x--------x------xx------x
    m   3 | --x------x-------------x--------x-
    p   :
    o   j | ----x--------x----------x--------
    n
    e   :
    n   n | ------x-------x-------x----------
    t

system       --x-xxx--xx-xxx--x---xx-xx-x---xx-----> time
                  x      x        x x
                                  x
```

x denotes a failure

Figure 1. Forms of Failure Data

IDENTIFICATION

Given the range of types of dependency described above it is
not surprising that there is no single identification method.
Different methods are needed to identify different types of

potential dependency. A number of qualitative engineering techniques are available for this and are described in Amendola [1], Ballard [2], Crellin et al [3], Edwards and Watson [4], Fleming et al [5], Games et al [6,7], Humphreys et al [8], Watson [9].

Here we consider data analysis techniques to aid the identification of CCF and TBF dependencies. The methods considered include simple plotting techniques, such as scatter plots to the use of more complex analytic tools, such as hazard modelling. Some of the methods will be overlapping. Underlying most of the models will be a stochastic characterisation of the failure process of the system. For non-repairable systems this may be either in terms of a distribution (eg. exponential, Weibull) or through more complex models such a competing risks. For repairable systems a wide range of models is available. Ascher and Feingold [11] suggest five different models which are frequently met in reliability studies. These are the Homogeneous Poisson Process (HPP), Non-Homogeneous Poisson Process (NHPP), Branching Process (BP), Renewal Process (RP) and the Superimposition of Renewal Processes (SRP).

CCF
The main statistical effect of CCF is to abnormally raise the rate of failure of an item. There are a number of methods which aim to identify statistically significant increases in the rate of failures, eg Walls and Bendell [17], Ansell and Phillips [18,19]. These methods are based on different models for the underlying failure process.

Walls and Bendell [17] describe a graphical technique which can be used to explore field data collected on a daily basis, or within some other time window, for potential dependent failures. These daily failure times for systems of identical or related components are described by a NHPP model with Weibull rate. This is based on the assumption that after repair the systems reliability remains essentially unchanged. The properties of this NHPP model are discussed in Walls and Bendell [17]. They fit this model to the data under the null hypothesis of no dependency. The data are examined for significant departures from the null model by constructing 95% confidence intervals about the estimated expected number of failures, obtained through simulation. Particularly high failure rates compared to the null model would lead to points falling outside the confidence interval and so indicate departure which may be due to dependence.

The environment model of Ansell and Phillips [18] may also be used to identify local changes in rates of failures which could be due to CCF. They assume the data arises from several repairable systems/components being run in the same time frame which are subject to a common stress which is dependent on time, say $S(t)$. Each repairable system/component forms a renewal process. This renewal process need not be identical and the components need not be assumed to be identical. An appropriately flexible form for S(t) needs to be selected. Any point for which S(t) rises significantly would require investigation to see if it is due to some CCF.

These methods outlined above have the advantage that they do not simply identify CCF via coincidental failures. Rather these methods allow for the possibility of delayed, unrevealed or unflagged failures due to common causes. On the other hand these approaches have the disadvantage that they assume the entire failure process can be described by a single model.

As usual it is important to exercise engineering judgment in interpreting statistically significant patterns indicated by the above approaches to decide whether the identified statistical dependencies are indeed practical dependencies such as CCF. There may be other reasons for statistically significant patterns. These include changes in the parameters or the form of the underlying model. There is also a (hopefully) small chance of a statistically significant pattern when the model is correct and no dependencies are present.

For other data forms it seems relatively easy to generalise from these two approaches. Using smoothing methods, based on either Fourier Transforms, amended Kalman Filtering Techniques or other methods, it may be possible obtain local failure rates which could then be compared to observed data as in the approach of Walls and Bendell [17]. Again it would be natural to take independence as the null model and departures from this should be investigated. An approach similar to that of Ansell and Phillips [18] would require specification of a dependency mechanism on top of the stochastic characterisation.

An advantage of the Walls and Bendell [17] model is that it is graphically based. It would be hoped that a generalisation based on some form of smoothing could also be developed as a graphical procedure. An alternative method might be to consider the present plethora of plotting techniques such as TTT-plots to see if certain abnormalities in plots can be traced to patterns associated with CCF.

There are other methods which can be used to detect clustered point events which might aid the identification of CCF when delayed failures occur. Hawkes and Oakes [20] consider self-exciting process which could be described a Poisson Cluster Process, see Lewis [21]. This type of model assumes an event locally creates other events according to some random process. The general type of process is a Branching Process. Estimation in such models may suffer from confounding unless the nature of the underlying process is well understood.

Georgiakodis and Jerwood [22] describe the potential use of discriminant analysis to monitor system reliability and to aid detection of common mode failures. As yet this approach has not been widely adopted and even Georgiakodis and Jerwood [22] illustrated their methods using medical data.

TBF
In the case of a single sequence of repeated failures for a repairable system the possible relationship in the TBFs are of interest. For example, is the lifetime of the ith component dependent on the lifetime of the jth component according to some mechanism?

A graphical approach can be adopted to investigate these TBFs. For example, plotting the cumulative number of failures against cumulative time on linear paper can highlight whether or not there is some trend in the data towards lengthening or shortening TBFs. Such trend may be taken as a form of dependency which can be attributed to environmental or procedural causes. Such a plot will also highlight other salient features of the data including rogue TBFs, clusters of failures etc. In the discussion of Ansell and Phillips [18] a number of alternative plotting techniques are outlined. If there is no trend or the trend has been removed then the inter-relationship between TBFs ,in general, k failures apart can be examined. A scatter plot of the lagged TBFs (ith TBF against (i+k)th TBF for k>0) could be used. The lag is progressively increased until a point is reached where the loss of information incurred by discarding points is too high. It is common not to progress beyond a lag of n/4, where n is the number of data points. Any systematic pattern would indicate statistical dependency. From the analysis of such reliability data Walls and Bendell [12] suggest that it can be difficult to characterise the underlying model for the patterns identified.

Obviously confirmation of patterns identified subjectively can be achieved using more formal methods. These may include trend tests such as Laplace and MIL-HNBK-219 , or tests for serial dependency based on the estimated lagged correlations, see Ascher and Feingold [11], Cox and Lewis [23], Bendell and Walls [24].

It should be mentioned that regression type approaches such as Logistic Modelling and Proportional Hazard Modelling may also be used to detect such dependency, see [19].

As usual it is important to confirm whether the identified statistical dependencies correspond to practical dependencies such as minimal repair, environmental effects etc.

Identification of dependency in the TBFs can be because it can be difficult to select an appropriate failure process. Ideally information is required at the lowest level (component), but in practice often is only available at system. In such cases spurious statistical dependencies are likely to be found.

There is also a need to search for dependency between the TBFs of two or more systems. However so far the work on the interaction between systems has only met with limited success.

The shock model approach of Marshall and Olkin [25], and the work of Downton [26] consider only the time to first failure of a number of components. It is hard to imagine easy generalisation of such a models to a general model for set of repairable systems.

The environmental model of Ansell and Phillips [18] as described above considers the interaction between systems through some common stress, but that may only be useful in particular circumstances. It assumes that each system is repaired. The degree of dependency will be seen through the changes in the estimated S(t). However there may be confounding with the estimation of the components own performance. Therefore it might be sensible to compare the estimates obtained when interaction is and is not present.

There are other models where dependency has been considered in a specific form. Smith [27] and Harlow, Smith and Taylor [28] have considered models where the failure of components transfer extra stresses onto other components within the system. These models are fairly simple being bundles of fibres, initially assumed to be identical in a series-parallel system. Problems associated with inference including identification have not been considered.

ASSESSMENT AND MODELLING

There are many problems associated with the complexity of modelling dependency. Even without dependency there are problems associated with the existing techniques for modelling system reliability. Most software is designed assuming Markovian behaviour, or at least Markovian to the first order. Given that it is not easy, and sometimes it is not possible, to produce theoretical results for failure distributions or availabilities we often resort to simulation, with all the drawbacks when dealing with very reliable components. The problem only increases when components are dependent.

The concentration on Markovian methods as employed in Fault Trees has lead many authors to adapting these models to incorporate dependency rather than developing new ones. This is particularly true for CCF. Hence many methods still concentrate on developing rate measures of dependency which are required by the software, rather than really considering the appropriateness of the model. The rates are based on engineering judgment and aggregated data. Typical of such approaches is the distributed failure probability method (DFP) proposed by Hughes [29]. A set of 'environments' are established using engineering judgment. Previously collected data is then disaggregate into these 'environments' and the rates calculated. The DFP method is very general and encompasses a wide range of other models such as the Beta Factor and the Binomial Failure Rate.

The main criticism of such models must be that the level of complexity rises rapidly without necessarily yielding a greater insight into the dependency. This is of particular concern given earlier comments about the inappropriateness of Markovian modelling. Obviously if the environments' in the DFP approach are based on sound engineering judgments then they ought to be incorporated into the model explicitly instead of introducing extra layers within the model without further analysis of the failure mechanism.

Similar criticism can be also levelled at the models such as Beta Factor and Binomial Failure Rate which assume a particular form of dependency to obtain parameters for the models. While the original concepts where developed for applications where the mechanism was apparent, these models have subsequently been transferred to areas where it is harder to argue their appropriateness.

Generally the form of dependency should reflect the physical context of the problem rather than being constrained by the availability of models. This applies equally to the system level as it does to the component level. Thus essentially we are advocating that the approach should be problem-led rather than technique-led.

Dependency identified in the TBFs can be described using , for example, NHPP models, clustering processes, time series methods, proportional hazards modelling. But there are problems. For the Ansell and Phillips [18] approach it can be difficult to describe the future stress, S(t), when modelling. Simulation would seem the only appropriate solution. In such circumstances great care would have to be taken over sensitivity analysis.

As indicated earlier there are limited probabilistic results available. The work of Smith [27] and Harlow et al [28] on 'Bundles of Fibres' gives some results about the asymptotic behaviour of series/parallel systems of identical components for particular forms of dependency. The results of Ansell and Bendell [30] concerning the characterisation of coherent systems as serial/parallel systems does not assume independence of the components and therefore may be helpful in generalising the ideas of Hughes [31] when modelling complex systems.

CONCLUSIONS

The paper has aimed to generalise the view of dependency from just CCF to other forms such as TBF. The latter has not received the attention it is due. The paper has attempted to give further clarification of the differences between these two forms of dependency and statistical methods of identification, assessment and modelling have been reviewed. Obviously the statistical approaches need to be incorporated into a sound engineering framework.

Dependency will continue to be a major concern at both the design stage and when assessing performance of systems. As the quality of data bases improve it may well be that further insight will be gained into dependency. Nevertheless there will be a continued need for work both on qualitative and quantitative aspects of dependency. There is obviously generally a need to improve our modelling capabilities of systems especially in the area of dependency.

REFERENCES

1. Amendola, A., Common cause failure analysis in reliability and risk assessment, Reliability Engineering, ed A. Amendola and A. S. De Bustamante, 1986, Kluwer Academic Pub. London, pp. 221-256.

2. Ballard, G.M., Depenent failure analysis in PSA, Proc IAEA Int. Conf. Nucl. Power Perf. & Saf., Viena, 1989, pp. 119-133.

3. Crellin, G.L., Jacobs, F.M., Smith, A.M. and Worledge, D.M., Organising dependent event data - a classification and analysis of multiple compoent fault reports, Reliab. Eng., 15, 1988, pp. 145-158.

4. Edwards, G.T., and Watson, I.A., A study of CMF, UKAEA Report no SRD-R-146, 1979.

5. Fleming, K.N., Mosleh, A., Kelley, A.J., On the analysis of dependent failure in risk assessment and reliability evaluation, Nucl. Saf., 24, 1983, pp. 637-657.

6. Games, A.M., Breewood, M., Amendola, A., Martin, P., Keller, A.Z., CCF investigation using the European reliability data system, Proc. 8th ARTS, 1984, B2/2/1-7.

7. Games, AM, Amendola, A, and Martin P, Multiple related failure events - risk, design maintenance and cost, 5th Nat. Reliab. Conf., Birmigham, 1985, 5B/4/1-7.

8. Humphreys, P., Games, A.M., Smith, A.F., and Worledge, D.H., Progress towards a better understanding of dependent failures by data collection, classification and improved modelling techniques, Proc Reliab '87, 1987, 2C/4/1-14.

9. Watson, I.A., Analysis of dependent event and multiple unavailability with particular reference to common-cause failures, Nucl Eng and Des, 93, 1986, pp. 227-244.

10. Virolainen, R., On CCF, statistical dependence and calculation of uncertainty; disagreement in interpretation of data, Nucl Eng and Des, 77, 1984, pp. 103-108.

11. Ascher, H., and Feingold, H., <u>Repairable systems reliability</u>, Marcel Dekker, London, (1984)

12. Walls, L.A., and Bendell, A., Exploring reliability data, <u>Qual & Rel Eng Int</u>, 1, 1985, pp.37-51.

13. Games, A.M., DEFEND - a dependent failure database, <u>Euredata</u>, 1989, pp.178-194.

14. Cross, A., and Stevens, B., Reliability data banks - friend foe or waste of time, <u>Proc Reliab '87</u>, 1987, 5C/5/1-15.

15. Gibson, I.K., McIntyre, P.J., and Witt, H.W., Aspects of a model to improve the reliability estimates of engineering components, <u>Proc Reliab '89</u>, 1989, 4Aa/2/1-14.

16. McIntrye, P.J., Gibson, I.K., and Witt, H.M., Addressing the problem of relevance of reliability data to varied applications, EureData, Siena, 1989, pp. 28-38.

17. Walls, L.A., and Bendell, A., Exploring field reliability data for potential dependent failures, <u>Proc Reliab '87</u>, 1989, pp. 4Ab/3/1-8.

18. Ansell, J.I., and Phillips, M.J., Practical problems in the statistical analysis of reliability data (with discussion), <u>Appl Stats</u>, 38, 1989, pp. 205-247.

19. Ansell, J.I., and Phillips, M.J., Practical reliability data analysis, 1990, <u>Rel Eng</u>, to appear.

20. Hawkes, A.G., and Oakes, D., A cluster process representation of self exciting process, <u>J. Appl. Prob</u>, 11, 1971, pp. 493-503.

21. Lewis, P.A.W., Branching Poisson process, <u>J. Roy Statist Soc, B</u>, , 1964, pp. 398-456.

22. Georgiakodis, F., and Jerwood, D., Application of multivariate techniques to monitor system reliability and detect common-mode failure, <u>Proc Reliab '87</u>, Birmingham, 1987, 2B/5/1-10.

23. Cox, D.R., and Lewis, P.A.W., <u>Statistical analysis of series of events</u>, 1966, Chapman Hall, London.

24. Bendell, A., and Walls, L.A., Exploring reliability data, <u>Qual Rel Eng Int</u>, 1, 1985, pp. 37-52.

25. Marshall, A.W., and Olkin, I., A multivariate exponential distribution, <u>J. Amer. Stats. Ass.</u>, 62, 1967, pp. 30-44.

26. Downton, F., Bivariate exponential distributions in reliability theory, <u>J. Roy. Statist Soc, B</u>, 32, 1970, pp. 408-417.

27. Smith, R.L., Limit theorems and approximations for reliability of load sharing systems, <u>Adv Appl Prob</u>, 15, 1983, pp. 304-330.

28. Harlow, D.G., Smith, R.L., and Taylor, H.M., Lower tail analysis of distribution of strength of load sharing systems, <u>J. Appl. Prob.</u>, 20, 1983, pp. 358-367.

29. Hughes, R.P., Distributed failure probability approach to dependent failure analysis and its applications, <u>EureData</u>, Siena, 1989, pp. 167-177.

30. Ansell, J.I., and Bendell, A., On the optimality of k-out-of-n:G systems, <u>IEEE Trans Reliab</u>, 31, 1982, pp. 206-210.

31. Hughes, RP, New concepts for systems analysis, <u>Proc Reliab '89</u>, Brighton, 1989, pp. 4Ab/1/1-9.

125

COMMON CAUSE FAILURE ANALYSIS OF A SUBSEA SYSTEM

JULIO C.C. BUENO
Petroleo Brasileiro S.A- PETROBRAS, Serviço de Engenharia
Av. Chile, 65, 11 floor, Rio de Janeiro, RJ,Brasil
DAVID J. SHERWIN
School of Engineering Production, The University of Birmingham
P.O Box 363, Birmingham B15 2TT

ABSTRACT

A risk analysis study was carried out to analyse the conceptual design of a subsea system for the Albacora's field. Albacora is an oil field located on the continental shelf off the coast of Brazil. The project is being conducted by Petroleo Brasileiro SA- PETROBRAS, owners of the field. Because subsea systems consists of several identical components and the environmental conditions are quite hostile, a risk analysis study of subsea systems should consider the possibility of the occurrence of common causes. In this paper, a model to assess the effects of common mode failure on subsea systems is suggested. Two other models are also used in the analysis: The Marshal-Olkin model and the model suggested by Henley and Kumamoto. The results of the application of these models on the conceptual design of the Template-Manifold of Albacora are shown. Some mathematical approximations which allow a general and easy estimation of the effects of common cause failures are suggested.

1- INTRODUCTION

A Common cause is "a single condition or event which causes multiple failures, or multiple basic events" (4). These failures are called common mode failures. Improper design, defects due to manufacture, improper operation and maintenance and natural dangers such as earthquakes, floods, hurricanes, etc are examples of common causes.

Because they are by nature rare events, sometimes common mode failures are not considered in Risk Analysis studies. In some cases, the predicted probability of failure of a system considering common mode failures can be significantly higher than

the probability predicted not considering the common causes
(5,7).

Some difficulties concerned with assessing the effects of common causes in Risk Analysis studies are :

- the identification of the common causes which should be
considered in the analysis;
- the estimation of the frequency of occurrence of the common
causes considered;
- the determination of the dependence relations among the e-
quipment under the effect of the common cause considered.

An important development of offshore systems, which requires
new perspectives in the oil industry, is concerned with the
development of subsea systems. Subsea systems represent a new
way to develop deep water reservoirs (water depths greater
than 180 m). Nowadays, more and more oil fields are being
discovered in deep waters. Risk Analysis studies are a useful
management tool in projects for subsea systems , mainly because
these projects require high investments and are, at the same
time, projects of high economic and safety risks. As the cost
of repair and maintenance in a subsea system is several times
the cost of a equivalent work carried out at on onshore site or
even on a platform, availability and reliability may be
crucial factors to the success of the project.

In the particular case of projects for subsea systems, a risk
analysis should consider the possibility of the occurrence of
common causes. Subsea systems are constituted by some identical
components. Furthermore, environmental conditions are quite
hostile: water depth greater than 180m and distance to the
coast often greater than 100 Km. So, the probability of occurrence of a common cause is higher than in the most of onshore projects or even other offshore projects .

A general difficulty in performing any risk analysis study
which is especially important in subsea systems projects is
getting accurate suitable failure data. As subsea systems were
developed only recently, historic data of systems which work in
similar conditions are not available. Therefore, in general,
data must be extrapolated from other offshore projects.

This work is part of a risk analysis study carried out to
analyse the conceptual design of a Template-Manifold. This
Template-Manifold will be applied in the development of the
Albacora oil reservoir, which is an oil field located on the
continental shelf off the coast of the Brazil. The project is
being conducted by Petroleo Brasileiro SA- PETROBRAS, owners of
the field.

The aim of this paper is to analyse some models for assessing
the effects of common cause failures on the availability of
subsea systems and to evaluate the importance of taking into
account common cause failures in risk analysis of these projects.

NOTATION

a - availability of the component (steady state);

c - common cause conditional failure intensity ;

n - total number of identical components in the system;

Pi- probability that the system is in state i at time t;

pi- derivative of P with respect to time;

q - unavailability of the component (steady state), q= 1 - a ;

Q - unavailability of the system at the steady state;

Q(t) - unavailability of the system at time t;

λ - conditional failure intensity;

μ - conditional repair intensity of failures caused by ordinary reasons;

μc - conditional repair intensity of failures due to common
cause

2- GENERAL DESCRIPTION OF THE CONCEPTUAL DESIGN OF THE TEMPLA-TE-MANIFOLD OF ALBACORA.

The main function of a Template-Manifold is to gather the oil produced by the wells and to send it through a riser to a Pro-duction Unit. A Template-Manifold is also used to conduct and distribute fluids (gas or water) which are injected into the wells in order to maintain the production of the reservoir.

The following are general features of the conceptual design of the Template-Manifold of Albacora:

- Number of wells : 10

- General dimension of the Template's structure: 25m x 16.3m x 4.3m.

- Manifold

The Manifold consists of four lines:

One Production line, used to gather the oil and gas produced from the Christmas-Tree to the Riser;
One Water injection line, used to supply the water which is injected into the wells;
One Gas lift line, used to supply the gas which is injected into the wells in order to reduce the weight of the fluid in the production column;
One production test line, used to conduct the production of

the   well under testing ;

- Christmas-tree:

Ten   Christmas-Trees are used in order to control the   production and the fluids injected into the wells. Each Christmas-
Tree consists of eight valves, two chokes and two   pressure
tranducers. The conceptual design also considers the possibility of drilling two satellite wells. In this case two Modules called Satellite Trees would be installed over the two
remote wells. The connections between the main Christmas-tree
and the Satellites-Trees are made by two flexible lines: one
to conduct the production and another to inject gas;

- Control system:

The control system consists of a primary system and a back-
up, called the secondary system. The primary system is a
multiplexed electrohydraulic control system. The secondary
system is a direct hydraulic control system. In the primary
system the wells are controlled by electronic signals transmitted through a umbilical cable. The system is fed by two
umbilicals: one electrohydraulic which transmits the electrical signal, feeds the hydraulic primary circuit and contains
five hydraulic lines to be used in the secondary system. The
other is only hydraulic. It is used as back-up to feed the
primary hydraulic circuit and contains five hydraulic lines
used in the secondary circuit. In the primary circuit each
Christmas-Tree is controlled by its own Control Module which
receives a particular signal. If the primary circuit fails,
the secondary circuit is actuated by directional valves,
located on the Control Module.

3- METHODS

Two models found in the literature were used to assess the effects of common cause failures on the availability of the
system: the Henley-Kumamoto model (4) and the Marshal-Olkin
model (10).

These two models are mainly applicable in cases where the mean
time to repair a failure of the system caused by a common cause
can be considered the same as the mean time to repair a failure
of system caused by different reasons. For instance, these
models are applicable if a common cause is originated by a
manufacture or a maintenance problem. However, they do not
distinguish a failure of the system caused by a common cause
from a failure of the system caused by different reasons.

These models are not completely suitable when a common cause
occurs and the mean time to repair is considerably longer than
the mean time to repair ordinary failures. In the case of
subsea systems, as the environmental conditions are quite hostile it is necessary to consider common cause failures originating from environmental problems such as earthquakes, hurri-

canes, big waves, etc. Under these conditions, to assume that the mean time to repair is significantly greater than in the ordinary situations is obviously more realistic. In order to consider this situation, a model called the Environmental Common Cause model was set up.

Because of the high cost and the inherent difficulties of the offshore work, the possibility of repairing only one equipment at a time was assumed in the Marshall-Olkin model and in the Environmental Common Cause model.

3.1 The Henley and Kumamoto Model

In this model, two situations are considered (Figure 1):

Situation A:
  - No common cause to time t and the system fails at t for ordinary reasons;

Situation B:
  - The last common cause occurs in time u - du to u, and no common causes from time u to t. The system fails at t for ordinary reasons.

Two Markov transition diagrams are set up . In situation A components are "up' at the initial stage and in situation B the components are "down" at the initial stage.

Figure 1. HENLEY AND KUMAMOTO MODEL

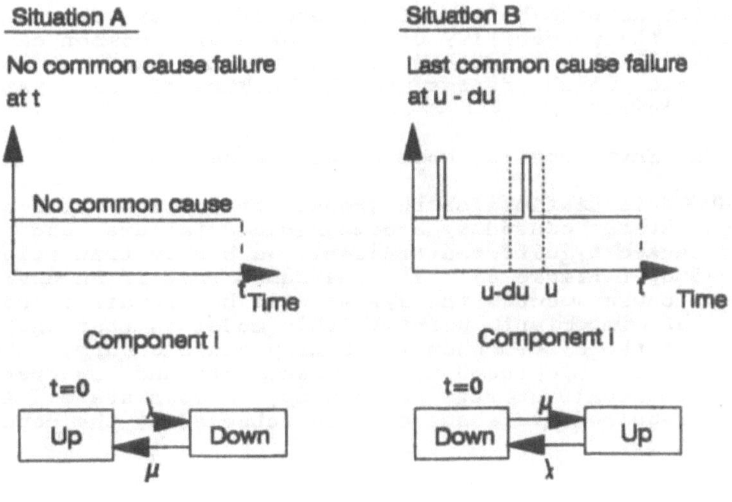

The following expression which estimates the probability that the system is failed at t is provided:

$$Q(t) = \exp(-ct) \prod_{i=1}^{n} \frac{\lambda_i}{\lambda_i + \mu_i} [\, 1 - \exp\{(\lambda_i + \mu_i)t\} \,]$$

$$+ \int_0^t c \exp(-cs) \sum_{i=0}^{n} [\, \frac{\lambda_i}{\lambda_i + \mu_i} + \frac{\lambda_i}{\lambda_i + \mu_i} \exp-(\lambda_i + \mu_i)s \,]\, ds$$

In the case of this subsea system, the components are identical. So, the development of the Kumamoto-Henley expression provides:

$$Q(t) = c \sum_{0}^{n} [\, \frac{\binom{n}{i} a^{i} q^{(n-i)}}{(\lambda + \mu)(n-i) + c} * (\, 1 - \exp-\{(\lambda + \mu)(n-i) + c\}t$$

$$+ \exp-ct \, (\frac{\lambda}{\lambda + \mu})^{n} \, (\, 1 - \exp-(\lambda + \mu)t \,)^{n} \,]$$

As can be observed, this model does not take into account the probability of a failure caused by a common cause at time t given that the system was operating before t.

3.2 Marshall-Olkin model

This model uses a Markov transition diagram (Figure 2) to estimate the probability of failure of all system. In order to consider the probability of occurrence of a common cause failure, all states which contain any element "up" are related to the state which represents the failure of the system by a transition rate c.

3.3- The Environmental Common Cause model (new)

In order to distinguish the repair transition rates between a failed state caused by a common cause failure and a failed state caused by different reasons, a Markov transition diagram was set up (Figure 3). It is assumed that if an environmental common cause occurs the system will be repaired completely. Thus, an important point in this model is that whatever the state of the system when the common cause occurs, the system goes at once to the 0 up, n down state and is returned by multiple repair at rate μ_c to n up, 0 down state. Any other repair required are assumed to be subsumed by the repair of the common mode.

131

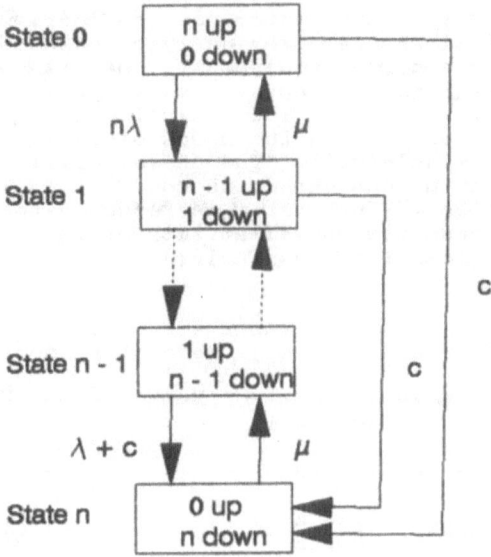

Figure 2. MARSHAL-OLKIN MODEL

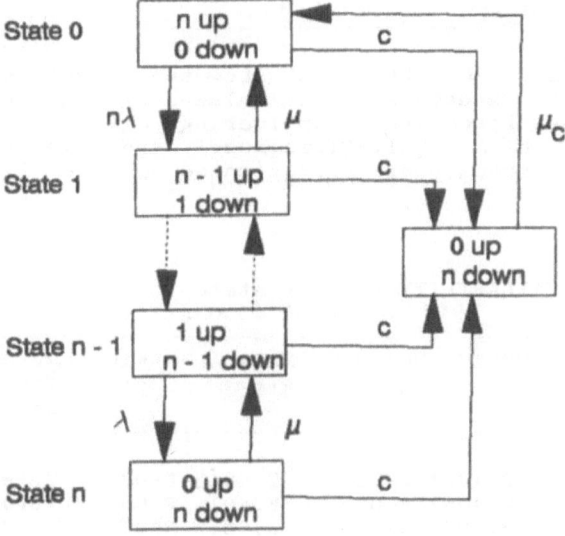

Figure 3. ENVIRONMENTAL COMMON CAUSE MODEL

4- RESULTS

The unavailability of the system not considering the effects of
common cause failures was previously calculated. Because the
wells share some common equipment, unavailability was calculated considering two aspects: a single well and the complete
Template-Manifold system. The unavailability of a single well
was estimated considering the modes of failures which could
only affect one single well. The unavailability estimated for
the complete system considered the modes of failure which could
affect all wells . The system was assumed to work with gas
lift. TABLE 1 shows the unavailability at the steady state as a
function of the mean time to repair.

TABLE 1
UNAVAILABILITY (Steady state); Mode of production: **gas lift**

MTTR (Hours)	Single well	Template-Manifold
24	5.5222 E-04	1.8219 E-04
48	1.1039 E-03	3.6439 E-04
72	1.6551 E-03	5.4660 E-04
96	2.2058 E-03	7.2882 E-04
120	2.7560 E-03	9.1106 E-04

TABLES 2A, 2B, 2C, 2D and 2E show the unavailability of the
system at the steady state calculated by the three models
previously described and considering different MTTR. The
results for the Marshall-Olkin model and the Environmental
Common Cause model were approximated by the finite Taylor
series expansion.

TABLE 2A
UNAVAILABILITY (Steady state); MTTR = 24h

c (1/hour)	HENLEY-KUMAMOTO	MARSHAL-OLKIN	ENVIRONMEN-TAL MODEL $\mu c=7200h$
1 E-08	2.4001 E-08	2.4013 E-07	7.1995 E-06
1 E-07	2.4001 E-07	2.4013 E-05	7.1992 E-05
1 E-06	2.4001 E-06	2.4012 E-05	7.1946 E-04

TABLE 2B
Unavailability (Steady state); MTTR = 48h

c (1/hour)	HENLEY-KUMAMOTO	MARSHAL-OLKIN	ENVIRONMEN-TAL MODE $\mu c=7200h$
1 E-08	4.8006 E-08	4.8054 E-07	7.1995 E-06
1 E-07	4.8006 E-07	4.8054 E-06	7.1992 E-05
1 E-06	4.8006 E-06	4.8051 E-05	7.1946 E-04

TABLE 2C
Unavailability (Steady state); MTTR = 72h

c (1/hour)	HENLEY-KUMAMOTO	MARSHAL-OLKIN	ENVIRONMEN-TAL MODEL $\mu c=7200h$
1 E-08	7.2013 E-08	7.2119 E-07	7.1995 E-06
1 E-07	7.2013 E-07	7.2119 E-06	7.1992 E-05
1 E-06	7.2013 E-06	7.2113 E-05	7.1946 E-04

TABLE 2D
Unavailability (Steady state); MTTR = 96h

c (1/hour)	HENLEY-KUMAMOTO	MARSHAL-OLKIN	ENVIRONMEN-TAL MODEL $\mu c=7200h$
1 E-08	9.6023 E-08	9.6209 E-07	7.1995 E-06
1 E-07	9.6023 E-07	9.6205 E-06	7.1992 E-05
1 E-06	9.6023 E-06	9.6197 E-05	7.1946 E-04

TABLE 2E
Unavailability (Steady state); MTTR = 120h

c (1/hour)	HENLEY-KUMAMOTO	MARSHAL-OLKIN	ENVIRONMEN-TAL MODEL $\mu c=7200h$
1 E-08	1.2004 E-07	1.2048 E-06	7.1995 E-06
1 E-07	1.2004 E-06	1.2030 E-05	7.1991 E-05
1 E-06	1.2004 E-05	1.2029 E-04	7.1946 E-04

5- DISCUSSION

On the basis of the results the following aspects can be pointed out:

5.1- Based on the results shown in TABLE 1, the probability of a failure of the system caused by a failure of all the wells together and not considering common mode failures is negligible when compared with the probability of a failure of the system caused by a failure of the equipments shared by the wells.

5.2- As can be seen in TABLE 2 the differences between the values found by Marshal-Olkin model and Kumamoto-Henley model is considerably large and approximately constant. The difference can be explained by the fact that Kumamoto-Henley model assumes implicitly the parallel repair of all components at the same time.

Using the expression developed in item 3.1, for the steady state, the Henley and Kumamoto model provides the following results:

$$Q = c \left[\sum_{i=0}^{n} \frac{\binom{i}{n} a^i q^{(n-i)}}{(n-i)(\lambda + \mu) + c} \right]$$

Given that, in this case, n = 10 and patterning a = 1 and q very small:

$$Q = \frac{c}{c + 10\mu}$$

So, in practical terms, at the steady state:

$$Q = \frac{c}{10\mu}$$

On the other hand, the Marshal-Olkin model provides the following expression for the last state:

$$pn = cP0 + cP1 + \ldots + cPn-1 + \lambda Pn-1 + \mu Pn$$

At the steady state $\dot{pn} = 0$. As P0 + P1 + ... + Pn-1 = 1 - Pn, then:

$$Q = Pn = \frac{c + \lambda Pn-1}{c + \mu}$$

It can be demonstrated that:

$$P_{n-1} = \frac{\mu}{\lambda + c + \mu} \, P_n \qquad\qquad (\, n \geq 1 \,)$$

In this particular case, $\mu >>> \lambda$ and c. So,

$$Q = \frac{c}{\mu}$$

which explain the differences found in the two models.

5.3- On the basis of paragraph 5.2, the following method can be used to estimate the availability of the system at the steady state, considering the Marshal-Olkin model. The approximation suggested can be useful if a computer program which provides a Taylor approximation is not available:

a- First approximation:

$$Q = \frac{c}{c + \mu}$$

b- Estimation of Pn-1

$$P_{n-1} = \frac{\mu}{\lambda + c + \mu} \, Q$$

c- Second approximation:

$$Q = \frac{c + \lambda P_{n-1}}{c + \mu}$$

d- Evaluation of the precision

$$r = \frac{Q}{P_{n-1}} \quad : \quad \frac{\mu}{\lambda + c + \mu}$$

if r > 1%, then go to b; if not, then stop.

In the case of the Environmental Common Cause model as the probability of failure of all wells together not considering common cause failures is practically negligible, the following expression can be used as an approximation of this model:

$$Q = \frac{c}{\mu c + c}$$

Regarding the state-of-art of failure data applied to subsea systems , which are mostly unavailable or do not exist , these approximations can easily and quickly provide a general idea of the importance of common cause failures to the availability of subsea systems, and their sensitivity to inaccuracy of the data.

5.4 As expected, the values found for the unavailability of the system, considering common mode failures, is strongly dependent on the values assumed for c and the model considered. TABLE 3 shows the maximum and minimum relations between the value of the unavailability caused by a common mode failure and the unavailability of the Template-Manifold caused by the equipment shared by all wells.

TABLE 3

Relative importance of common mode failures

Model	Minimum importance MTTR= 120 Hours c= 10-8	Maximum importance MTTR= 24 Hours c= 10-6
Henley-Kumamoto	1.3176 E-04	1.3174 E-02
Marshall-Olkin	1.3224 E-03	1.3180 E-01
Environmental model	7.9159 E-03	3.9490 E 00

The definition of the value of c to be used is not only a technical task but also a management decision. As known, an offshore structure is usually required to resist a 100 year wave (1.14 E-06 hours-1). On the other hand, the "Guidelines for Safety Evaluation of Platform Conceptual Design" (6) as-sumes that events for which frequencies are less than 1 E-04 /year (1.14 E -07 hours-1) represent residual risks. The value of 1 E-07 hours might be a good suggestion to quantify the effects of common cause failures at this stage. As can be observed, in this case the value found by the Environmental Common Cause model is 39% above the unavailability of the system if common causes are not considered.

6- CONCLUSION

6.1- Common mode failures are an important aspect to be consi-dered in a Risk Analysis study of a subsea system. Depending on the the common cause conditional failure intensity and the model assumed the total unavailability of the system can be significan-tly affected by common mode failures. In the case of the concep-tual design of the Template-Manifold of Albacora, the unavailabi-lity of the system is increased about 39% if the probability of occurrence of the common cause is assumed to be 1.0 E-07/year and Environmental Common Cause model is used .

6.2- Despite the inherent difficulties, a gross estimation of the effects of common mode failures can be easily performed if some approximations are assumed. As, at the moment, failure data of subsea systems are also mainly estimated, the results obtained can be considered plausible.

6.3- The effects of common cause failures estimated by the Henley and Kumamoto model was significantly lower than the effects of common cause failures estimated by Marshal-Olkin model because the Henley and Kumamoto model implicitly assumes the repair of all equipment at the same time . As , in the case of subsea systems, the resources are expected to be limited the use of the Henley and Kumamoto model can underestimate the effects of common cause failures concerned with the availability of the system.

6.4- In the case where a common cause failure implies in a time to repair significantly greater than the normal mean time to repair the system, the Henley and Kumamoto model and Marshall-Olkin model are not completely suitable. The model suggested, called the Environmental Common Cause model, estimated the unavailability of the system as substantially greater than the other models.

7- REFERENCES

1- HUGHES, R.P. . A New Approach to Common Cause Failures, Reliability Engineering, 1987, 17, p211.

2- FLEMING, K.W. and al. A Systematic Procedure For the Incorporation of Common cause Events into Risk and Reliability Models. Nucl. Eng. and Design. 1986, 93, p245.

3- DHILLON ,B.S. and SING, C. . Engineering Reliability: New Techniques and Application. John Wiley and Sons, U.S.A., 1981. ISBN-0-471-0514-8.

4- HENLEY E. and KUMAMOTO, H. . Reliability engineering and risk assessment. Prentice Hall, U.S.A., 1981. ISBN-0-12-772251-6.

5- Mc CORMICK, N.J. . Reliability and Risk Analysis. Academic Press, U.S.A., 1981. ISBN-0-12-482360-2.

6- NORWEGIAN PETROLEUM DIRECTORATE. Guidelines for safety evaluation of the platforms, 1981.

7- HAGEN, E.W. . Common -Mode/Common-Cause failure: A Review. Annals of Nuclear Energy, 1980, 7, p509

8- JOLLY, M.F. and WRAATHHALL, J. . Common-Mode Failures in Reactor Safety Systems. Nuclear Safety, 1977, 18(5), p624.

9- APOSTOLAKIS, G.E . The effect of a certain class of potential common mode failures on the reliability of redundant
systems. Nucl. Eng. design., 1976, 36, p123.

10- MARSHALL, A.W. and OLKIN, I. . A multivariate exponential
distribution. J. Am. Statist. Assos. . 1967, 62, p30.

DEPENDENT FAILURES ANALYSIS :- DEFEND DATABASE

I.K.GIBSON
R.H. MATTHEWS

Safety & Reliability Directorate,
AEA Technology, Culcheth, Warrington,
Cheshire WA3 4NE, UK.

ABSTRACT

The objective of this paper is to report progress on the development of
the DEpendent Failures ENgineering Database (DEFEND). This PC-based
database has been developed to support Dependent/Common Cause failure
analysis within safety and reliability studies.

The paper describes existing failure event data sources and the
extraction, structuring and coding of information which has been assessed
as desirable to facilitate dependent failures analysis. Details of the
present status of the DEFEND database is presented together with a brief
review of some of the data currently held within the database.

INTRODUCTION

In recent years many models have been developed appertaining to dependent
failures analysis. Difficulties have been encountered with the
application of such models and the incorporation of dependent failures
analysis within safety and reliability studies. These difficulties are
perceived not to be entirely due to a lack of accurate models. They can,
however, be attributed to the deficiencies of the supporting data.
Relevant data may not be readily available, or of acceptable quantity or
quality, or in a usable format to be effectively used by the theoretical
models. DEFEND has been developed to address this problem.

The development of the DEpendent Failures ENgineering Database (DEFEND)
has been undertaken to attempt to provide a means of presenting available
data in an upgraded and systematic manner within a PC based database. The
desired objectives being, that the maximum amount of relevant information
is extracted from the available source data and that the database
information can be utilised in support of as many different dependent
failures models as practicable.

SOURCE DATA

Existing data which could possibly be utilised within the DEFEND database
for dependent failures analysis was not originally collected specifically
with the purpose of supporting such analysis studies. Thus the development
of the DEFEND database has had to overcome the difficulties in using such
data and thus additional review and coding methodologies have had to be
developed.

Source data falls into two broad categories:-

1. Event/incident reports

This type of data usually consists of a descriptive text record of what occurred during an unusual or non-routine incident or event. This data is usually collected for Regulatory Authority or management safety investigations purposes. Examples of such data within the nuclear industry are Licensee Event Reports (LER) and Nuclear Power Experience Reports (NPER). Event reports of this type can vary in complexity from a laconic summary of a simple independent failure of a particular component to complex event scenarios pertaining to numerous related component failures consisting of both independent and dependent failure sequences.

2. Component failure event data

This is the most common form of data available, consisting of individual component failure records. This information is usually collected by the plant staff for maintenance or reliability objectives. Ostensibly, all the events recorded are independent thus in order to detect dependencies within the data a systematic method of searching and sorting data has to be applied. Such methods have been proposed [1,2] so that failures with specific common features and attributes can be bracketed and investigated to establish if they contribute a dependent failure. An exercise has been undertaken [3] which has utilised the dependent failures identification methods from standard component failures data sources mentioned above. The dependent failures identified have not yet been incorporated into the DEFEND database to date.

The remainder of this paper reflects the work performed to date on data extracted from the descriptive event/incident type of reports, more specifically the LER & NPER data sources.

EVENT REPRESENTATION

As for the component failure event data the detection of dependent failures within the descriptive event/incident reports such as LER/NPER requires a structured investigation and presentation methodology. Los Alamos Technical Associates (LATA) [4] developed a scheme based on cause-effect logic whereby each component unavailability can be expressed in terms of a cause and its effect on a component state. For example, (Fig 1)

141

Fig 1 SIMPLE CAUSE–EFFECT LOGIC DIAGRAM

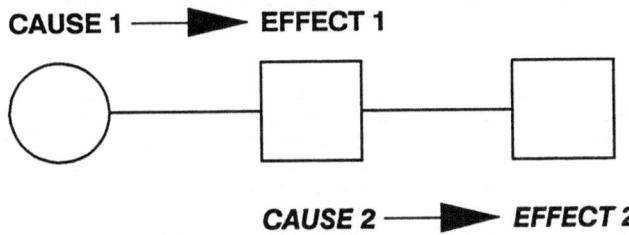

The root cause of the event [CAUSE 1] (denoted by the circle) has the effect [EFFECT 1] of making a component (denoted by the box) unavailable. The failure of this component causes [CAUSE 2] the failure of another component [EFFECT 2]. By this methodology an overall event description within a particular LER/NPER report can be dissected and expressed in a consistent manner. Several cause–effect logic chains may be present in each NPER/LER report. The logic expressed by these diagrams can be fairly complex and can represent a single cause having multiple effects and extensive cascades of component failures. (Fig 2)

Fig 2 MORE COMPLEX CAUSE–EFFECT LOGIC DIAGRAM

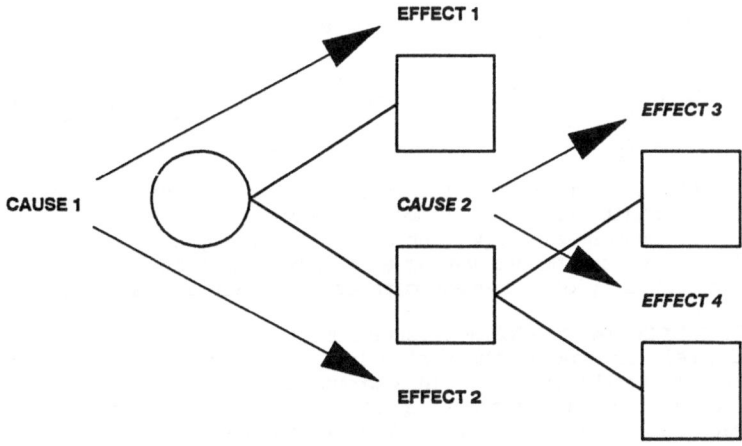

Note that a dependent event is an event that causes more than one effect. This can be a 'cause' with more than one effect (eg CAUSE 1 giving EFFECT 1 and EFFECT 2) (See Figure 2). That is a branching process. Alternatively it can be a cascade process (see figure 1) (CAUSE 1 gives rise to EFFECT 1 which in turn gives rise to EFFECT 2).

The LATA methodology has been adapted and modified by a number of research teams and the variants have been analysed and compared at SRD[5]. Pickard Lowe & Garrick (PLG) in their version described events on a component by component basis and this methodology has been adapted for the DEFEND database. The structure of each record in the database consists of a "top" form wherein descriptors of the overall event are recorded eg; where it occurred, its effect on the plant system etc. Linked to this form are items relating to the individual components involved in the scenario detailing such attributes as for example its failure modes, or the cause of its unavailability.

DATABASE INFORMATION

At present 46 data attributes are entered within the database for each event description and the components involved. The coding schemes developed for attributes such as root causes, modes of failure etc; have been adapted from existing widely used data coding schemes. It is evident that a balance has to be achieved between the ability to fully describe an event and difficulty created when too many codes are available thus increasing problems of interpretation.

The attributes identified at the outset of the development of the database were those assessed to be required to support many of the current dependent failures models and reflected the extent of the information that what was felt should be possible to extract from existing data sources.

The attribute information extracted for each event falls into a number of groupings -

   1. **SOURCE** - attributes designed to record the source and type of the source information and any cross references to any previous analysis or review of the event information.

   2. **LOCATION** - this grouping relates to the identification of the plant, plant system and components involved in the event. This group of attributes are important as it is usually this information that is used by by the analyst as a primary "sort" to assess the applicability of the event data to the analysis under study.

   3. **DESIGN INFORMATION** - ie attributes such as; the number of channels in system, the voting logic, minimum number of channels for system operability etc. This group of attributes are not normally recorded in failure event reports however such information is material for many dependent failures models.

   4. **QA INFORMATION** - although within the coding exercise it is attempted to reduce the subjectivity of the reviewer's interpretation of events it is impossible to remove it completely. In addition, the

analyst utilising the data may interpret certain aspects in a different way to the original reviewer. Thus to aid consistency and the ability to re-assess certain reviewer's interpretation certain QA information is included. Data recorded encompasses eg; reviewer identification, assumptions made by reviewer in coding the event etc.

5. EVENT DESCRIPTION - attributes designed to describe the event itself eg; effect on plant and system, status of plant at time of incident, the root cause or trigger of the event, corrective actions taken etc. Included in this grouping are attributes more specifically related to dependent failures analysis eg;

A) coupling mechanisms:- ie; the reason which propagated a trigger cause to affect multiple channels eg; same hardware, same procedures etc.

B) defences against initial trigger event and coupling mechanisms:- ie; different hardware, location or procedures which in the reviewers opinion may provide a defence against the event.

6. COMPONENT DESCRIPTION - attributes relating to the components involved in the event eg; component type, function and failure mode etc. Each component involved in the event needs to be uniquely referenced such that the overall cause-effect logic can be built up.

CURRENT DATABASE STATUS

DEFEND is a PC-based database set up for IBM compatible PCs and has been developed using an relational database software package.

To date over a 1000 events have been entered into the database. 474 events have been classified as "common-cause" dependent failures events. The other failures can be classified into categories such as linear or branched cascade events, single cause/single component failures; etc. The database allows the analyst to define what he considers to be a "dependent" failure.

The entry/coding of further data has been curtailed at the present time as the data already in the database is being assessed to see if any improvements can be made. The aim of the present review exercise is to; identify any areas of weaknesses in the data and to try and apply the data to support dependent failures models such as the Trigger Coupling model[6]. Any improvements required or additional information that are identified at this stage can thus be implemented in the further versions of the database.

The database at present includes a number of standard reports and the usual data extraction and viewing facilities. A menu driven search routine allows the extraction of data which is relevant to the study area of the analyst. The search routines are carried out on a level by level basis and the search information is stored at each stage such that if a particular search criteria is found not to detect suitable data then the same group of sorted data can be further searched by another criteria.

The data base structure is being continually developed and improved to provide additional standard reports/data extraction as they are indicated desirable by the users of the database.

EXAMPLES OF DATA

In an attempt to illustrate the type and scope of the information held within the current database, outlined below are a small number of examples of the data that can be extracted and some preliminary insights gained:

A) COUPLING MECHANISMS

As stated previously, the coupling mechanism is defined as an explanation why a single trigger event has a multiple effect on a number of components or system trains. In other words the feature of "sameness" that couples the components or trains in a common cause event. Nine overall groupings of coupling mechanisms are coded in the database (these overall groupings are further sub-divided in the database information). The occurrence of the various groupings for the coupling mechanism for the common cause events within the database are shown in Table 1

TABLE 1 COUPLING MECHANISMS

	%
MANAGEMENT & SUPERVISION	3.3
PROCEDURES	20.7
HUMAN ACTIONS	16.6
EQUIPMENT	8.7
QUALITY ASSURANCE	1.4
LOCATION	2.2
ENVIRONMENT	6.0
HARDWARE	39.9
TIMING	1.4

Thus it is seen that the database information indicates that the data reviewers have identified the major coupling mechanism as Hardware ie the trigger event affects a number of similar types of component hardware. Other coupling mechanisms which appear to be quite significant are Procedures and Human Actions. It is also seen from more detailed study of the coupling mechanism data within the database that over 50% of the coupling mechanisms assigned to procedures and human actions are connected with maintenance activities.

Another attribute that is contained in the database is the reviewers perceptions on the defences which could be utilised to prevent the coupling mechanism applying and preventing the trigger event affecting multiple components and channels. The reviewers assigned defences against the coupling mechanisms are shown in Table 2:

TABLE 2 DEFENCES AGAINST COUPLING MECHANISMS

	%
MANAGEMENT & SUPERVISION	4.5
PROCEDURES	32.6
HUMAN ACTIONS	3.8
EQUIPMENT	4.9
QUALITY ASSURANCE	22.3
LOCATION	0.1
ENVIRONMENT	1.1
HARDWARE	25.8
TIMING	0.8

It is observed that the reviewers have considered that different and improved maintenance/operating and QA procedures are important defences against common cause events. This result may not be too surprising but it is interesting that nearly a third of the events are assessed to possibly have had an adequate defence provided by different and improved procedures. It is also interesting that changes to procedures appears more important than the more obvious defence of establishing diversity in component hardware. (It should be noted the assumption has been made by the reviewers that the defence of different hardware was feasible for all cases ie; no account was taken for cost and practicability aspects for the option of introducing "diverse" hardware).

B) ROOT CAUSE OF TRIGGER EVENT

The root cause of the trigger event is also classified in the DEFEND database. The top level the groupings were as shown in Table 3.

TABLE 3 ROOT CAUSE OF TRIGGER EVENTS (TOP LEVEL CODES)

		COMMON CAUSE EVENTS %	OTHER EVENTS %
DESIGN	(D)	18.8	9.7
MANUFACTURE	(F)	4.6	1.4
CONSTRUCTION	(C)	4.2	3.4
MAINTENANCE	(M)	15.7	14.8
OPERATION	(O)	10.7	6.3
ENVIRONMENTAL STRESSES	(E)	35.8	40.8
OTHERS + UNKNOWN		10.3	23.5

It is observed that Environmental stresses is the most commonly assigned root cause. This finding is to be expected as this grouping contains such items as electrical and mechanical failure, material interaction (wear, erosion/corrosion) etc.

If we now examine the data at the lowest level of coding the most common assigned root causes are as shown in Table 4:

TABLE 4 ROOT CAUSE OF TRIGGER EVENT

		No. OF ENTRIES		
		TOTAL	COMMON CAUSE EVENTS	OTHER EVENTS
DESIGN ERROR OR INADEQUACY	(D20)	183	42	141
UNKNOWN	(U00)	150	85	65
MECHANICAL FAILURE	(E100)	82	46	36
DEFECTIVE CAL. PROCEDURES	(M23)	71	9	62
DEFECTIVE MAIN. PROCEDURES	(M20)	69	33	36
WEAR	(E96)	68	36	32
ELECTRICAL FAILURE	(E40)	67	40	27
CONSTR/INSTAL/COMMISSION ERROR	(C10)	66	24	42
MATERIALS INTERACTION	(E90)	61	7	54
DEFECTIVE OP. PROCEDURES	(O20)	59	18	41
SET POINT DRIFT	(E46)	58	47	47

It is observed that certain root cause assignments of the coders/reviewers
are much more common within common cause events than in other types of
events eg; design errors, defective calibration procedures, defective
operating procedures. This fact may have been expected as causes such as
defective calibration procedures could be envisaged to have a strong
impact on common cause dependent failures ie if these are wrong then
obviously a number of components/redundancies are likely to be affected.

Although detailed analysis of the database information has still to be
carried out some insights can be obtained from the preliminary form of
review illustrated above. Data of this nature can substantiate possible
knowledge eg; what defences against dependent failures should be
addressed. Obviously the basic design of the plant has to analysed to
protect against dependent failures but the data has indicated that it is
important that procedures are carefully investigated to alleviate against
dependent failures.

DISCUSSION

Numerous other preliminary reviews of the data within the DEFEND database
are currently being carried out. This activity is giving the database
developers valuable information on what is in fact achievable from
existing data and identifying the inadequacies within the source data of
the event\incident type utilised to date. For example, the LER/NPER source
data is very sparse on information on the effect of the fault on the
system or plant and thus in many cases this has had to be inferred. In
addition, the LER/NPER data is variable in content, some reports are very
brief giving minimal information while others give detailed descriptions.

In addition, it is evident, that many of the fields within the database
have a large degree of subjectivity placed within them by the analyst eg
the root cause of the event, the coupling mechanism, etc. Currently

exercises are being carried out to assess the variability in the assessments of various coders/reviewers. Another area being studied is the consistency of a particular reviewer in coding similar events.

QA/review procedures are currently being undertaken on the data and the suitability of the various coding structures are being re-assessed particularly in the areas where subjectivity is involved within the interpretation. It is hoped that alteration of the codes and the structure may assist in the improvement in the consistency of the data. Overall the difficulties encountered in coding the data are those that could be envisaged with the analysis of secondary data ie data not originally recorded for the purpose it is now being used.

Preliminary analysis of the data is continuing and insights into the data obtained. The database is about to be utilised to investigate the usefulness of the data in support of dependent failures models, specifically by the use of data within the trigger-coupling model[6] currently under development within SRD.

SUMMARY

The DEpendent Failures ENgineering Database (DEFEND) provides a computerised record of failure events which can be utilised within dependent failures analysis. The data is stored in a systematic manner and aims to provide a consistent review and analysis of the event reports from various data sources.

To date over 1000 failure reports have been reviewed and entered into the database. Currently the content of the database is being re-examined. This exercise is planned to:- highlight any inadequacies within the data; to gain insights from the current data; and to assess its usefulness in supporting current dependent failure models.

Difficulties encountered have been mainly concerned with the level of information from current data sources, and classifying and coding this information to the level that is thought desirable to support the analysis models. In addition, problems have been encountered ensuring consistency and reducing subjectivity in the interpretation and therefore the coding of the events.

The DEFEND package provides a menu driven enquiry facilities including a number of "standard" search routines/reports.

The data and database are actively being extended and modified to:-

i) cover a wider range of plant systems/components

ii) include data from additional sources.ie; incorporation of dependent failures identified from component failure data sources.

iii) reflect the needs of various modelling techniques.

iv) respond to additional requirements of any user.

148

REFERENCES

1. Games, A.M., Some aspects of Common Cause Failure Analysis in Engineering Systems, PhD Thesis, University of Liverpool, October 1986.

2. Games, A.M. et al, Exploitation of a Component Event Data Bank for Common Cause Failure Analysis, International ANS/ENS Topical Meeting on Probabilistic Safety Methods & Applications, San Francisco, Feb 1985.

3. Analysis of Data from Joint UKAEA/CEGB Dependent Failures Collection Exercise. To be published.

4. Los Alamos Technical Associates, A Study of Common Cause Failures, Phase 2, A Comprehensive Classification System for Component Fault Analysis EPRI NP-3837, Project 2169-1, Interim Report, June 1985.

5. Humphreys, P. et al, Progress towards a better understanding of Dependent Failure by Data Collection, Classification & Improved Modelling Techniques, Reliability 87 Conference, Birmingham, April 1987.

6. Ballard G.M., Dependent Failures Analysis in PSA, IAEA International Conference on Nuclear Power Performance & Safety, Vienna, Austria, October 1987.

SOFTWARE IMPACT ON SYSTEM RELIABILITY

IAN C. PYLE
SD-Scicon plc
Abbey House, Farnborough, Hants   GU14 7NA

ABSTRACT

There is currently much concern about the best way to take into account the possibility of systematic faults in a computer-based system. Techniques are available to determine the reliability in the face of random faults (principally hardware), but it has been found difficult to bring the concept of "software reliability" into the framework of those techniques. This paper suggests an approach based on the random occurrence of operational situations combined with possible systematic faults arising from residual (undetected) defects in software or hardware design. The approach is illustrated by three case studies of different kinds of system.

INTRODUCTION

Conventionally, the reliability of a system is determined from its hardware design, taking account of replication of components to provide redundancy. There is no room in such a structure for design defects. In a computer-based system, there is the possibility that mistakes have been made during the development, and the certainty that omissions have been made during the pre-delivery checking. These facts give grounds for scepticism about the correctness of delivered software, but do not immediately tie up with the (time-dependent) reliability of the software-intensive system. We can equate quality with the likely density of residual errors in the delivered software, but we need an additional factor to link it with the expected rate of random failures in the hardware.

During operational use, the variety of operational situations grows with the passage of time, within a domain which is very large and in practice impossible to test exhaustively. Thus there is a time-dependent probability that the actual data will cause execution of a path or condition in the software that was not tested. Such a path may or may not contain an error: the likelihood of it being faulty is unlikely to be better than that for the parts of the software that were tested before delivery. A condition that was not thought to be significant when the softwre was developed would similarly not have been tested. The present approach takes these possibilities into account, as an additional mode of failure, which is random because the actual occurrence of such an unextepcted situation is not at a predictable time. Software failures can thus be combined with the hardware reliability elements.

The idea of interpreting software reliability in this way has been used in several studies recently. The actual details are not available for publication, but their gist is expressed in the three case studies which follow. The numbers provided are illustrative only. The simplest case studied is a message-processor, where we can estimate the values for the three terms in the formula. Another is a development of a command and control system, which has gone through several stages of evolution, using increasingly sophisticated development technology. The third is a distibuted, high-availability system, in which the hardware reliability is determined by traditional methods, but the system reliability is determined from that by systematic analysis of the failure modes and recovery strategies. In each case the analysis gives us information about the reliability to be expected, and what too look for in further analysis to provide better estimates.

THREE-TERM FORMULA

The contribution to system reliability of systematic (principally software) faults is the rate at which, during operational conditions, a failure occurs as a result of a systematic fault which exists in the system. We can express this as a basic three-term formula:

$$S = pD * pV * pC$$

where
   pD is the rate of arrival of stimuli (operational data, events);
   pV is the proportion of stimuli whose responses are not verified before operation;
   pC is the density of faults in the software (as created)
and
   S is the rate of operational data causing execution of a path that contains a fault.

Note that these three terms reflect activities at significantly different periods relevant to the system: its operational use; its pre-operational checking (testing, verification and validation); and its invention.

The error density pC depends on the software development technology and quality control before verification. Sources of such faults are (a) deficient or ambiguous specification; (b) unjustified assumptions; and (c) bad logic in the development. The faults are likely to occur in any part of the design, so can be estimated by determining the proportion of software faults discovered during the pre-operational testing, on the assumption that the faults are equally likely to be made in the checked as in the unchecked parts.Verification gives both an estimate of their density, and the feedback for exposure of a proportion of them.

The unverified proportion pV depends on the coverage of the checking before deployment. A proportion of paths in the program are exercised by a particular test set, which we call pV_s: the static proportion unchecked. Various techniques are available for determining them, such as Test Execution Ratios (Hennel et al 1989). However, it is important to note that the distribution of operational situations is far from uniform, so that the frequently-occurring situations exercise a small proportion of the paths (and a large proportion of the paths correspond to rarely-occurring situations). Because of this, the proportion of *situations* checked is very much higher that the proportion of *paths* covered:

$$pV \ll pV_s$$

We must interpret the specified rate of working to take this into account. By testing over a period of time (applying data selected randomly in the valid and invalid domains) we distort the distribution to improve the coverage of the less common situations. The result of the combination is that the software contains a (static) proportion of undetected faults: $pV_s \times pC$.

The operational rate pD for the system is independent of the software or hardware (except that response times and turn-around delays can slow down the rate at which stimuli are handled).

We have here an interaction between the anticipated operational situations and the testing strategy. If an operational situation is recognisable distinct, then the partitioning of the test cases should have included a test which executed the same software and demonstrated acceptable behaviour. The converse of this condition, when an operational situation leads to behaviour which is *not* acceptable, indicates that the partitioning is inadequate, and that something about the operational situation is different from the tests *without that difference being realised before the situation occurred operationally.*

The above analysis applies for a single item of software, without considering its design in any greater detail. As with hardware, a more significant analysis can only be carried out after the system has been designed. The impact of software fault-tolerant design, with exception detectors and handlers, is to qualify the above analysis for individual software units, and indicate how the static measures for each unit should be combined according to the pattern of data and situations which it has to handle.

Checking

The checks applied to software before delivery may include static analyses as well as dynamic tests. Verification can be considered as the software analogue of accelerated life testing in hardware. Whereas it is necessary to find ways of increasing the probabilistic rate of random errors in hardware (by, for example, running the unit at a higher temperature) to establish the raw failure rate, in the case of systematic errors this is no problem: if the error happens once, it will happen (reliably) every time, and whatever happens with a particular situation will always happen in that situation. Thus the situations that arise frequently need be tested only once, giving time for testing the situations that arise only rarely during operational use. However, the characteristic of software-based systems (and modern VLSI hardware) is that exhaustive testing is infeasible: the number of possible states and data combinations is astronomical, and the variety of behaviour is too great for exhaustive testing.

Static analysis seeks to escape from this dilemma by identifying *regions* of the data space in which the behaviour is analytically analysed, rather than testing at single data points, and confirming that the behaviour determined by the software is expected throughout the region. This certainly improves the coverage of the checks, but still leaves open the possibilities of mistakes in the partitioning in the data space (in other words, in the conditions used to discriminate between different actions), and in the behaviour of the underlying hardware/software infrastructure.

If a particular situation is included in the set of tests carried out before operational use, we can presume that the resulting behaviour is correct (because if it did not do so, it exposes a fault in the software, which can be corrected before operational deployment). The domain of test data inputs must take account of all situations significant for the activity of the system. The proportion of operational situations that are exercised during pre-operational checking is (usually) intended to be completely representative, since it cannot be exhaustively complete. Bougé (1983) has analysed the implications of representative testing.

Operational conditions generate data and situations at a certain rate, covering a range dependent on the elapsed time of exposure. When an operational situation arises that is outside the verified domain (in other words, a situation that was not anticipated when the verified situations were prepared), it will execute software which has the original susceptibility to error, without the benefit of the verification exposure. Thus we get a time-dependent probability, which can be combined with the hardware failure rates.

Case 1: Message processor

A simple system to which this analysis has been applied is a message processor, in which messages received over a data link are checked according to the relevant protocol before being passed to the local destination for action. The configuration of the system was initially decided on the basis of hardware reliability prediction: the messages were passed along two parallel channels, and compared before disposal. The problem was to assess the reliability of the complete system including software (and implicitly to establish constraints on the software development methods to be used). Regrettably, it is significant and typical of such developments that considerable effort had already been put into the hardware design before software design was contemplated. This imbalance meant that details about the hardware reliability were based on an established structure, whereas those about software were not.

Thus we have to treat the software for the system as a single undifferentiated item, not yet designed or analysed in any more detail. The technology to be used for the software development was based on natural language specification, programming in Ada and comprehensive checking (in other words, with no prototypes to confirm the protocol and no formal specification of the intended behaviour.) The specification listed all the message types to be handled, giving the rules of the protocol for their checking and disposal for interpretation. It also gave the expected message rate (i.e. the channel capacity of the data link), but no indication of the relative frequencies of the different message types. The checking of the software was to be by a combination of three techniques: independent verification and validation, selected static analyses, and testing with random messages over a period of time before delivery.

On this basis, what can we usefully say about the likely correctness of the delivered software, and the reliability of the message processing system? According to the theory presented above, we look for the contributions from the three terms in the formula.

For the software creation in Ada, with its strong type consistency rules, package specifications and recompilation control, we can expect significantly better quality than with previous programming languages. There should be fewer occurrences of undetected bad logic in the development, but no change to the incidence of faults due to deficient or ambiguous specification, or to unjustified assumptions. The error density in delivered software is reported to reduce to $3_{10}-3$ per line using Ada [ ]. Thus the proportion of software that is probably faulty before checking is

$$pE = 3_{10}-3$$

The software checking, by three contrasting techniques, largely deals with the problems in the natural language specification and software/hardware interaction. The static analyses were selected to identify which sets of messages are handled by the same path in the software, to confirm (by the independent V&V) that the partitioning was consistent with the specification, and (by the tests with random data) that the proper behaviour was achieved for some messages in each set. Thus there is planned to be checking against deficient or ambiguous specifications, and confirmation of (at least some of) the assumptions. However, none of these techniques give any assurance of completeness, but their omissions would not be correlated.

In the given specification, there were about 100 message types, each with about 10 distinct conditions for processing. The planned testing includes every one of the message types and every distinct executable part of the software, but not all combinations of validity checking, or of hardware reconfiguration. The system is thus to be tested with data that cover the full variety of operational inputs: 100 message types (all covered), with two different actions for each (accept or reject the message) dependent on the message context (tested with representative situations for each action), and the variety of reasons for which a message may be acceptable or not. We estimate the test coverage as the reciprocal of the variety we have been able to foresee: since we can envisage 1000 possibilities (at this level of analysis), the gaps in our coverage are likely to be of order 1/1000:

$$pV_s = 10^{-3}$$

This figure provides a basis for quality control, as the target for unit testing. Combining the above two terms, we predict the residual error density in the delivered software as $3_{10}-6$ after checking. But because the distribution of stimuli is non-random, we presume at least an order of magnitude better than this: say

$$pV = 1_{10}-4.$$

The operational message rate is 1000 messages/hr. If the message types and contexts were randomly distributed, the testing would have missed 1 message per hour, and the probability rate for the software handling that message incorrectly would be $3_{10}-3$ per hour. But the non-rndom distribution implies that unexpected situations (of input data outside the tested domain) arise at

$$pD = 10^{-1} \text{ /hr}$$

Putting these values into the original formula (or, equivanlently, combining the rate of unexpected situations with the residual error density), we obtain

$$S = 1000 \times 10^{-4} \times 3_{10}-3 \text{ /hr}$$
$$= 10^{-1} \times 3_{10}-3 \text{ /hr}$$

$$= 3_{10}-4 \text{ /hr}$$

i.e. on average one failure from this source per 3333 hours.

This figure was combined with the hardware failure rates to give the total system reliability.

154

Case 2: Command and control system

The particular CCS concerned here is now being re-implemented for the third time, and we want to take advantage of the information obtained from the failures encountered during trials and operational use in its first and second implementations. Software design has not yet started, so we wish to estimate the likely reliability using a particular development style.

The first implementation was 200k lines long. During trials before acceptance 45 significant software errors were detected; in operational use, 24 different software errors were detected in 3 replicates over 2½years. Most of the errors occurred when the loading was heavy.

The second implementation, of a subset of the facilities, was only 40k lines long. This was released for limited use when 8 software errors were known; one new software errors was detected during four years of service in 9 replicates, and two different software errors were detected by other means. In addition, many specification changes were instituted in the light of trial use.

The third implementation is 400k lines long. During trials before acceptance 9 significant software errors were detected; no further errors have been detected during two years of operational use in 2 replicates.

For these systems, we do not separate the terms for creation and checking errors: the residual error density of the software delivered for trials or operationl use is estimated by the ratio of errors detected to size: $2_{10}-4$ per line in the first two systems, $2_{10}-5$ per line in the third (reflecting improved production technology).

The observed operational error rates are taken to indicate the rate at which operational situations arise outside the anticipated domain, and expose a fault which already exists in the software. Most operational situations outside the expected domain get handled satisfactorily, but 3 per year for each replicate in the first system do not. On the basis of the above fault density, the rate of unexpected situations is about 2 per hour in each replicate.

The second implementation was tested in a wider expected domain, so that the rate of unexpected situations was reduced to 2 per day per replicate.

The third implementation, assuming a worst case of still 2 unexpected situations per day, but $2_{10}-5$ probability of them reaching a residual software error, would thus fail at a rate of $4_{10}-5$ per day, which is $1.5_{10}-2$ per year. or MTBF of 67 replicate-years.

Thus we obtain a value for the softwrae component of reliability to combine with the hardware reliability values, which the raw observation did not.

Case 3: High availability system

This is a system which is well-developed and currently undergoing pre-operational testing. The design of the hardware and software is well-established. The failure modes are derived from its configuration, and failure rates estimated from quasi-operational use during the testing period.

Each distinct unit is considered as a possible source of failure, and each kind of failure that has a distinct repair strategy is analysed. Software failures may contribute directly, but most significantly by attempted automatic repairs that do not work. We list below the principal failure modes and repair strategies in which software is involved, and the resulting contribution of the software to the system unreliability (through residual errors in the software becoming exposed by operational conditions).

155

Failure mode	Recovery strategy	MTBF (sec)	Repair time	Repair fail	Software Contribution
Link loss	divert traffic	10^5	10^{-1}	10^{-1}	10^{-6}
Single user loss	log-on another terminal	10^6	10^2	10^{-2}	10^{-8}
Users' data lost	use back-up	10^4	10^2	10^{-1}	10^{-5}
Node fails	re-con figure	10^6	10^2	10^{-1}	10^{-7}

The positive contribution of the software to the system reliability is the improvement it permits the full configuration to have over the basic unreliability of the hardware. However, the residual faults in the software reduce this improvement to the repair failure probability divided by the MTBF (unless that is definitely because of hardware failures). We have taken the repair failure probability as indicating $pV \times pG$, and the MTBF as the inverse of the rate pD.

As a result of this analysis, the loss of users' data was seen to be significant, and steps taken to reduce the proportion of repairs to the database (from the back-up) that did not work.

CONCLUSION

Software is perfectly reliable – it may not do what had been intended, but it can be relied on to do the same thing in the same circumstances (within bounds for non-deterministic programs). Thus any perceived failures are in our expectations: software controls a plant to behave in a prescribed way, but being formal it does not take account of all aspects of use (which a responsible person would). If we expect a software-controlled system to cope with unpredicted situations, the fault is in our expectation, not in the software. This is the basis of what we call software (un)reliability.

Standards have a modest part to play in this context – mainly in setting acceptable probabilities for system failure, and by promulgating techniques that give appropriate confidence in the product. No amount of software checking can eliminate failures entirely.

While the evidence presented here does not actually confirm the hypothesis of the formula for software reliability, the analysis has shown that software reliability can be accomodated in calculations of system reliability by bringing the three factors together. Estimates of its value depend on our ability to predict the shape of the distribution of operational situations, as well as on the density of residual errors in the software. The approach gives no information about the likely *severity* of the failures that may occur. Neverthless, this appears to be an important step in integrating the reliability analyses of software and hardware in systems.

Acknowledgement

The ideas described here are based on work done at SD-Scicon, but do not necessarily reflect the official views of the company.

REFERENCES

Hennell M.A., Hedley D., and Riddell I.J.: Program analysis and systematic testing; in "High-integrity software", *ed* C.T. Sennett, ISBN 0-273-03158-9, Pitman 1989.

Bougé L.: A proposition for a theory of testing – An abstract approach to the testing process; DAIMI PB-160, Computer Science department, Aarhus University, Denmark (May 1983).

Carré B.: Program analysis and verification; in "High-integrity software", *ed* C.T. Sennett, ISBN 0-273-03158-9, Pitman 1989.

QUANTITATIVE & QUALITATIVE RELIABILITY ASSESSMENT OF SYSTEMS INCLUDING SOFTWARE

PAUL HELYER & COLIN DAVIES
Technology Group,
Aerosystems International
Scientific House, 40-44 Coombe Road, New Malden, Surrey, KT3 4QF

ABSTRACT

Established reliability modelling and analysis techniques are not readily extended to address whole systems which include software - it is difficult to incorporate software elements of a system within reliability block diagrams and no means of predicting the reliability of software elements has become established. An approach has been developed based on the technique of functional thread analysis - borrowed from software structured analysis and design methods - which allows reliability modelling and analyses of entire software intensive systems to be initiated from early in the system development.

INTRODUCTION

Software plays an evermore significant role in complex systems, yet reliability analysis techniques generally focus on hardware element of systems. It has been recognised that techniques are required that may be used as design drivers - providing input into the design and specification processes using only the information known at the time.

This paper discusses some aspects of an approach to the problem of performing appropriate reliability analyses on software intensive systems.

After a brief outline of some aspects of the problem of performing reliability analyses on software intensive systems, the paper proceeds to identify and discuss various features of the proposed approach. Finally some indications are provided as to the future activities expected to follow from the current work.

158

PROBLEM

The purpose in performing reliability analyses is to affect the end product (through its specification, design or production) such as to improve its reliability. To achieve maximum benefit from such analyses, they should be able to provide input throughout the whole of a system's life cycle and relating to all parts and levels of the system's specification, design and production.

Many reliability analysis methods require a considerable degree of detail to be available before they can be applied; this generally results in little, or no, reliability feedback being available to assist in the highest levels of the design. Furthermore, existing methods tend to address either software or hardware; rarely considering the reliability of the overall system.

Although the use of the negative exponential model of reliability has become established practice for hardware and various methods provide estimates of the hazard rate to be used for hardware elements of a system, no established method exists for predicting the reliability of software elements of a system. It is generally accepted that there are a large number of factors that affect the reliability of software (and, indeed, hardware). The pace of technological change has contributed to the problems of identifying and quantifying the effect of the numerous factors affecting reliability. Some mechanism needs to be identified to take account of the impact of new technology or techniques on the reliability of parts of a system with, possibly, no more than intuitive information about how reliability may be affected.

Software intensive systems generally perform multiple functions and may continue to provide significant functionality even in the event of a failure. In considering the reliability of a system it may be simplistic to consider the system as "two state" - either working or failed -since there may be a whole range of degraded modes of operation. Conventional reliability modelling methods do not readily distinguish between failures which cause loss of a single function and those which cause loss of multiple functions.

Reliability analyses are concerned with more than simply how often a system is expected to fail or even what proportion of functions is expected to survive a mission. The action of a system in the event of a failure may be of the highest importance and should be considered from the outset. Not only should failure mode analyses (eg. fault tree analysis and failure modes and effects analyses) be initiated early enough to impact the high level design decisions, but they should address the whole system - hardware, software and the interaction between them. In order to support these analyses, some means of deducing the likely or potential failure modes of various system elements needs to be developed.

SOLUTION

Overview

The approach to the solution of the problem of applying reliability methods to systems including software discussed within this paper has the following key features:

- a technique for producing a system functional model which may be initiated early within the system life cycle and applied to the hardware and software parts of a system;

- the use of high level working estimates of the reliability of elements of a system to enable the reliability characteristics of the system architecture to be examined before the detailed information about the system design is available;

- the identification of fundamental categories of failure cause and suggestions as to basic hazard rate models which may assist the production of element reliability estimates;

- the introduction of two new reliability parameters which will assist in modelling the degree of system degradation which may result from individual failure;

- the use of lists of generic failure modes for various system element types which, together with the application of the functional model, will assist in the performance of failure modes analyses.

Each of these features is described in more detail below.

Functional Thread Analysis

The functional model of the system which forms the basis for subsequent reliability analyses, is based on the use of functional thread analysis. Functional thread analysis is used within structured analysis and design techniques and is supported by most, if not all, such structured methods.

A functional thread is the sequence of actions through a system, subsystem or element that relates a single input to a single output. A functional thread may be dependent upon the state of other functions or elements of the system.

The functional model of the system covers all the functional threads of the system, identifying the various functional elements (processes, dataflows, stores, controllers) within the system. Individual elements may be subsystems in their own right and may be further subdivided in a hierarchical manner.

The use of functional thread analysis, within structured methods, starts with the analysis phase of the life cycle. At such time the system model will be a logical model, that is the elements of the model will be defined only in terms of their functions (actions on the data flowing through the system). As time progresses, not only will the model be modified as lower levels of detail are addressed, but also as physical features of the system (that is the manner in which the logical model is to be implemented) become defined.

Hardware and software are treated identically within the model: logical models will not even identify whether functions are implemented in software or hardware.

The use of functional thread analysis, coupled with suitable structured analysis and design methodologies, is seen as being beneficial to the system specification and design as well as the effectiveness of system testing, all of which will, in turn, contribute to the reliability of the end system; however, it is the use of functional thread analysis in reliability modelling and failure mode analyses which is the concern of this paper.

Conventional reliability block diagrams may be used to indicate the elements within each functional thread and any redundancy between them (redundancy within functional threads may only be identified when some physical rather than purely logical information is put into the model). The reliability block diagrams will also indicate elements of a system which provide the environment necessary for the functional thread to operate, but do not in themselves form part of the functional thread concerned. The environmental elements of the reliability block diagrams will relate to such aspects as the provision of electrical power, the hardware processor upon which a software process resides, the data highway where a particular data flow occurs etc.

Figure 1 shows an example functional model in the form of a data flow diagram for a hypothetical system. Figure 2 shows the reliability block diagrams which have been produced for the three functional threads within the system.

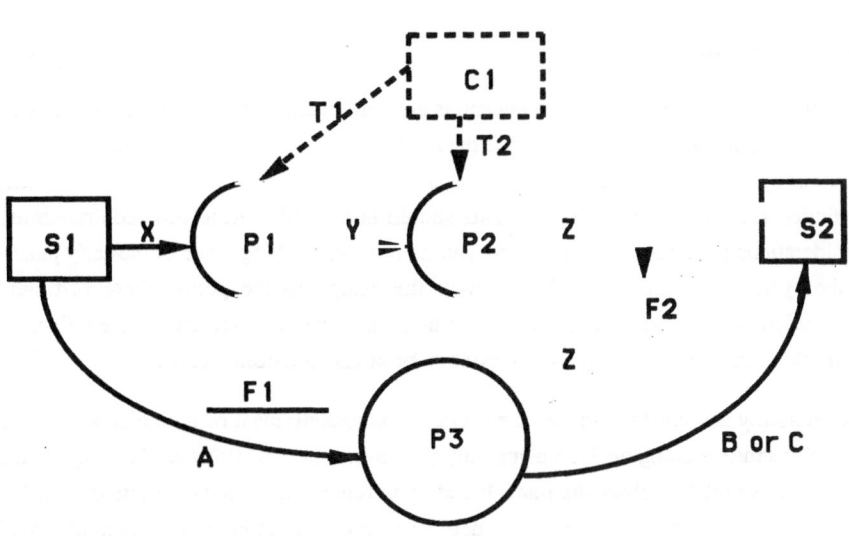

Figure 1. Example of a functional model of a hypothetical system.

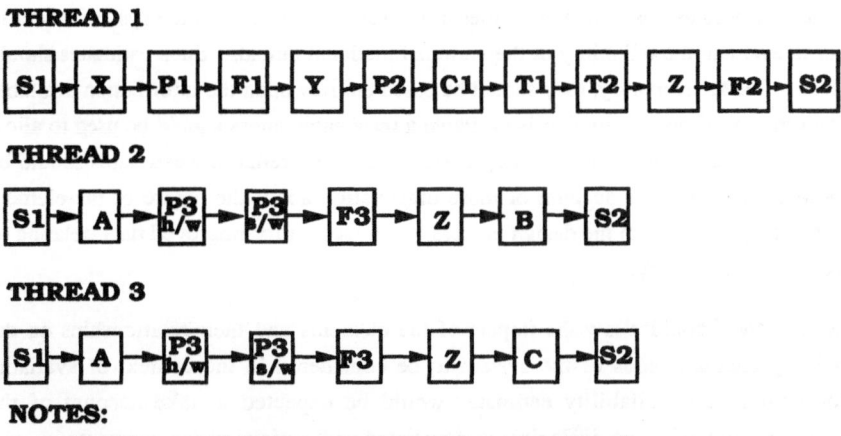

THREAD 1

$S1 \rightarrow X \rightarrow P1 \rightarrow F1 \rightarrow Y \rightarrow P2 \rightarrow C1 \rightarrow T1 \rightarrow T2 \rightarrow Z \rightarrow F2 \rightarrow S2$

THREAD 2

$S1 \rightarrow A \rightarrow \frac{P3}{h/w} \rightarrow \frac{P3}{s/w} \rightarrow F3 \rightarrow Z \rightarrow B \rightarrow S2$

THREAD 3

$S1 \rightarrow A \rightarrow \frac{P3}{h/w} \rightarrow \frac{P3}{s/w} \rightarrow F3 \rightarrow Z \rightarrow C \rightarrow S2$

NOTES:

1) **Both Hardware and Software aspects of P3 are considered**

2) **The threads are interrelated via F2**

Figure 2. Reliability Block Diagrams for the three functional threads within the
hypothetical system

Reliability Estimates

Early in the life of a system, when decisions are being made which will dictate the basic architecture and functionality and hence the fundamental form (and inherent reliability) of the final system, the detailed information necessary to permit use of many reliability prediction models is not available. This should not be allowed to preclude reliability considerations to feature in the decision making processes taking place in the early phases of the system development. To progress the design to the point where sufficient information is available to allow models to be used is clearly ineffective - the reliability feedback occurs far too late in the life cycle to be of use as a design driver.

The reliability associated with the basic system architecture must be modelled at the time the architecture is being decided using only the information available at that stage of the life cycle. Reliability block diagrams based upon functional threads identified at such a time may form the basis for such a model. The functional threads concerned may well not be completely defined (that is, some of the elements of the thread may still need to be further decomposed before they can be fully specified) and the elements of the thread may be logical rather than physical, nevertheless some of the reliability characteristics of the prospective system may already be examined.

It is necessary to estimate the reliabilities of the various elements of the system model in order to evaluate the reliability of the various functional threads. Such estimates should be recognised as temporary, they should be based entirely upon the information available at the time and if no information is available a reasonable guess should be used to allow the method to continue. As time progresses the element reliability estimates should be reviewed, either to take account of more information about the nature of the element concerned resulting from the design process, or to account for improved data relating to the element's reliability.

The estimates should allow the impact of the elements and their relationships on the reliability characteristics of the system to be considered in the context of available information. The reliability estimates would be expected to take account of the postulated complexity and difficulty etc. associated with various system elements.

Element reliability estimates may be made by a number of methods according to the information available. A combination of the "similar equipment" and "comparative complexity methods - comparing the reliability of an unknown element to that of some element of known reliability - is considered to be one of the best sources of quick reliability estimates of reasonable accuracy. It is important, however, to recognize the reasons for differing reliabilities associated with different elements.

Causes of Failure

In order to assist in the production of suitable estimates of the reliability of system elements, consideration is directed towards various causes of failure.

It is postulated that any failure may be attributed to one of four possible causes:

- the specification is incorrect;
- the design is not consistent with the specification;
- the production is not consistent with the design;
- the applied environment is not as specified.

Each of these causes in turn may have underlying causes such as: human error, lack of understanding or knowledge of some physical process; constraints applying to the system; etc.

It is suggested that each of the causes should be considered when determining reliability estimates. Different factors will be apparent in assessing the reliability associated with each cause.

The negative exponential distribution is suggested as a basic model of the reliability associated with each of the first three causes, not because it is considered to be a good fit, but because of its simplicity and the intuitive acceptance of the concept of the "constant hazard rate" which stems from it.

The basic model proposed is:

$$R = e^{-f(P) \cdot d}$$

Where

R is the reliability of the aspect of the element concerned (either specification, design or production).

P is the set of system element parameters considered in the model.

d is some function of the duration or period of use.

f   is some function of P which may be termed the hazard rate of the element.

$$f(P) = \frac{M(m_j) \cdot B(b_j)}{T(x_j)}$$

Where

B   is a basic hazard rate for the element which is a function of various scale and complexity attributes, b_j, of the element.

M   is a modifying factor which is a function of various qualitative and descriptive attributes (expressed as factors, m_j) of the element which detract from, or contribute towards, its reliability.

T   is a modifying factor reflecting the reliability growth of the element resulting from experience and improvements made to it. T is a function of various measures of age, x_j.

For each of the first three categories of failure cause, identified above, attributes may be identified against each of the functions B, M and T which can be used to assist in producing estimates of the reliability of various elements within a system. It is possible that models can be established, or customised, to provide reliability estimates for elements from the attributes that are known about them. The models would require a number of default values to apply to attributes which are "not yet known". Both the default values and the models themselves could be tailored to suit particular circumstances, latest thinking on reliability significant factors etc.

The reliability of the applied environment appropriate to an item should be considered separately. It will depend upon the reliability of other elements either internal or external to the system. In some cases these "environment failures" may be considered as special cases of specification failures since they may be considered as failures to capture all the requirements.

To quantify the reliability attributed to the environment of a given element, it is necessary to identify those parts of the system and the system's environment which form the environment of the element concerned.

System Reliability Measures

Conventional reliability modelling considers two system reliability parameters:

- logistic reliability - the probability of defect free operation - possibly expressed in terms of a mean time between defects;

- function reliability - the probability of a function being provided for a given period or purpose.

For a multi-function system the reliability of each function may be modelled separately. If the system can be classified as either working or failed solely in terms of the combination(s) of functions provided by the system, then a single model of the functional reliability of the system can be produced. In general, however, for a multi-function system, there will be a range of degraded modes of operation.

Conventional system reliability models do not distinguish between, say, element failures which cause complete loss of all system functions and those which cause loss of a single function only. Two new measures are introduced to assist is the design of systems will maximum functionality being available at the end of a mission or period of interest. These new measures have been termed the Reliability Index and the Independence Index. They are intended to be used to supplement the more conventional logistic and functional reliabilities described above, in the early stages of a system's life cycle.

The Reliability Index is related to the proportion of a system's functionality expected to survive a mission. It is evaluated by modelling the total system functionality in terms of the system functional threads as if the functional threads were independent. (ie. if there is no redundancy between functional threads in a system, the Reliability Index will be the product of the reliabilities for the functional threads.) The index does not represent a probability since an element which is part of two distinct threads will be "counted twice". However, this feature allows the Reliability Index to emphasise the effect of elements which contribute to more than one function.

The Independence Index is a measure of the proportion of system functions which is expected to survive an element failure. It is calculated by averaging the proportion of system functional threads which would survive loss of each system element. The simplest method is to calculate the averages purely according to the number of elements; a refinement of the method is to weight the average according to estimates of the failure rate of the elements.

A further refinement to both the reliability and the Independence Index is to apply weightings to each functional thread according to some criteria reflecting its relative importance to the system's usefulness.

Failure Mode Analyses

The system functional model resulting from functional thread analysis may also be of use in the performance of failure mode analyses (fault tree analyses and failure modes and effects analyses) on systems including software - particularly to support the conduct of such analyses during the early stages of the system life cycle.

Checklists may be produced which describe possible failure modes of various types of system element in general terms. By consideration of the particular role(s) of elements with respect to the functional threads to which they contribute, it should be possible to deduce a "customised" set of failure modes applicable to the elements within a specific system. These may then be used for the basis of a failure modes and effects analysis.

When determining the effects of a given failure mode, it may be useful to identify the functional thread(s) by which the effects would be manifested (together with the environment necessary to support the thread). This will aid the identification of areas of incomplete or incorrect analysis where the failed element under consideration is also required to function for the postulated effects to be manifested. (For example a processor failure may not be diagnosed by Built In Test (BIT) software if the BIT software runs on the same processor.)

Consideration given to the relationships between functional threads and to common environmental elements appearing in distinct functional threads may be of assistance in the identification of single point failures or in identification of instances where an apparently minor failure may have far reaching end effects.

It is considered that it is through distinct functional threads sharing common environments that some software faults may result in wide ranging, unpredictable effects.

Summary

Functional threads are identified and are used to describe a system. The functional threads may be defined in a hierarchical manner and will become better defined as time progresses. Reliability models are produced for each functional thread to a level of detail appropriate to the information available at the time.

Working estimates of the reliability of the elements, as identified within the functional threads, are produced, taking into account the information that is available and considering potential causes of failure and factors which may affect them.

Various reliability parameters for the system are then calculated which may be used to examine the inherent reliability characteristics of the system architecture. The characteristics considered will include the extent to which an element failure will affect the total system functionality and the proportion of functions expected to survive a mission or period of operation.

The identified functional threads and their elements may also be used to enable or assist the performance of failure mode analyses at an early stage in the system life cycle.

THE FUTURE

Draft working procedures have been produced to describe both the reliability modelling and the failure modes analyses methods. It is intended that a pilot study will be conducted to try out the procedures retrospectively on a real project. The pilot project has the primary purpose of ensuring the the the procedures are practical and that they are suitably documented. It is anticipated that working changes will need to be made to the draft procedures during the pilot exercise.

Should the pilot exercise show the procedures to be practical and useful, then a secondary aim of the pilot project will be the generation of material to be used for training exercises in the application of the procedures.

The models suggested for producing estimates of element reliabilities may be developed. Indeed, use of the models to guide educated guesses will assist in the collection of data necessary to produce better models. It has been suggested that individual organisations or projects may produce customised sets of parameters, factors and default values to be used with the models. Such sets would be peculiar to certain situations (eg. users, operating environment, particular manufacturers or designers etc.) and would reflect particular experience or concerns.

It is anticipated that the procedures would benefit from the use of software tools. Ultimately it is hoped that such tools could be integrated with structured analysis and design tools as part of an overall integrated product support environment.

CONCLUSION

Some aspects of an approach for conducting reliability analyses on software intensive systems have been discussed. The approach is compatible with various structured analysis and design methodologies and may potentially be automated within an integrated product support environment.

The approach enables reliability analyses to support the early, high level decisions which define the system architecture and its inherent reliability. It provides a method for considering the reliability merits of a system based upon what is known or what may be guessed at the time decisions need to be made.

The approach considers hardware and software elements of a system equally, recognising that it is the reliability of the end system that matters rather than the reliability of hardware or software parts of it.

The approach includes suggestions for models of the hazard rate contribution associated with various causes of failure. These models may be customised to account for the experience and concerns of different practitioners. The models may also be used as a means of modelling the intuitive impact of new techniques or methods on the reliability of elements of the system.

Two new parameters are suggested which distinguish between failures which cause loss of a single function and those which cause loss of multiple functions. These parameters may be of particular use in comparisons between design solutions before the system architecture is finalised.

Preliminary failure modes analyses may be conducted, providing reliability input into early design decisions, by using checklists of potential failure modes for various types of system element and considering the functional mechanisms within the system through which end effects may be manifested.

Overall, it is considered that the proposed approach to reliability analyses will assist in bringing the reliability analysis process closer to the specification, design and production process that "creates" the reliability of the end product.

ACKNOWLEDGEMENTS

The approach discussed within this paper has been developed by Aerosystems International under contract to the European Space Agency.

THE ROLE OF STATIC ANALYSIS IN SOFTWARE RELIABILITY ASSESSMENT

Manfred Kersken

Gesellschaft für Reaktorsicherheit (GRS) mbH
Forschungsgelände
8046 Garching, FRG

ABSTRACT

Static analysis as a part of software reliability assessment in general aims in the detection of faults and anomalies, in the preparation of tests and in the collection of metrics for the qualitative and quantitative description of software attributes. It is shown that in addition to these objectives the demonstration of compliance between functions which are implemented in a program and those which are specified (in the software requirements specification) can be achieved by means of static analysis.

INTRODUCTION

The three principles for the development of software which is required to fail only to a specified minimum probability are fault-avoidance, fault-removal and fault-tolerance. Fault-avoidance is achieved during the constructive task of establishing software along a predefined life cycle according to good software engineering practice. Fault-removal is the analytic task of finding and correcting faults which are remaining in spite of a careful software construction. Fault-tolerance aims in detecting failures during the execution of a program system including a following recovery procedure and subsequent failure-masking. Thus failures are tolerated which occur in spite of the constructive and analytic measures taken against faults.

As concerning the assessment of software reliability, licensing bodies and associated companies often receive software for certification only after it is already coded, these institutions can participate in V&V via assessing the V&V procedure of the producer, which is said to be passive V&V here. If, however, an assessor likes to take part actively in V&V he is left with the analytical techniques of fault-removal only. The main analytical techniques are static analysis and testing. For an independent assessor it is often impossible to establish test cases for software which was developed by others - in spite of the fact that good software should be designed to be testable. Static analysis is remaining as a flexible and powerful means for the assessment of software.

OBJECTIVES OF STATIC ANALYSIS

Static Analysis is often regarded as a means for obtaining a coarse overview of the software system like e.g. a calling hierarchy of subroutines, some representation of module interconnections and/or data dependencies.

Fig. 1 shows a general scheme of software development and static analysis.

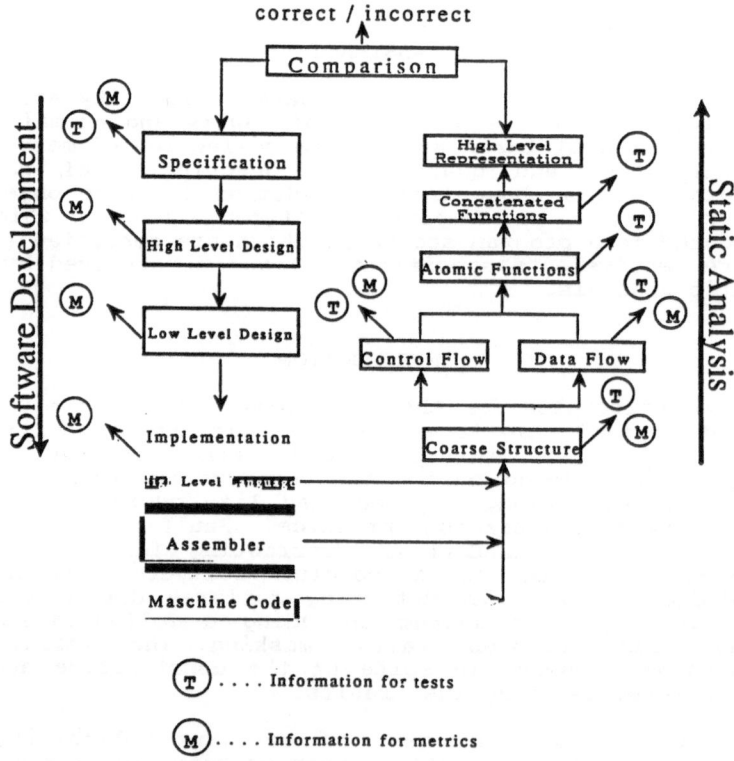

Figure 1. Software development and static analysis

The left-hand column is a simplified software development route (ignoring the verification steps at each level) which ends in the implementation of a program being represented as either a high level language, assembly language or machine code. Some highly safety critical applications require start of analysis from machine code as the ultimate form of a program. This is to avoid a rigid verification and validation of complex software like a compiler. A tool which supports this approach will be described later.

The column on the right-hand side shows a possible approach of static analysis. In a first step the coarse structure (e.g. calling hierarchy) is identified, afterwards the more detailed structure in terms of control flow and data flow. Up until this level static analysis is well supported by computerized tools. The next step is based on a common view of control flow and data flow, i.e. what happens with the data when a distinct path or set of paths is traversed. This is the identification of "atomic functions" which are (not well defined) small action, e.g. read an array, perform some linear operations on it and write the result in another array. These functions are then concatenated to bigger ones, etc. until all the program functions are represented in a way which allows comparison with the specified functions. If e.g. the specification contains logic diagrams, then the high level representation derived by static analysis should also be a logic diagram. During this procedure information for the preparation of tests and for the extraction of metrics can be obtained. Thus the four main objectives of static analysis are:

1. Collection of product metrics
2. Preparation of tests
3. Identification of faults and anomalies
4. Demonstration of compliance of implemented functions with specified ones

COLLECTION OF PRODUCT METRICS

The first step within static analysis is a meaningful partition of the source code to be analyzed. Partitioning includes the identification of the coarse structure of a program, e.g. in terms of subroutines, modules, functions, etc. as well as a breakdown into smaller units which form the basis for the representation of the control flow (e.g. basic blocks, sequential parts, linear code sequence and jump,...). This in turn requires the scanning through the source code instruction by instruction. Scanning as well as the identification of the coarse structure can provide for a lot of primitive metrics like no. of modules, no. of lines of code, no. of specific instructions, no. of operands/operators, etc. from which more sophisticated metrics can be computed (Halstead [1], McCabe [2]....). These metrics may be used to draw conclusions (compute correlations) to software quality attributes. The relationship between quality attributes and metrics, however, is still a matter of research.

PREPARATION OF TESTS

During the process of static analysis as shown in Fig. 1, the analyst gains a deep insight into the programs coarse structure (calling hierarchy, module interconnection) and fine structure (control flow). Also the functions which are performed by small units of code up until concatenated ones (realistically up until 50 to 100 third generation language instructions) are now known.

With this knowledge functional as well as structural tests can be prepared.

IDENTIFICATION OF FAULTS AND ANOMALIES

Several classes of faults can be found by means of static analysis. During preparation and performance of control flow analysis typing errors and code which is not reachable is recognized. Concerning assembler programs the use of wrong register and flags is indicated which may result in control flow as well as data flow errors.

Data flow analysis is concerned with the usage of variables; i.e. variables which are referenced but not defined (error message) or defined but never referenced (warning). Mode checking reveals anomalies in the mode of variables and constants in assignments and expressions and in the number and mode of formal and actual parameters of CALL statements (e.g. to subroutines and functions).

DEMONSTRATION OF COMPLIANCE OF IMPLEMENTED FUNCTIONS WITH SPECIFIED ONES.

The most ambitious objective of static analysis is the demonstration of compliance of the actually implemented functions with those which are fixed in the program specification (see Fig. 1). In this approach a reverse development procedure from a low level program representation (machine code, assembly language, high level language) to a high level one is involved. For a large part of this procedure no computerized tool support is available, i.e. it is performed purely mentally. This implies that the approach is a) error-prone and b) time consuming, thus in general it will only be applied to small programs. Both disadvantages can be drastically reduced if the program to be analyzed is well structured and readable. Readable in this context means that the code itself should be readable. It needs not necessarily to be well commented; in many cases it will even be delivered without any comments to the analyst in order to maintain independence between the developer and the analyst to avoid common errors. Another powerful possibility of reducing the error-proneness of the mental approach of reverse development is to specify test cases on the basis of the knowledge of control flow, data flow and functional behavior of the analyzed program. Such a test will be shown in the following example.

EXAMPLE OF STATIC ANALYSIS

The following example which is taken from a core protection system shows how static analysis of a piece of assembler code is performed in a purely mental way. In this procedure the source code is converted instruction by instruction into a graphical representation called the Control Flow and Data Flow Diagram CDD (Fig. 2), clearly illustrating control and data flow. The source code is split into sequential parts SP's starting either at the entry point of the module or at a label or after a branch.

In Fig. 2 the symbol SPi, i = Ø,...,22, denotes the SP s of the module and the lines between the SP s represent the control flow. On the right-hand side of the SP's columns are reserved for the register of the computer GØ,...,G7, for the constants K used in the module, and for the input and output area E and A respectively.

The data flow is recorded in this part, i.e. the right hand side of the SP s whereby the symbols denote:

X read access
O write access
▨ write access, where the previous content of the register/storage location is relevant, e.g. arithmetic or logic instructions.
→ direction of control flow or data flow

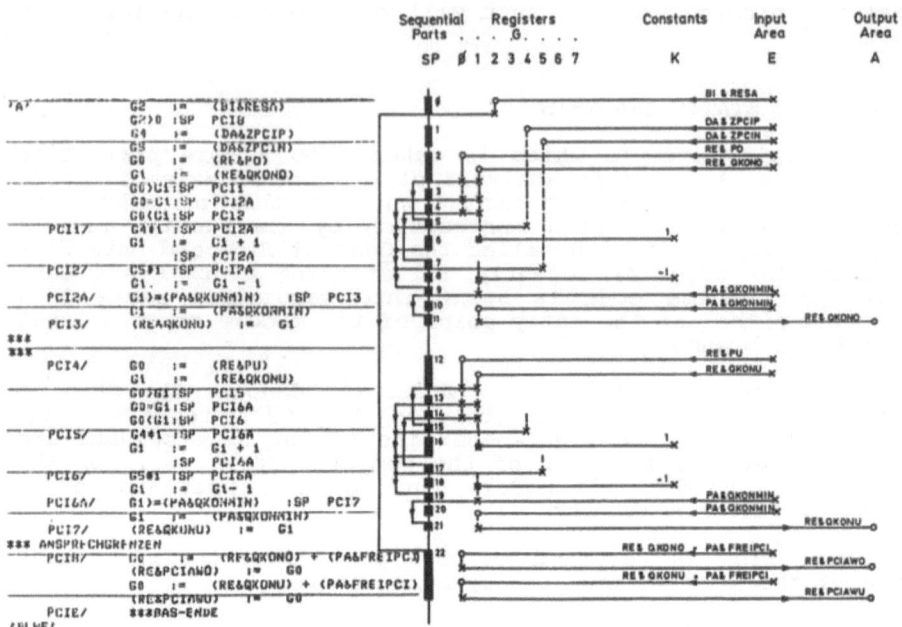

Figure 2. Code example with control flow and data flow diagram CDD

Fig. 3 shows the control flow in a manner more suitable for path analysis. Without detailed evidence it can be started that the number of paths in SPØ,22 is 14x14+1=197. The code, however, reveals some of them as not logical (shown in brackets), i.e. they are never used due to the branching conditions. The branching conditions in SP2,3,4 are GØ ≥G1, GØ = G1 and GØ < G1; thus, a jump is identified, within SP2,SP3 or SP4 and a transition SP4→SP5 is impossible. It is also noted that SP1,11 and SP11,21 reveal the same structure. This is explained by the same computations performed for the upper and lower part of the reactor core.

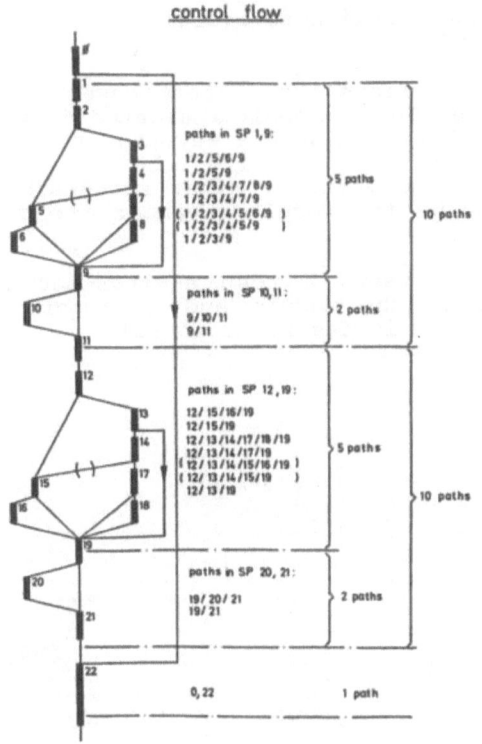

control flow

paths in SP 1,9:

1/2/5/6/9
1/2/5/9
1/2/3/4/7/8/9
1/2/3/4/7/9
(1/2/3/4/5/6/9)
(1/2/3/4/5/9)
1/2/3/9

5 paths

paths in SP 10,11:

9/10/11
9/11

2 paths

paths in SP 12,19:

12/15/16/19
12/15/19
12/13/14/17/18/19
12/13/14/17/19
(12/13/14/16/19)
(12/13/14/15/19)
12/13/19

5 paths

paths in SP 20,21:

19/20/21
19/21

2 paths

0,22

1 path

10 paths

10 paths

Figure 3. Control flow for path analysis

On behalf of the information from path analysis the number of test cases for checking the whole structure of SPØ,22 may be reduced to:

    10 test cases for SP1,11
    10 test cases for SP11,21
     1 test case for path SPØ SP22

i.e. 21 test cases to check the correctness of the analysts work.

Based on the knowledge about control flow and data flow from CDD and some additional information from the code, test cases may be specified for the module under consideration which allow testing of the structure and the functional properties as well. The basic testing criteria refer to the functional properties of the module. Additional tests with a path co-verage growing with increasing importance of the module - with respects to the contribution to overall software reliability - should be performed.

The table in Fig. 4 lists the 21 test cases stated above for the example module. The operands indicated in the head of the table have to be chosen in such a way that the operators OP in the branching conditions receive the logical values ($>$, $\leq, \geq, =, \neq$) of the table entries. After successful execution of the individual path, the result given in the column "action" must appear.

Of course for such tests a test environment must exist to enable execution of the module and to report test results. This test environment is described in [3].

path - no.	path	SP4 RESA OP 4	SP 2,3,4 PO OP QKONOold	SP5 ZPCIP OP 1	SP7 ZPCIN OP 1	SP 9 QKONOnew OP QKONMIN	Action	
#	S4/22	>					$<QKONO_{new}^0> * <QKONO_{old}^0>$	
1	S1/2/5/9/11	≤	>	≠		>	$<QKONO_{new}> * <QKONO_{old}>$	
2	S1/2/3/9/11	≤	=			>	"	
3	S1/2/3/4/7/9/11	≤	<		≠	>	" * "	
4	S1/2/5/6/9/11	≤	>			>	" * $<QKONO>_{old}$ +1	
5=2	S1/2/3/9/11	≤	=			>	" * $<QKONO>_{old}$	$<PCIAWO> *$
6	S1/2/3/4/7/8/9/11	≤	<		=	>	" * $<QKONO>_{old}$ -1	$*<QKONO>_{new} * <FREIPCI>$
7	S1/2/5/9/10/11	≤	>	≠		<	" * $<QKONMIN>$	
8	S1/2/3/9/10/11	≤	=			<	" * "	
9	S1/2/3/4/7/9/10/11	≤	<		≠	<	" * "	
10	S1/2/5/6/9/10/11	≤	>	=		<	" * "	
11=8	S1/2/3/9/10/11	≤	=			<	" * "	
12	S1/2/3/4/7/8/9/10/11	≤	<		=	<	" * "	

(upper part of core)

path - no.	path	SP4 RESA OP 4	SP 12,13,14 PU OP QKONUold	SP15 ZPCIP OP 1	SP17 ZPCIN OP 1	SP 19 QKONUnew OP QKONMIN	Action	
13	S11/12/15/19/21	≤	>	≠		>	$<QKONU_{new}> * <QKONU_{old}>$	
14	S11/12/13/19/21	≤	=			>	" * "	
15	S11/12/13/14/17/19/21	≤	<		≠	>	" * "	
16	S11/12/15/16/19/21	≤	>	=		>	" * $<QKONU>_{old}$ +1	
17=14	S11/12/13/19/21	≤	=			>	" * $<QKONU>_{old}$	$<PCIAWU> *$
18	S11/12/13/14/17/18/19/21	≤	<		=	>	" * $<QKONU>_{old}$ -1	$*<QKONU>_{new} * <FREIPCI>$
19	S11/12/15/19/20/21	≤	>	≠		<	" * $<QKONMIN>$	
20	S11/12/13/19/20/21	≤	=			<	" * "	
21	S11/12/13/14/17/19/20/21	≤	<		≠	<	" * "	
22	S11/12/15/16/19/20/21	≤	>	=		<	" * "	
23=20	S11/12/13/19/20/21	≤	=			<	" * "	
24	S11/12/13/14/17/18/19/20/21	≤	<		=	<	" * "	

(lower part of core)

Figure 4. Test table for the example module

TOOLS FOR STATIC ANALYSIS

As already mentioned the process of static analysis can be based on tools up until the step exhibiting control flow and data flow. Some tools provide for more than control flow and data flow but in most cases these additional features are only applicable under severe restrictions of the code. The control flow is represented by many tools similar as shown in Fig. 3, as they use a graph-theoretic approach to break down the source code to be analyzed. Thus the sequential parts of Fig. 3 are represented as nodes of a graph and the lines between the sequential parts become directed arcs of a graph.

THE FORTRAN ANALYZER FANAL

The FORTRAN-Analyzer FANAL is used for the analysis of control flow and data flow. Fig. 5 shows the structure of FANAL.

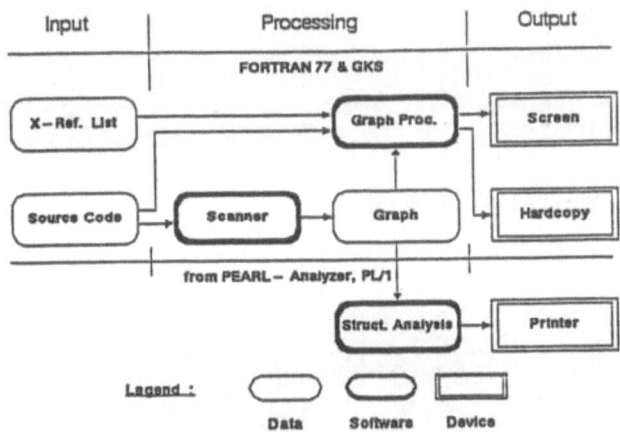

Figure 5. The structure of FANAL [4]

The main interfaces are:

- A text string (source code) as input.
- A cross reference list (normally produced by a compiler) as input.
- The representation of a directed graph in form of a data structure as output.
- Dialogue input.
- Some graphical representation of directed graphs as a user interface.

FANAL provides possibilities for an interactive traversal of the control flow graph to show the reachablility of nodes (forward, backward, both directions) for chosen parts of the graph. Fig. 6 is an example of a control flow representation.

178

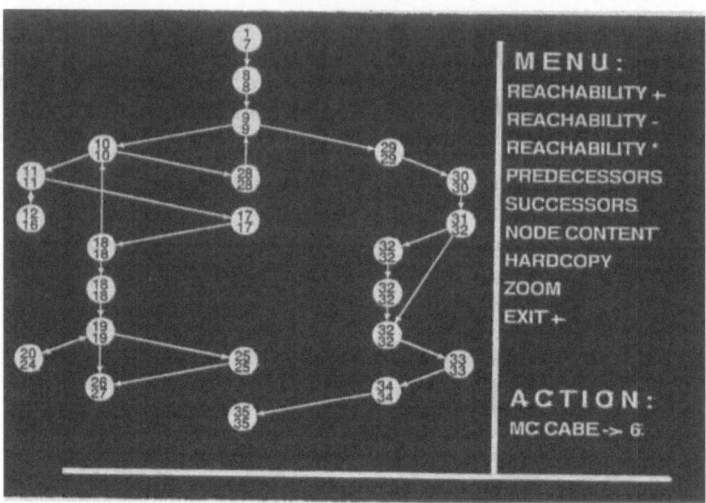

Figure 6. Control flow graph

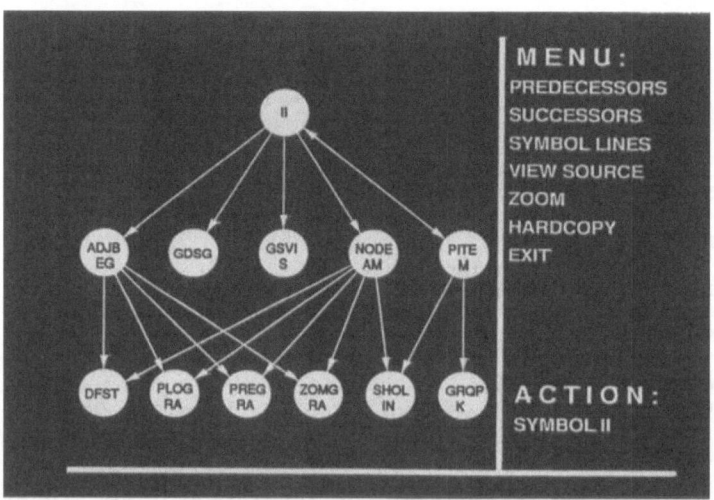

Figure 7. Data flow graph of the variable II

Concerning data flow, the tool enables the representation of data dependencies. A typical question is: What variables are

used to generate the value of a distinct variable. The answer
of FANAL to this question for a variable II is shown in
Fig. 7.

THE COMMON ASSEMBLY LANGUAGE ANALYZER STAN

The common assembly language CAL is defined on the basis of
the assembly language of the 8-bit processor Z80, 16-bit
processors MC68000 and INTEL 8086 and 32-bit processors VAX
and IBM Series-1. These different processors, which are
commonly used in many applications give hope, that CAL can
also cover the instruction sets of further microprocessors.

Assembly languages manipulate objects such as memory locations, registers, flags, ports etc. So in CAL all these
objects are treated as variables, which can be modified or
referenced. Assembly instructions often have implicit effects
on e.g. registers or flags. In CAL all these implicit effects
are represented explicitly by assignment statements to the
corresponding objects. Therefore each processor-specific
assembly instruction is represented in CAL by a set of one or
more expressions explicitly describing all the effects of the
assembly instruction. Thus all modifications of objects are
shown by assignment statements. This results in a uniform
handling of all objects and their usage can be analysed using
data flow analysis methods. CAL is described in detail in [5].

STAN converts the CAL-Code of each routine into a directed
graph. The nodes of the graph are the blocks (linear sequences) of CAL statements while the edges describe the
possible transfer of execution control between the blocks.
Structural analysis of this graph identifies all loops
inclusive their nesting. A subgraph is attached to each loop.
The resulting subgraph hierarchy reveals the structure of the
CAL-code (fig. 8)

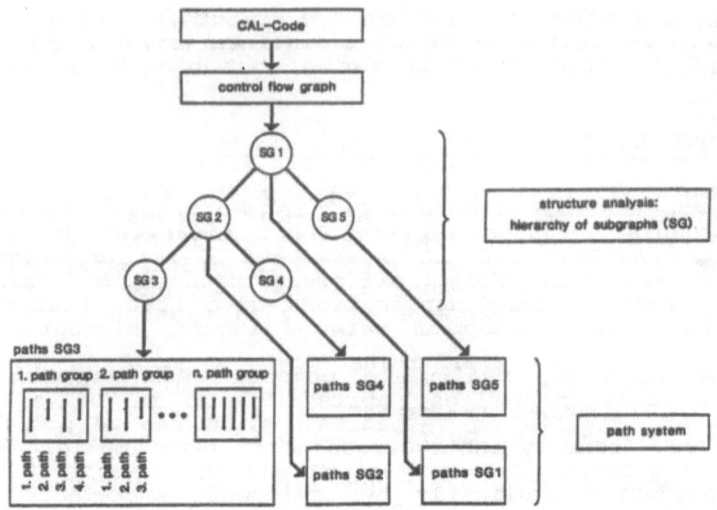

Figure 8. Scheme of the generation of program structure and path system

For each subgraph of each routine a print-out of the following details is given:

- type of the subgraph
- level in hierarchy
- number of nodes (blocks and sub-subgraphs)
- the start block
- the exit blocks
- the return or stop blocks
- the latch blocks (blocks with an arc back to the start block)
- the loops contained in the subgraph.

It is also possible to generate plots of these subgraphs.

During structural analysis STAN also generates some sub-ordinate results, such as: labels that are not used or irreducibility of a routine graph, one of the most serious violations of the rules of structured programming.

The set of all path of a graph which is in general infinite for practical programs has to be described in a finite form by a formal system. STAN generates such a system (Chomsky-2-Grammar) for the description of all path of a control flow graph. The Chomsky-2-Grammar consists of a finite set of path-sections and of a set of rules describing how to combine these path-sections. It is possible to address distinct paths thus supporting the generation of path predicates and test data.

Moreover a minimal C1-coverage is computed to ensure that the analyzed code can be tested by a minor number of test runs. C1-coverage is the smallest possible set of paths with the characteristic that it contains each edge of the control flow graph at least once.

Just as control flow analysis deals with uncovering of flow of control in a given program, the data flow analysis deals with definition-usage of variables in programs and tries to locate improper use of variables. In CAL, registers, ports, memory locations and flags are treated as variables and effects of instructions are expressed as definitions or references of these variables. The two types of anomalous situations dealt with by data flow analysis are:

- the use of uninitialised variables
- unused definitions of variables

Certain "gross" data flow anomalies can be discovered simply from the variable cross-reference for the program under consideration. If the cross-reference shows only references to a variable but no definitions then this is clearly a case of the use of an uninitialised variable. Similarly unused definitions can be detected. However, this approach is not always adequate. If for example cross-reference shows some definitions and some references for a variable it is not clear if each reference is preceded by a definition. It is then necessary to take into account various paths in the program and establish that irrespective of the path taken through the program a definition always precedes any reference to a given variable. Such an analysis is done by Reaching Definitions Analysis.

The reaching definitions analysis is carried out for each procedure or routine in the given program. The method combines information about variable references and definitions obtained from cross-reference with control flow graphs of routines obtained during structural analysis. The approach used by STAN is an iterative one, which works on the unreduced control flow graph of a routine as against some methods which require graph reductions and special node orderings. For a more detailed description of the data flow analysis of STAN see [6].

CONCLUSIONS

Static Analysis is used in general for the purposes of iden- tifying faults and anomalies, collecting metrics and preparing tests. In this paper it has been shown, that the more am- bitiuns objective of demonstrating the compliance of the actually implemented functions in a program and the specified ones can also be achieved by static analysis. This is done by reverse program development from a low level code representation of a program to some high level representation which allows comparison with the specifications. As the approach is hardly supported by tools, specific test cases should be specified to ensure that errors in the reverse development are detected.

REFERENCES

[1]   Halstead, M.H.
       Elements of Software Science
       Elsevier, New York, 1977

[2]   McCabe, T.J.
       A complexity measure
       IEEE Trans. on Software Engineering,
       Vol. SE-2, No. 4, Dec. 1976.

[3]   Kersken, M.; Rietzsch, L.; Mertens, U.
       Qualification of a computer system for the limitation
       of power density in a reactor core
       Proc. COMPSAC 84, Chicago, Nov. 7-9, 1984
       IEEE Computer Society

[4]   Puhr-Westerheide, P.
       FANAL-Description (in German)
       Internal Paper
       Gesellschaft für Reaktorsicherheit (GRS) mbH
       Forschungsgelände, 8046 Garching, 1989

[5]   Dahll, G. et al
       Tools for standardised software safety assessment
       Report of the OECD Halden Reactor Project HWR-211
       (May 1987)

[6]   Maertz, J.; Dhodapkar, S.D.
       Data flow analysis of common assembly language pro-
       grams by means of static analysis tool STAN
       Hardware and software for real time process con-
       trol,   (Eds. J. Zalewski, W. Ehrenberger)
       North Holland, 1989

TIME SEQUENTIAL RELIABILITY ASSESSMENT OF ELECTRICAL AUXILIARY SYSTEMS

R.N.Allan   and   T.O.Inga-Rojas

Electrical Energy and Power Systems Group
UMIST
Manchester, UK

ABSTRACT

This paper describes models and evaluation techniques for performing time sequential reliability evaluation of electrical auxiliary systems. All the models and techniques have been tested in a computer code. Some assumptions were introduced in order to arrive at a compromise between practical usefulness and objective assessments of the results. The basic approach combines different techniques including the use of network modelling, minimal path tracing techniques, minimal cut set analysis to identify initiating events, the event tree approach to assess restoration of supply and minimal cut set techniques to assess the guaranteed supply.

INTRODUCTION

Electrical auxiliary systems comprise broadly three subsystems: the main, the guaranteed and the essential supplies. The main supply feeds loads and drives which depend on the type of fuel used. If failures occur in this supply, the generating unit may be either forced into a derated state or tripped. The guaranteed supply feeds loads which are expected to continue to function with a supply of guaranteed quality, e.g. during large system disturbances, unit starting or shutdown. The essential supply system comprises standby emergency sources and drives, which are generally used for the safe operation and post trip cooling of nuclear stations and for black starting of fossil fired stations. A reliability assessment of all three subsystems is necessary.

A reliability assessment has different characteristics according to whether aspects of safety or adequacy are considered. These two aspects involve the analysis of modes of failures, causes of failures and consequences. It is important that adequacy and safety be evaluated as precisely as possible so that alternative designs may be compared with each other and with specific reliability targets. The approach used should not make unrealistic assumptions and approximations and should permit many possible systems and operating policies to be compared. Sensitivity studies should also be performed to identify weak areas in the system and to assess the impact of inaccurate component reliability data.

In adequacy assessment, the evaluation of limiting state reliability indices provides measures of unit and station availability. Typical examples of these indices are: expected failure rate, average outage duration and annual outage duration.

In many safety assessments, the system to be examined becomes available or operational to perform a specific task during a specified time following some system changes or disturbances. This implies that, depending on the task being performed, only particular subsystems of the overall plant have a significant role to play. Many aspects in electrical auxiliary systems such as: starting up and running standby emergency supplies, load shedding, load reconnection are events which entail operational procedures based on reliability and safety requirements. The quantitative assessment involved is time sequential and requires an estimate of the system probability of failure which is a function of: the switching actions carried out, components failure characteristics, number of tasks performed and system configurations. The average failure rate associated with a particular subsystem and unit can however provide useful input to this safety assessment procedure.

Many methods for system safety and risk assessment have been developed in recent years and applied to complex industries such as: nuclear, chemical and aerospace [1-9]. However, there is a need for further development of techniques for time-sequential analysis particularly those that provide efficient computational algorithms. This paper describes models and evaluation techniques developed at UMIST in conjunction with the CEGB for assessing the reliability of time sequential events of electrical auxiliary systems. The techniques are based on network analysis [10-11] and event tree methods [10,12,13] to assess the likelihood of recovering supply after a system failure and then to maintain that supply for a specified period without further failure.

SYSTEM DESIGN CHARACTERISTICS

Many designs are possible for the electrical auxiliary systems and final choices are based on factors such as; safety, capital costs, running costs and reliability. The main elements in all systems include:
a) Sources of supply- These are of two types. The first are the generating units and the electrical grid system. The second comprises standby generators and batteries which are used during emergency conditions.
b) Busbars - These are points in the system which interconnect supplies and loads. These include main supply busbars, essential supply busbars and guaranteed supply busbars. Each satisfies the function described previously.
c) Loads - These can be divided into four classes. Class 1 are essential loads which are normally connected and remain standing on the board in the event of loss of supply. Class 2 are non essential loads normally connected on the board but shed and left disconnected in the event of loss of supply. Class 3 are essential loads normally connected on the board, shed in the event of loss of supply and reconnected after supply on the board has been restored. Class 4 are essential loads normally on standby mode but connected on the board in emergency situations.
d) Other electrical equipment. These include all other network components such as transformers, cables, circuit breakers and protective devices.

SYSTEM OPERATIONAL CHARACTERISTICS

The operation of electrical auxiliary systems can be described as a series of control actions taken in order to maintain supply at each busbar in the system. These control actions, manual or automatic, are ruled by decision making processes based on operating modes of the system, available information, safety requirements, engineering knowledge, experience and intuition. Appropriate control actions are required to respond to the various states in which the system can be found. These are:

a) Normal operating state. In this state, the system is operated in such a way that supply to all the loads and drives is satisfied. The main objectives of control actions are to provide continuous supply.

b) Restorative operating state. This state arises when supply to one or more busbars has been lost usually following the occurrence of an abnormal event.

c) Emergency operating state. This state comes about when supply is restored and must be maintained for a specified but limited period of time.

PROBLEM FORMULATION

The principles of the design and operational characteristics described in previous sections must be taken into full account in formulating the problem and defining acceptable solution methods. The probabilistic approach described in this paper is centred on the development of models and techniques that recognise the different stages which take place in the time sequence of events. These stages are as follows:

a) The initiating event: this stage identifies failures which cause initial loss of supply at the load point of interest. It also represents the beginning of the time sequential analysis, and its occurrence in the time scale of events is assumed to take place at zero time.

b) The restoration process: this stage is concerned with the ability of the system in re-establishing supply within a short period of time defined as the mission time T1. The restoration process implies the acceptable loss of supply due to the occurrence of the initiating event.

c) The guaranteed supply process: this stage is concerned with the ability of the system in maintaining the guaranteed supply. This supply must be maintained continuously for a short period of time defined as the mission time T2. This stage follows the successful restoration of supply.

d) The continued availability of supply: this stage is concerned with the availability of the system in supplying power for a period of time T3. During this period, the load point of interest can tolerate down states of time duration T4 separated by up states of time duration T5.

These stages are shown in Fig.1. This paper considers each of these

TABLE 1 Minimal Paths to Busbar S3

path	status	nodes and branches	source
1	n/c	S3 S2 S1 G1 10 8 1	G1
2	n/c	S3 S2 S1 G1 10 9 1	G1
3	n/o	S3 S2 L3 10 4	L3
4	n/o	S3 S2 L2 10 3	L2
5	n/o	S3 S2 S5 L1 10 16 2	L1
6	n/o	S3 S2 S1 G2 10 8 5	G2
7	n/o	S3 S2 S1 G2 10 9 5	G2
8	n/o	S3 S2 S1 S4 W1 8 17 6	W1
9	n/o	S3 S2 S1 S4 W1 9 17 6	W1

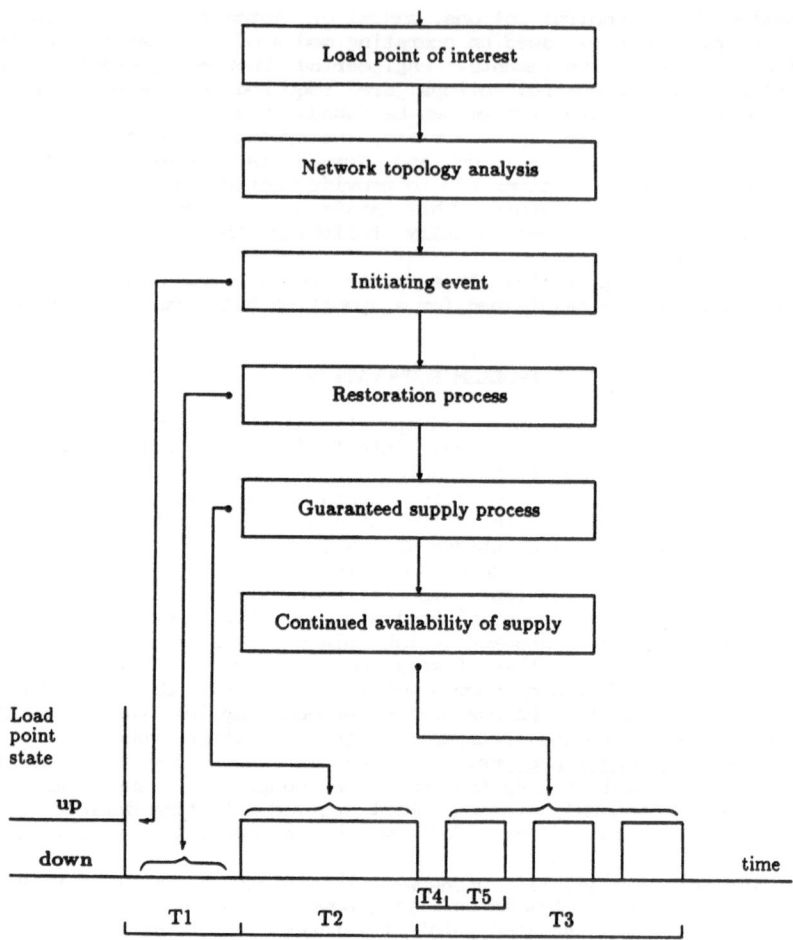

Figure 1 Summary of the time sequential assessment

stages except the last one, i.e. the continued availability of supply.
Therefore, the basic characteristic of this probabilistic approach is
concerned with identifying initiating events that lead to system failure
and subsequently with the evaluation of the likelihood of re-establishing
supply and then of maintaining supply for predetermined mission times T1
and T2 respectively.

NETWORK TOPOLOGY ANALYSIS

In order to quantify the sequential events in terms of reliability
measures, a qualitative assessment of the system is first required. This is
achieved in terms of the minimal paths that exist between the sources of
supply and the load point of interest. These minimal paths define that part
of the system whose intended function is to supply power under normal

TABLE 2   Initiating Events for Busbar S3

event failure mode   available n/o paths

1	S2	none
2	S3	none
3	B11	none
4	B12	none
5	T4	none
6	S1	3 4 5
7	G1	3 4 5 6 7 8 9
8	T1	3 4 5 6 7 8 9
9	B1	3 4 5 6 7 8 9
10	B7, B9	3 4 5
11	B7, T3	3 4 5
12	B7, B10	3 4 5
13	T2, B9	3 4 5
14	T2, T3	3 4 5
15	T2, B10	3 4 5
16	B8, B9	3 4 5
17	B8, T3	3 4 5
18	B8, B10	3 4 5

operating conditions, and are used to identify failure events leading to loss of supply at the load point of interest. They also define the various existing alternatives of supply to initiate a restoration process and identify the operational procedures to be carried out in order to recover supply. Finally they can be used to define that part of the system being operated during the emergency conditions and to identify the failure events which would cause a subsequent loss of supply.

The method used to determine these minimal paths and to identify those that are normally closed (normal operating condition) and those that are normally open (standby condition) is fully described in Ref [10,14].

The normally open paths represent the various ways that supply can be restored using standby sources. These paths are generally ordered quite arbitrarily by a computational algorithm. However the restoration process must follow required operational practices and the standby sources (normally open paths) must be considered in the appropriate order. Computationally therefore, the techniques must permit either specified default ordering of these normally open paths or allow the user to be able to select/order the paths according to required operational practices.

INITIATING EVENT

The loss of supply to a load point of interest operating under normal conditions can occur in many different ways. Each unique way is referred to as an initiating event. These represent the failure modes of a component or a combination of components. There are a number of established methods for deducing these failure modes. One of the most efficient is to deduce the minimal cut sets associated with the minimal closed paths found during the network topology analysis. A technique for deducing these minimal cut sets is described in Ref [14].

An additional qualitative aspect must also be deduced whilst identifying the initiating events from the minimal cut sets. This aspect is whether the loss of supply following each initiating event can be restored

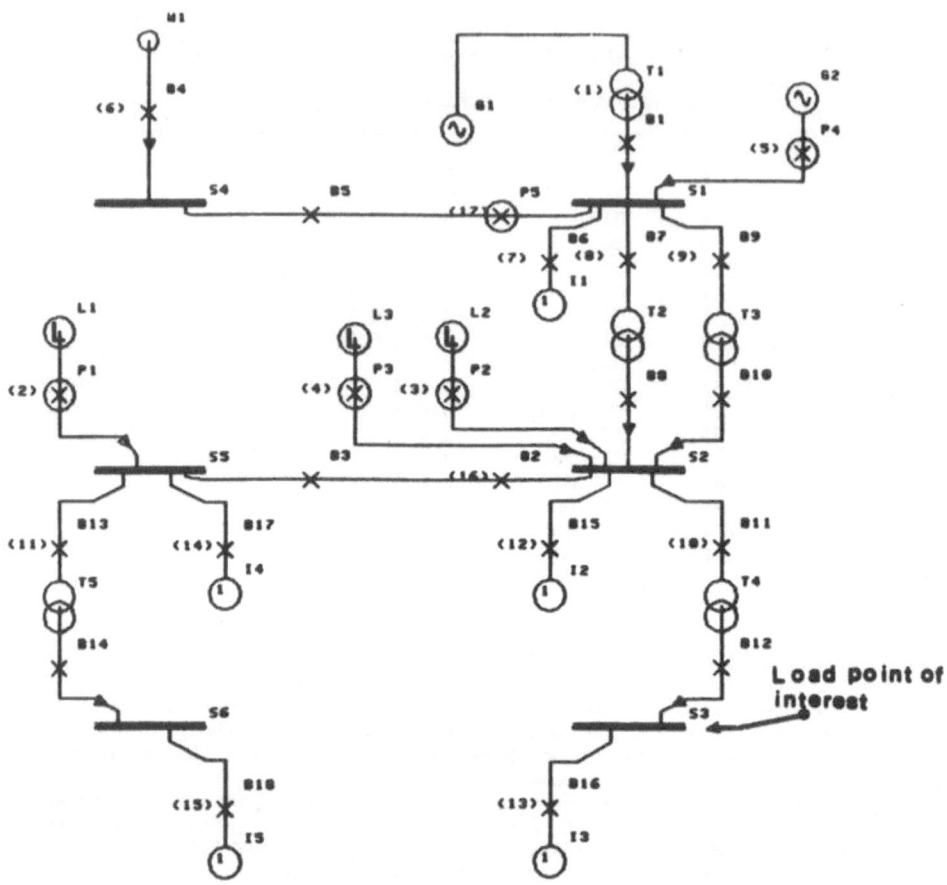

Figure 2   Example network

using standby sources (normally open path) or whether the failed component(s) must be repaired or replaced. This aspect can also be established from the list of minimal paths [14,15]. Only restoration using standby sources is considered in this paper.

The technique is illustrated by analysing busbar S3 of the system shown in Fig.2. The minimal paths are listed in Table 1 and the initiating events in Table 2. It is seen that there are 18 initiating events of which 5 involve repair or replacement. These events are not considered further. The remaining 13 events involve the use of alternative sources associated with the normally open paths. These failure modes are of paramount importance and are considered further in the modelling of the restoration process.

It can also be seen that several of the initiating events are associated with the same restoration process. In the present example, two groupings of the restoration processes can be identified; Group 1 being paths (3,4,5) and Group 2 being paths (3,4,5,6,7,8,9). In order to ensure an efficient computational process, it is essential to create a systematic approach which groups those initiating events having the same restoration

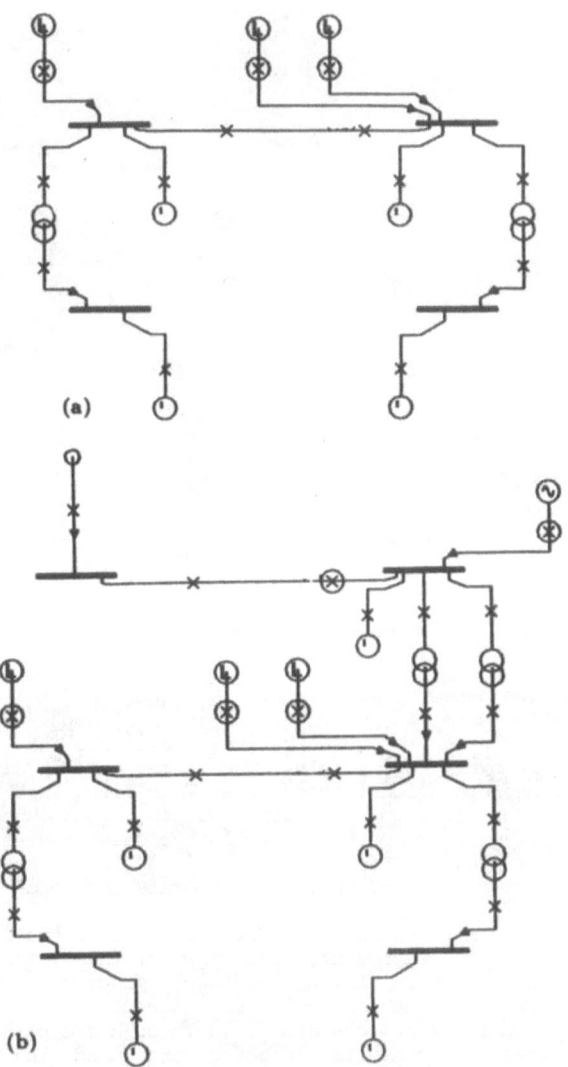

Figure 3 Configurations for groups of intiating events

process and simulating the restoration of that group only once. In the present example, the two relevant groups create the topological configurations shown in Fig.3.

RESTORATION PROCESS

Identifying Normally Open Paths

After deducing the initiating events the time sequential assessment proceeds with the analysis of the restoration process. This implies the

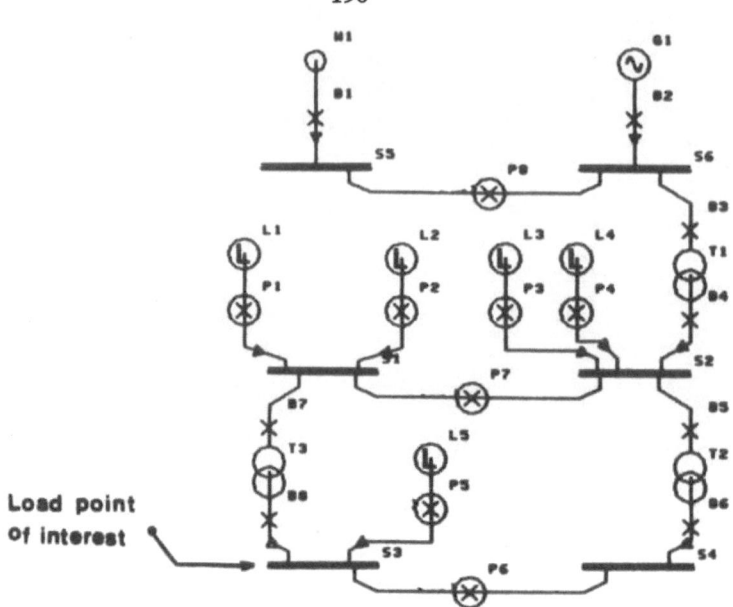

Figure 4   Test network

consideration of two aspects. First, it is necessary to determine the different possible ways in which restoration of supply can take place. Second, there is a need to quantitatively evaluate the corresponding probability of recovery of supply. The former case is directed towards system operation and operational procedures. Therefore, switching actions, starting up standby emergency supply sources, load shedding and load reconnection are of major concern. In the latter case, however, reliability characteristics of the system are determined from the statistical information available on the failure to operate and repair cycles of the components.

Busbar S3 of the network shown in Fig.4 is used as the example in the present analysis. This represents that part of the system which is left after the occurrence of an initiating event or a group of initiating events causing loss of supply at the load point of interest.

There are seven input sources connected through n/o paths to this load point. Each of these sources is selected and used in the following sequential order; 1st, standby source L5, 2nd, standby sources L1 and L2, 3rd, standby sources L3 and L4, 4th, the electrical grid system represented by source W1 and 5th, the generating unit represented by source G1. Although this order may not reflect a practical situation, it shows the basic characteristics of the techniques. Based on this order, the n/o paths deduced are presented in a diagrammatic form in Fig.5. This figure shows eleven n/o paths and each represents an alternative for restoring supply to load point S3.

In general, n/o paths comprise one or more n/o branches. These are presently restricted to a maximum of two n/o branches per path. This gives four categories, all of which exist in Fig.5.
1: n/o path defined by one n/o branch. Associated with this n/o branch are a standby source and its n/o breaker, (paths 1-3).

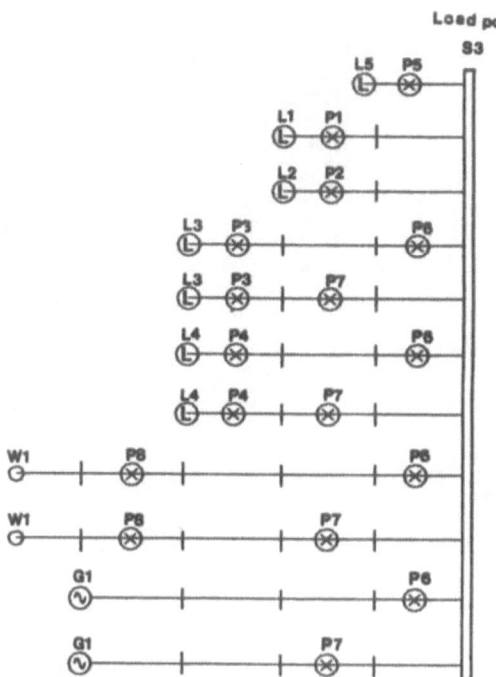

Figure 5   Representation of the n/o paths for busbar S3

2: n/o path defined by two n/o branches. The first one is similar to category 1, but the second one is defined by a n/o breaker connected between two busbars, (paths 4-7).

3: n/o path defined by one n/o branch. This n/o branch is defined by a n/o breaker connected between two busbars. An input source in this n/o path is normally operating (paths 10-11).

4: n/o path defined by two n/o branches. Both of these n/o branches are similar to category 3, (paths 8-9).

In all cases, the sequence of events involves starting up the standby source (if necessary) and closing the breakers as a sequential process. This requires a set of detailed and comprehensive algorithms which are described in detail in Ref [16].

The main objective in defining a n/o path in terms of categories is that each n/o path must be independent. As shown in Fig 5, this objective is not necessarily achieved at this stage of the analysis. Therefore, a systematic way of reducing the n/o paths is needed.

The first step is to compare paths in the same category. If dependence exists, the source of dependence is eliminated by appropriately combining the dependent paths. This gives equivalent n/o paths (enp) in each category. When this technique is applied to the network of Fig.4, the 11 n/o paths of Fig.5 are reduced to the six equivalent paths shown in Fig.6. However, it can be seen in Fig.6 that there are still some situations in which more than one n/o branch is included in different equivalent n/o paths. Therefore an extension of the analysis is required, involving a systematic combination of equivalent n/o paths of different

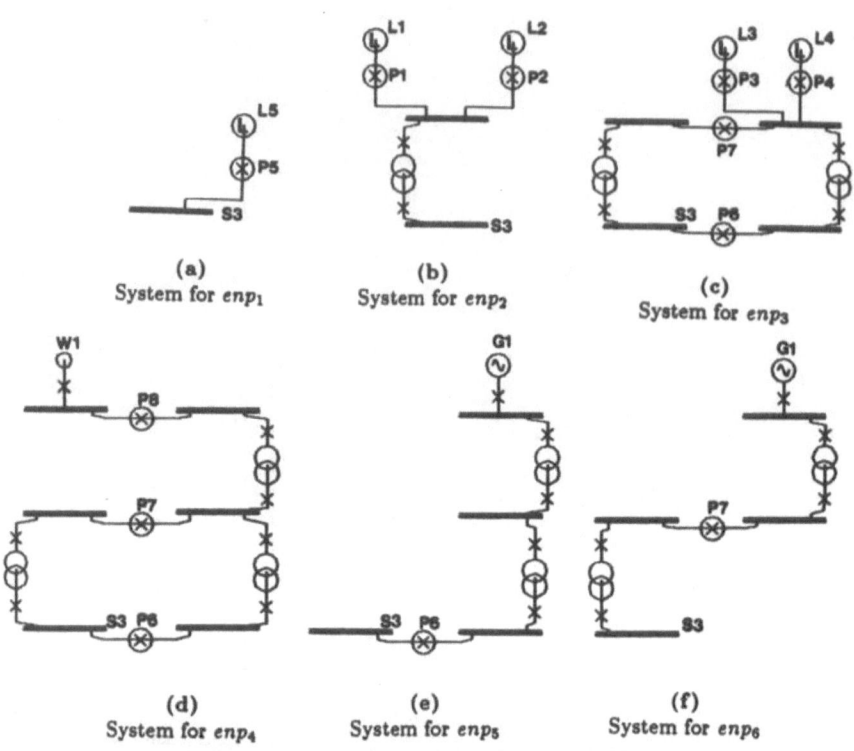

Figure 6   Configurations for each equivalent n/o path

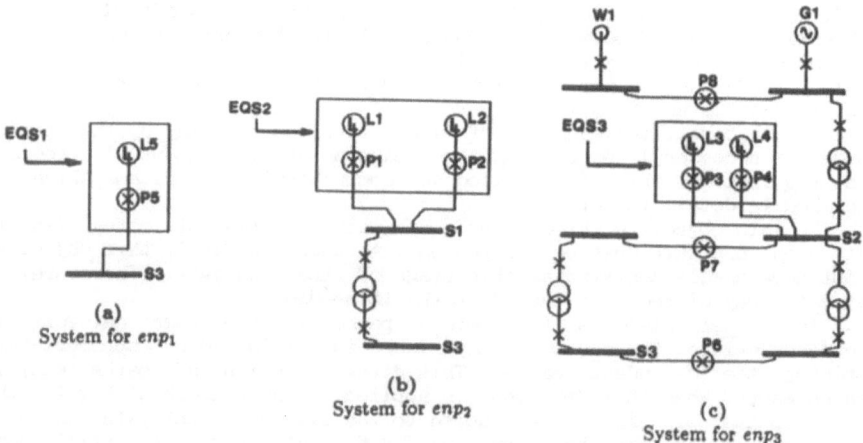

Figure 7   Reduced configurations for n/o paths

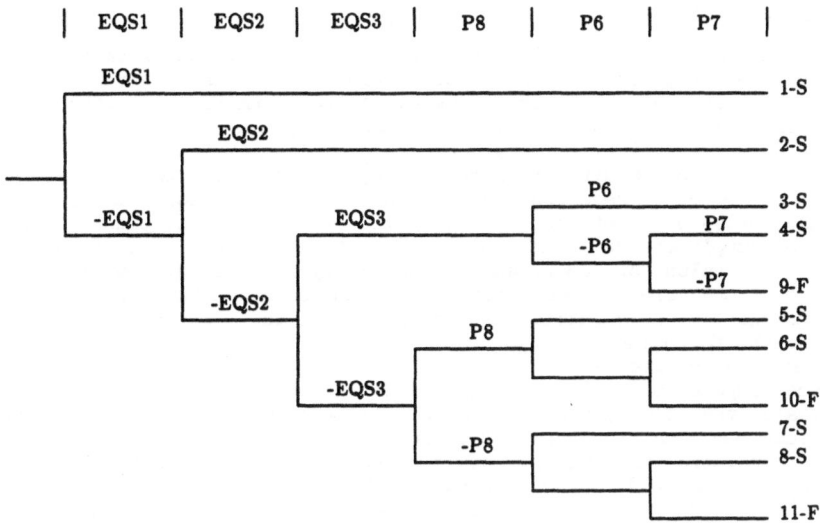

Figure 8 Event tree

categories. This second reduction considers the combination of equivalent n/o paths having at least one n/o branch in common. This is a more complex procedure because all combinations of categories must be considered. A detailed algorithm for achieving this is given in Ref [16].

When this technique is applied to the test network of Fig.4, the number of effective equivalent configurations is reduced further to the three shown in Fig.7.

When more than one standby source is connected to the same busbar, they can be defined in terms of an equivalent source. In reliability terms, these equivalent sources are represented by the concept of majority vote systems (m-out-of-n). In the present example, there are three equivalent sources; EQS1, EQS2 and EQS3. These are also shown in Fig.7.

Event Tree Analysis

The network analysis described above reduces the complexity of the topological configuration of a system to a level where the deduced equivalent n/o paths define systems which can be treated separately. This process can be followed by an event tree analysis in order to ascertain first the various sequence of events that can occur, the outcome of each sequence, i.e. whether it leads to success or failure, and finally the likelihood of each sequence and thus of success or failure in terms of probability.

The construction of an event tree must recognise the events that can take place and the sequence in which they occur. In the case of electrical systems these events include starting up standby sources and closing n/o breakers. There are several techniques available for constructing these event trees. Applying the concept of reduced event trees [10] to the present example, gives the reduced event tree shown in Fig.8.

Quantitative evaluation

A quantitative analysis involves the evaluation of the probability of restoring supply at the load point of interest. This analysis requires the list of event sequences deduced in the event tree analysis together with the reliability characteristics of the components involved in the restoration process, i.e. standby sources (diesel generators, gas turbines, batteries) and n/o circuit breakers. Also, as defined previously, the restoration process requires the definition of a mission time T1 during which the series of events determined by the event tree take place in order to recover supply at the load point of interest. This specified time permits the value of occurrence probability to be evaluated from a knowledge of the relevant probability distribution and parameters.

Consequently, a quantitative analysis of the restoration process considers the following steps.

1: define mission time T1.
2: deduce components involved in restoration process.
3: evaluate the time dependent probability (success or failure to operate for each component in Step 2. The probability for an equivalent source also considers the effect of m-out-of-n systems when more than one source is connected to the same busbar. This step can also consider the underlyingprobability distribution associated with the components in the restoration process and therefore the values of probability can be time dependent.
4: evaluate the probability for each event sequence. This probability is calculated from the product of the appropriate probability of the components included in each event sequence as evaluated in Step 3.
5: from the results of Step 4, the sum of the probabilities of the event sequences leading to success and the sum of the probabilities of the event sequences leading to failure represent the probabilities of success and failure for restoring supply at the load point of interest for the mission time T1.

Application

An application of this reliability evaluation considers the probability of restoring supply at busbar S3 in the test network of Fig.4. It is assumed that the probability of failing to start up for all the standby sources is equal to 0.1 and the probability of failing to close for all the n/o breakers sources is equal to 0.05. These values are considered fixed in this example and therefore the mission time T1 does not have any effect. In practice however, these probabilities would be calculated from a knowledge of T1 and the relevant distributions. Therefore they would be time dependent and would tend to decrease with increased time T1.

This gives the success and failure probabilities for the equivalent sources EQS1, EQS2 and EQS3 as shown in Table 3 from which the

TABLE 3 Probability Evaluation for Equivalent Sources

equivalent source	system components	success criterion	probability success	failure
EQS1	Fig.7a L5/P5	1-out-of-1	8.55×10^{-1}	1.45×10^{-1}
EQS2	Fig.7b L1/P1 L2/P2	2-out-of-2	7.31×10^{-1}	2.69×10^{-1}
EQS3	Fig.7c L3/P3 L4/P4	1-out-of-2	9.79×10^{-1}	2.10×10^{-2}

195

TABLE 4   Probabilities of the Restoration Process

event sequence (Fig.8)	outcome	probability
1	S	8.55×10^{-1}
2	S	1.06×10^{-1}
3	S	3.63×10^{-2}
4	S	1.81×10^{-3}
5	S	7.40×10^{-4}
6	S	3.70×10^{-5}
7	S	3.89×10^{-5}
8	S	1.95×10^{-6}
9	F	9.54×10^{-5}
10	F	1.95×10^{-6}
11	F	1.03×10^{-7}

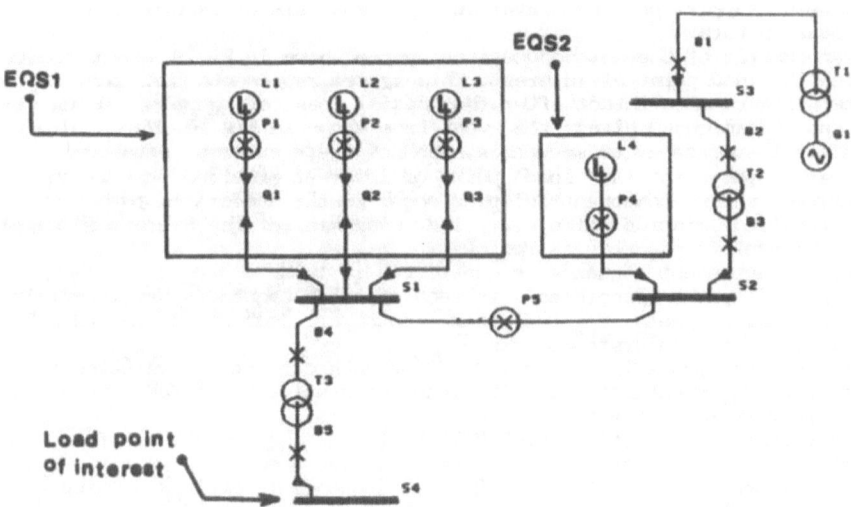

Figure 9   Test network

probabilities of each event sequence and the overall success/failure can be evaluated. These are shown in Table 4.

MAINTAINING A GUARANTEED SUPPLY

Technique

After the supply to the load point of interest has been restored, reliance is placed upon the standby emergency sources in order to guarantee supply during the emergency operating state defined by mission time T2. Further failures during this time are not to be tolerated. The successful event sequences deduced during the restoration process are mutually exclusive and can be treated separately. Therefore, the modelling of the

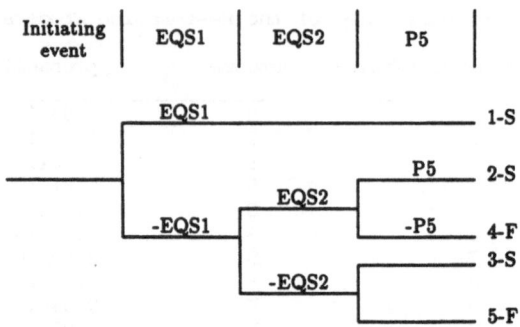

Figure 10 Event tree for restoring supply to busbar S4

guaranteed supply process involves the analysis of each of these event sequences in turn.

An example of the method uses the system shown in Fig.9 where busbar S4 is the load point of interest. This system represents that part of the system which remains intact after the initial loss of supply. Using the previous techniques gives the event tree shown in Fig.10. This indicates there are 3 success event sequences. Each of these success sequences will maintain supply to the load point of interest provided that it becomes operational and no component failures occur in the associated paths during the specified mission time T2. The analysis of the guaranteed supply process therefore proceeds as follows:

- each success event sequence is considered in turn
- the minimal paths associated with each of these sequences are identified
- the minimal cut sets of these minimal paths are deduced. These are shown in Table 5 for the system of Fig.9
- the failure probability during T2 of each cut set is evaluated from a knowledge of probability distributions and statistical parameters of each component in the cut set
- in the case of the failure probability of the equivalent sources, the required m-out-of-n operational status is also needed
- the failure probability of each event sequence is evaluated using basic probability theory [10] by summating the failure probabilities of each cut set
- the overall failure probability is evaluated by conditionally summating the values of the failure probability of each event sequence, i.e. weighting them by the likelihood of occurrence. These conditional probabilities are the event sequence probabilities evaluated during the restoration process.

Application

Consider busbar S4 of the system of Fig.9, event sequence 1 shown in Fig.10 and the corresponding cut sets shown in Table 5. Assume a mission time T2 of 3hr, the data shown in Table 6 and that a 2-out-of-3 criterion is needed for ESQ1. Also consider two operational procedures (OP) associatd with ESQ1:
a) OP1: all units which start are loaded and kept running
b) OP2: of those units which start, only 2 (= m) are loaded and kept running.

Using the techniques presented in this and the previous sections gives

TABLE 5   List of Cut Sets

cut	cut sets for event sequence		
	1	2	3
1	S4	S4	S4
2	S1	S1	S1
3	EQS1	S2	S2
4	B4	B4	S3
5	T3	T3	G1
6	B5	B5	B4
7		P5	T3
8		EQS2, S3	B5
9		EQS2, G1	P5
10		EQS2, T1	B2
11		EQS2, B1	T2
12		EQS2, B2	B3
13		EQS2, T2	T1
14		EQS2, B3	B1

TABLE 6   Reliability Data of Components

component	probability of failing to start or to close	failure rate while operating, f/yr
generator L1	0.03	26.28
generator L2	0.04	26.28
generator L3	0.05	26.28
open breakers P	0.01	0.02
closed breakers B		0.04
cables Q		0.10
busbars S		0.005
transformers T		0.110

NB: times to failure are assumed exponential

the results shown in Table 7 for both the restoration process and the guaranteed supply process.

CONCLUSIONS

This paper has described the development of models and efficient evaluation techniques for performing time sequential reliability evaluation of electrical auxiliary systems. All these models and techniques have been tested in a computer code. The techniques are performed in the following steps.
a) Network modelling of the system to be analysed using path tracing techniques and minimal path analysis
b) Identification of failure events using minimal cut set analysis of the minimal paths
c) Modelling and evaluation of the restoration of supply using an event tree approach and probability evaluation techniques
d) Modelling and evaluation of the guaranteed supply using minimal cut set techniques applied to the remaining system.

TABLE 7   Overall Reliability Assessment

operational procedure	restoration success	failure	guaranteed supply success	failure
OP1	9.93×10^{-1}	7.05×10^{-3}	9.90×10^{-1}	9.73×10^{-3}
OP2	9.93×10^{-1}	7.05×10^{-3}	9.75×10^{-1}	2.49×10^{-2}

This procedure, although very simple in concept becomes extremely complex when applied to realistic problems. Some assumptions were introduced in order to arrive at a compromise between practical usefulness of the developed techniques and objective assessments of the results obtained. The basic approach adopted was the combination of different techniques in order to give the most suitable solution to the time sequential reliability problem.

REFERENCES

1. Apostolakis,G.E., "Mathematical Methods of Probabilistic Safety Analysis", California Univ, Los Angeles, USA (1974).
2. Fussell,J.B., Burdick,G.R. (eds), "Nuclear Systems Reliability Engineering and Risk Assessment", SIAM (1977).
3. Snaith,E.R., "Reliability Evaluation of an Electrical Supply System for a Nuclear Power Station", IEE Conference on Reliability of Power Supply Systems, IEE Conf Pub 148, 1977, pp.1-6.
4. Apostolakis,G., Garriba,S., Volta,G. (eds), Proceedings of the NATO Advanced Study Institute of Synthesis and Analysis Methods for Safety and Reliability Studies, Plenum Press (1980).
5. McCormick,N.J., "Reliability and Risk Analysis", Academic Press (1981).
6. Henley,E.J., Kumamoto,H., "Reliability Engineering and Risk Assessment", Prentice-Hall (1981).
7. Henley,E.J., Kumamoto,H., "Designing for Reliability and Safety Control", Prentice-Hall (1985).
8. European Federation of Chemical Engineering, "Risk Analysis in The Process Industries", EFCE Publications Series No 45 (1985).
9. Wang,J., Modarres,M., Hunt,R. "Probabilistic Risk Assessment: A Look at The Role of Artificial Intelligence", Nuclear Engineering and Design, Vol 106, 1988, pp.375-387.
10. Billinton,R., Allan,R.N., "Reliability Evaluation of Engineering Systems: Concepts and Techniques", Plenum Publishing (1983).
11. Billinton,R., Allan,R.N., "Reliability Evaluation of Power Systems", Plenum Publishing (1984).
12. Yellman,T.W., "Event-Sequence Analysis", Proceedings of The Annual Reliability and Maintainability Symposium, IEEE, 1975, pp.286-291.
13. Yellman,T.W., "Event-Sequence Analysis vs. The Fault Tree", Proceedings of The Annual Reliability and Maintainability Symposium, IEEE, 1981, pp.446-451.
14. Allan,R.N., Billinton,R., De Oliveira,M.F., "An Efficient Algorithm for Deducing the Minimal Cut and Reliability Indices of a General Network Configuration", IEEE Trans on Reliability, R-25, 1976, pp.226-233.
15. Allan,R.N., Billinton,R., De Oliveira,M.F., "Reliability Evaluation of the Auxiliary Electrical Systems of Power Stations", IEEE Trans, PAS-96, 1977, pp.1441-1449.
16. Inga-Rojas, T.O., "Time Dependent Reliability Evaluation of Electrical Systems", PhD Thesis, University of Manchester, 1989.

CANADIAN ELECTRIC POWER SYSTEM RELIABILITY ASSESSMENT - PERFORMANCE AND PREDICTION

R. BILLINTON
Power System Research Group
University of Saskatchewan
Saskatoon, Sask., S7N 0W0, Canada

M. OPRISAN, I.M. CLARK
Canadian Electrical Association
Suite 580 - One Westmount Square
Montreal, Quebec, H3Z 2P9, Canada

ABSTRACT

The Canadian Electrical Association (CEA) Equipment Reliability Information System (ERIS) was initiated in 1975 and consists of a comprehensive procedure for reporting generation and transmission equipment performance and developing reliability parameters from the basic data. These parameters are the basic data used in reliability assessment of generation and transmission systems. The CEA-ERIS data base will be briefly described in this paper and related to the predictive reliability assessment performed by Canadian electric power utilities.

The CEA has also initiated the Electric Power System Reliability Assessment (EPSRA) procedure which is designed to provide data on overall power system performance. This procedure is in the process of evolution and at the present time contains systems for compiling information on Bulk Electricity System disturbances, Bulk System Delivery Point performance and Customer Service Continuity statistics. This paper will describe the EPSRA procedure and illustrate some of the information generated. The EPSRA and ERIS data bases provide the basic information to quantitatively assess the past performance of the bulk electricity system and the basic component data required to predict future performance.

INTRODUCTION

Overall reliability evaluation of an electrical power system can be divided into the two basic elements of past performance assessment and future performance prediction. This paper illustrates the procedures which have been instituted by the Consultative Committee on Outage Statistics (CCOS) of the Canadian Electrical Association (CEA) to provide uniform and consistent reporting procedures for assessing past system performance and providing the basic equipment data for predicting future performance.

Effective quantitative assessment of power system reliability requires, in addition to other basic information, suitable mathematical models and realistic failure and repair statistics for the relevant system components. The failure and repair statistics are normally obtained from past operating data of comparable components and subcomponents in the system. Data collection is therefore an integral and indispensable part of quantitative power system reliability analysis. All the major electric power utilities in Canada participate in a single data collection and analysis system called the Equipment Reliability Information System (ERIS) of the Canadian Electrical Association (CEA). CEA started collecting data on generation and transmission outages in 1977 and since then has published a number of reports in both areas [1,2].

CEA has also initiated an Electric Power System Reliability Assessment (EPSRA) procedure which is designed to provide data on the past performance of an overall power system. This procedure is in the process of evolution and at the present time contains systems for compiling information on bulk system disturbances, bulk system delivery point performance and customer service continuity statistics. In the latter area, these data have been collected and published for many years by the Distribution Section of the Engineering and Operating Division of CEA. This responsibility has now been assigned to CCOS. The bulk system performance assessment procedures are now established and data should be available in the near future.

The term "reliability" as applied to power systems has a very wide range of meaning. The reliability concerns can be divided into the two general categories of system adequacy and system security, as shown in Figure 1. System adequacy relates to the existence of sufficient generation, transmission and distribution facilities within the system to satisfy customer load demand. Adequacy is, therefore, associated with static system conditions without consideration of any system disturbances. System security, on the other hand, relates to the ability of the system to respond to disturbances arising within the system.

The bulk of the work done on predictive reliability evaluation has been in the area of adequacy assessment. There has been some work done in the security domain such as operating reserve assessment and probabilistic evaluation of transient stability but the number of available publications in the area of quantitative security assessment is very small compared to those in the adequacy domain [3,4,5,6].

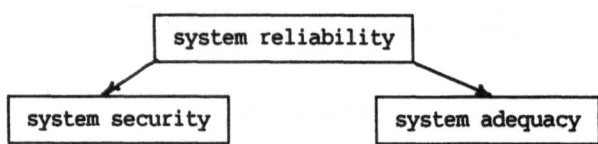

Figure 1. Subdivision of system reliability.

A power system can be broadly categorized by three functional zones containing generation, transmission and distribution facilities. These three zones can be combined to provide a practical and consistent framework for predictive reliability evaluation containing three hierarchical levels [7]. Hierarchical level one (HLI) is composed of only the generation facilities. Hierarchical level two (HLII) includes both the generation and the transmission facilities. Hierarchical level three (HLIII) encompasses all three functional zones. The functional zones and hierarchical levels are shown in Figure 2. There is a long history of reliability application at the HLI level, where the primary objective is the determination of the adequacy of the total system generation to satisfy the total system load demand.

HLII evaluation is still at its infancy and while there is considerable published material [3,4,5,6] available, there is still no consensus regarding requirements, techniques and indices. Evaluation at the HLIII level is a major task and if done at all is accomplished by using the HLII indices as input to the distribution functional zone.

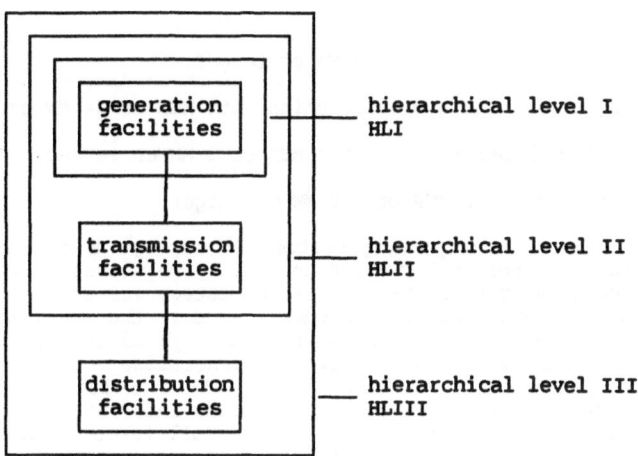

Figure 2. Hierarchical levels in power system reliability evaluation.

The Equipment Reliability Information System (ERIS)

The basic structure of the ERIS is shown in Figure 3.

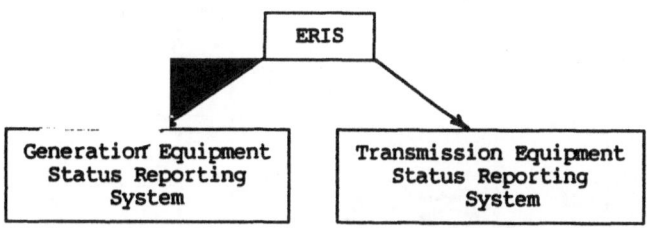

Figure 3. Basic components of the ERIS.

The ERIS data base does not contain any information on distribution system equipment. This is obviously a very important aspect of utility performance, and work is now in progress to initiate the collection of distribution component outage data. The ERIS data base therefore provides the basic data required to perform quantitative reliability assessment at HLI and HLII.

The following is a brief description of the generation and transmission equipment reporting systems with particular emphasis on their ability to support predictive reliability assessment.

Generation Equipment Status Reporting System

The generation data reporting system covers commercial generating units in Canada of the following types and sizes.

1. Combustion turbines with Maximum Continuous Ratings (MCR) of 1 MW or larger.

2. Fossil units with MCR of 60 MW or larger.

3. Pumped storage and hydraulic units with MCR of 24 MW or larger.

4. Internal combustion units with MCR of 1 MW or larger.

5. Nuclear units with MCR of 200 MW or larger.

The generation data reporting system [1] employs a continuous state monitoring procedure for each generating unit. The states in which a unit can reside have been grouped into the eleven categories shown in Figure 4. These states are numbered 11 through 16 for the available states (operating or capable of operation) and 21 through 25 for the unavailable (outage) states. These 11 states provide a generalized eleven-state model for each generating unit. States 12, 13, 15 and 16 provide a three dimensional aspect, as these states also include information on the extent of the unit derating. State 21 which is the full forced outage state has an "outage type" associated with it which designates the outage as a sudden forced outage, an immediately deferrable forced outage, a deferrable forced outage or a starting failure.

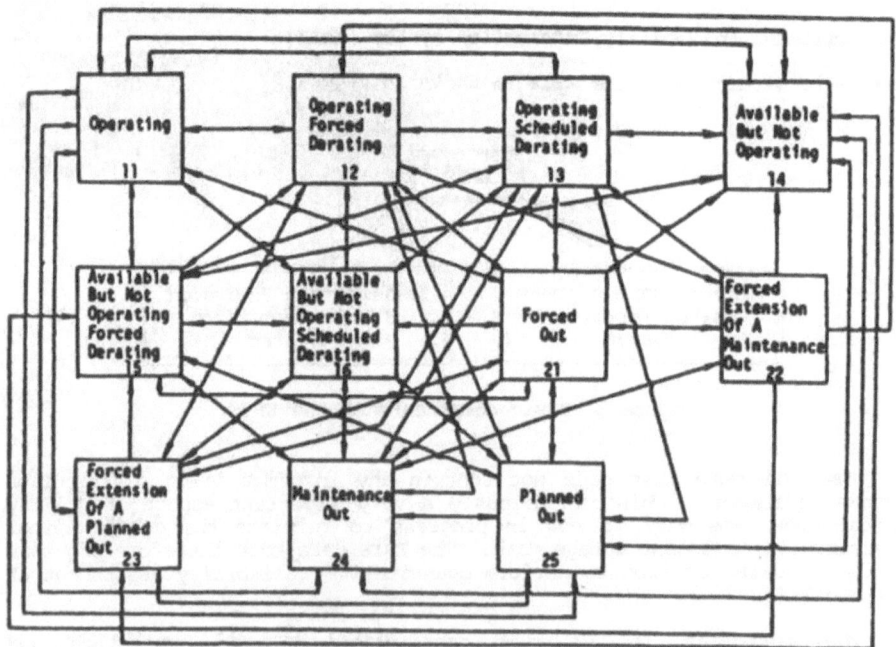

Figure 4. Basic state space diagram for a generating unit.

The time spent in any of the states can be easily calculated from the recorded data and important generating unit parameters such as the forced outage rate (FOR) and the derating adjusted forced outage rate (DAFOR) can be computed using the relevant state residence times. The continuous state monitoring approach also provides detailed information on transition rates associated with the model. These transition rates are essential elements in frequency and duration (F&D) analysis and in multi-state modeling of generating units.

Adequacy Assessment at Hierarchical Level I

Generating capacity adequacy evaluation is a common procedure in most major utilities throughout North America. Recent surveys have shown that virtually all Canadian utilities now use probabilistic rather than deterministic criteria to evaluate the adequacy of installed and planned generating facilities. The basic probabilistic techniques and the ability of ERIS to provide the required data can be briefly summarized as follows.

Loss of Load Expectation [8]

Loss of load expectation (LOLE) is the most commonly used generation adequacy index and is normally defined as the expected number of days in a year in which the load will exceed the generation. The basic generating unit parameter used in the calculation of the LOLE is the unit FOR. This is sometimes replaced by the DAFOR for large generating units which tends to give a pessimistic appraisal of the unit and therefore of the system adequacy. As a result, there is a growing tendency to utilize multi-state generating unit models. The ERIS data base provides all the basic generating unit information required for the calculation of the conventional LOLE index.

Loss of Energy Expectation [8]

Several Canadian utilities use loss of energy expectation (LOEE) as their HLI adequacy index. This parameter is the expected energy that will be curtailed due to inadequate capacity. As in the calculation of a LOLE index, either two, three or multi-state generating unit models can be used to compute the LOEE. All the basic generating unit data for the LOEE technique can be readily obtained from the ERIS data base.

Frequency and Duration Analysis [8]

The LOLE indicates the expected number of days or hours that load will be curtailed. The LOEE on the other hand estimates the expected amount of energy curtailed. These indices do not provide any indication of the number of times that load will be curtailed nor the extent of the curtailment and the average duration of such a curtailment. This information can be obtained from a frequency and duration analysis. The additional parameters required for such an analysis are the transition rates associated with the appropriate generating unit model. The ERIS data base contains all the generating unit state information and therefore it is possible to obtain transition rate data for any degree of complexity.

Security Assessment At Hierarchical Level I

All Canadian utilities are concerned with security assessment within their own system and in the interconnected configuration within which their own system resides. These assessments are done primarily using methods which utilize deterministic criteria. Some work has been done in the area of operating reserve assessment and probabilistic evaluation of transient stability. The ERIS data base can provide almost all the basic data required in a probabilistic operating reserve assessment.

Transmission Equipment Status Reporting System

The CEA-ERIS defines transmission equipment as those equipment rated 110 kV and above and associated with the transmission of electric power. They include synchronous and static compensators and also shunt reactors and capacitors connected on the tertiaries of transformers rated 110 kV and above. Forced outages of transmission lines have been categorized into line-related outages and terminal-related outages. The transmission line performance statistics are expressed on a "per 100 kilometer-year" (km.a) basis for line-related outages and on a "per terminal-year" (t.a) basis for terminal related outages. Outages are also classified according to their duration. Outages of less than one minute are classified as transient outages and those of one minute and longer are called sustained outages. Figure 5 illustrates some of the basic information regarding transmission line forced outages which can be extracted from the ERIS data base.

The reporting system on transmission equipment is limited to forced outages, their duration, primary cause and subcomponents involved, if any. Other parameters necessary for reliability analysis such as repair or replacement times of any component can be easily computed from the recorded data. Transmission lines are exposed to the weather and therefore are susceptible to adverse weather conditions such as lightning strokes, gales, typhoons, snow and ice. The failure rates in these kinds of weather conditions are higher than those in normal weather. Although the CEA data collection system provides for recording the number of failures in adverse weather, it does not monitor the duration of such

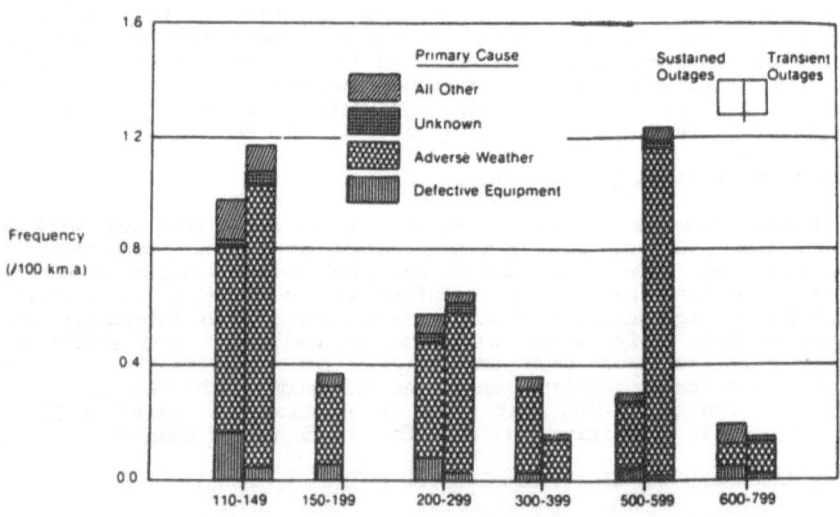

Figure 5. Frequency of line-related forced outages of Canadian transmission lines in 1982-86.

weather conditions. Adverse and normal weather failure rates can not therefore be obtained directly from the ERIS data base. The statistics shown in Figure 5 are for the last five years of data supplied to the CEA. It can clearly be seen from Figure 5 that adverse weather is a major contributor to both sustained and transient line-related forced outages.

This section has described the salient features of the CEA-ERIS transmission equipment data base. The system can provide the required data to support the basic two-state models required in conventional composite generation and transmission system adequacy evaluation. There are many aspects of advanced system models which cannot be supported by the ERIS transmission equipment data base, i.e. adverse and normal weather data and dependent failure parameters. Many of these are system and location specific and must be provided by the system analyst. The data base, however, provides the basic data required for predictive transmission system reliability evaluation.

Electric Power System Performance Assessment (EPSRA)

The basic structure of the EPSRA is shown in Figure 6.

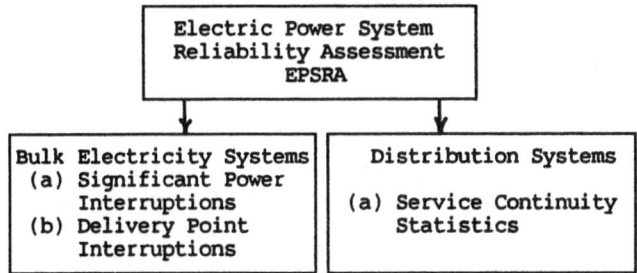

Figure 6. Basic structure of EPSRA.

The bulk electricity system (BES) parameters provide valuable HLII data. The service continuity statistics collected at the distribution system level provide overall HLIII indices. The following is a brief description of the salient features of the EPSRA components.

BES - Significant Power Interruptions

This is a relatively simple reporting procedure in which each participating utility provides annual data on the frequency and severity of significant power interruptions on its system [9]. Typically these events will involve wide spread customer interruptions or localized power interruptions of an extended duration. It is expected that these events will occur infrequently. Disturbance severity is defined as the unsupplied energy in an event and is measured in MW - minutes. This is transformed into "System Minutes" by dividing the unsupplied energy by the annual peak system load.

The disturbances are grouped by severity as follows

Degree 1 - an incident with a Severity from 1-9 System Minutes
Degree 2 - an incident with a Severity from 10-99 System Minutes
Degree 3 - an incident with a Severity from 100-999 System Minutes.

The basic definition of a significant power interruption is that it is an event which originates on the BES and is of Degree 1 or higher severity.

The reporting process was initiated January 1, 1985 and the first report has now been released by the CEA–Consultative Committee on Outage Statistics. An overall summary of the data is shown in Figure 7. These results are based on 20 utility years of data for the 1985 and 1986 periods.

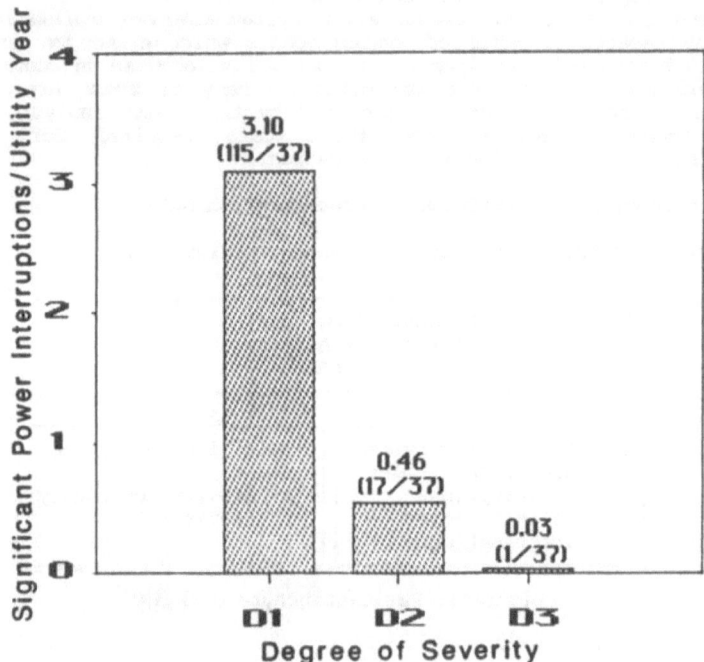

Figure 7. Overall BES disturbance statistics.

BES – Delivery Point Interruptions

This reporting procedure was initiated on January 1, 1988 and is intended to provide a centralized source of BES delivery point performance data at the national level which will allow utilities to compare their performance with that of other Canadian utilities. The measurement system will focus on the collection of all delivery point interruptions due to problems within the BES and therefore will include both adequacy and security considerations. The interruptions will be divided into momentary and sustained events. It is expected that this reporting procedure will provide some very important information for Canadian utilities and will be an important area of discussion in the CEA Power System Reliability Subsection.

A BES delivery point is defined as the interface between the BES and the distribution system. A delivery point is considered to be interrupted when its supply voltage is interrupted due to a problem with the busbar or due to problems within the BES.

The measurement system will focus on the collection of all delivery point interruptions arising from these problems. It includes interruptions with a duration of less than one (1) minute which may be restored by automatic reclosure facilities (Momentary Interruptions) and those of one minute or more (Sustained Interruptions). Delivery point interruptions caused by problems originating on the customer or distribution system are excluded, unless the BES fails to limit their impact to other delivery points.

As previously noted, a delivery point is considered to be the point of supply where the energy from the BES is transferred to the distribution system or the retail customer. This point is generally taken as the low voltage busbar at step down transformer stations (the voltage is stepped down from a transmission or sub-transmission voltage which may cover the range of 50 – 750 kV to a distribution voltage of under 50 kV but above 2 kV). For customer-owned stations supplied directly from the transmission system, this point is generally taken as the interface between utility-owned equipment and the customers equipment.

An interruption to a delivery point which has a duration of less than one minute is assigned zero duration for computational purposes. These are interruptions generally restored by automatic reclosure facilities, which are of very short duration (of the order of a few seconds).

An interruption of supply voltage to a delivery point which has a duration of one minute or more is designated as a sustained interruption.

In the case of sustained interruptions, three quantities are considered to be important, namely:

(a) Delivery point sustained interruption frequency.

(b) Interruption duration of BES supply voltage to the delivery point.

(c) Interruption duration of customer load supplied from the delivery point.

Generally (b) and (c) will be the same but some exceptions will exist and should be identified, as the impact of the interruption will differ. For example, in most Canadian utilities, the distribution network from a delivery point is radial and an interruption to the delivery point will result in all load supplied from it being interrupted. There may, however, be some situations where load is not interrupted such as where a distribution system is operated as a meshed network or where load is restored by an alternative path while the supply failure to the delivery point is repaired.

The delivery points can be categorized as multi-circuit or single circuit facilities.

A multi-circuit delivery point is one which is supplied from the BES by more than one circuit such that the interruption of one circuit does not cause a delivery point interruption. A single-circuit supplied delivery point is one which is supplied from the BES by one circuit whereby the interruption of that circuit will cause an interruption to the delivery point.

The transmission voltage levels before transformation to the delivery point are categorized into four levels or ranges.

```
Level 1    50 -  99 kV
Level 2   100 - 199 kV
Level 3   200 - 299 kV
Level 4   300 - 750 kV
```

It is believed that the designations of single circuit and multi-circuit delivery points together with the voltage level considerations will provide important utility (delivery point) characteristic data. In order to provide performance statistics which reflect these characteristics, the reported data must include the distinctions.

The data collected by this reporting system will be reported annually by CEA in tabular and graphical form. The format will be such that it will allow participating utilities to compare their delivery point performance relative to each other and to the overall national average.

Service Continuity Statistics

Service continuity statistics have been collected by many Canadian utilities for over twenty years. These data were compiled by the Distribution Section of the CEA Engineering and Operating Division. The responsibility for the procedure was assigned to CCOS in 1986 and the 1986 report was prepared by CEA staff. The report contains individual utility statistics and overall Canadian data [10]. Table 1 shows the basic service continuity statistics on a national basis for the 1987 period.

TABLE 1
Basic Canadian service continuity statistics
1988 Period

System Average Interruption Frequency Index	= 4.39 int/syst-cust
System Average Interruption Duration Index	= 6.58 hrs/syst-cust
Customer Average Interruption Duration Index	= 1.50 hrs/cust
Index of Reliability	= 0.999250

These indices are HLIII parameters and provide a valuable indication of customer service. The report also provides a breakdown of the primary causes of interruption. Figures 8 and 9 illustrate the contributions to the number of customer interruptions and the customer hours of interruption respectively due to the various primary causes.

In order to provide a more consistent basis for comparison between participating utilities, the overall data has been divided into three regions.

```
Region 1 - Urban utilities.
Region 2 - Integrated utilities containing urban and rural areas.
Region 3 - Large utilities contained in Region 2 but with separate
           reporting areas.
```

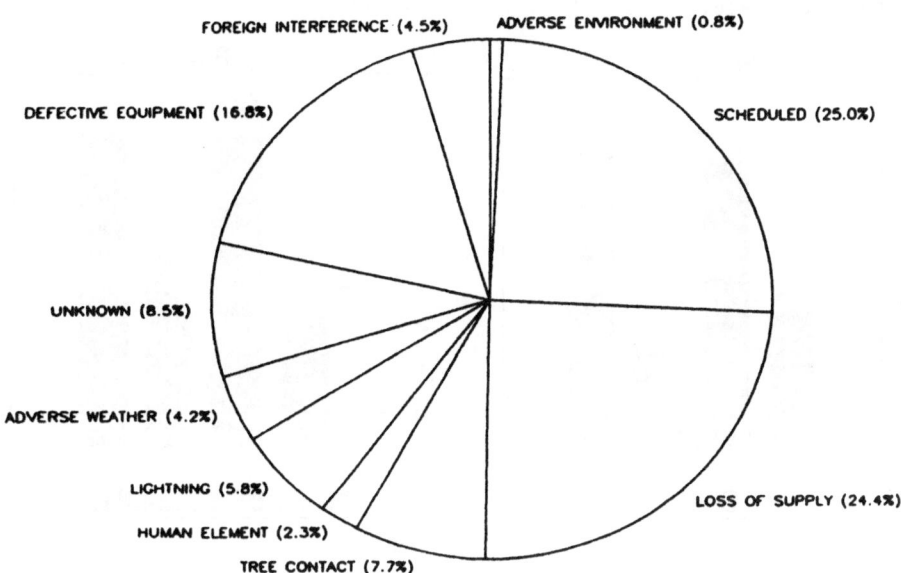

Figure 8. Percentage of customer interruptions due to the primary causes.

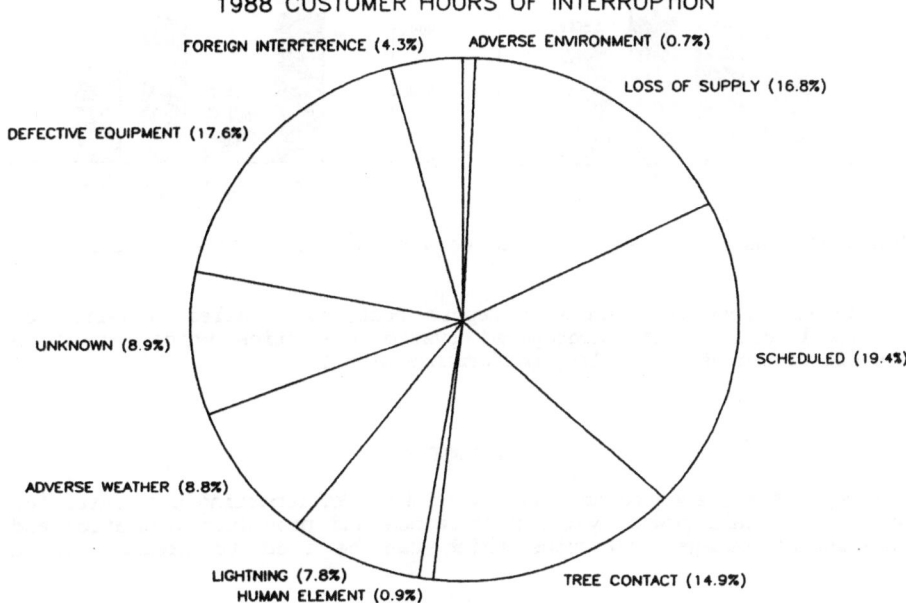

Figure 9. Percentage of customer hours of interruption due to the primary causes.

Figure 10 shows the basic system indices for the three regions and the overall data base.

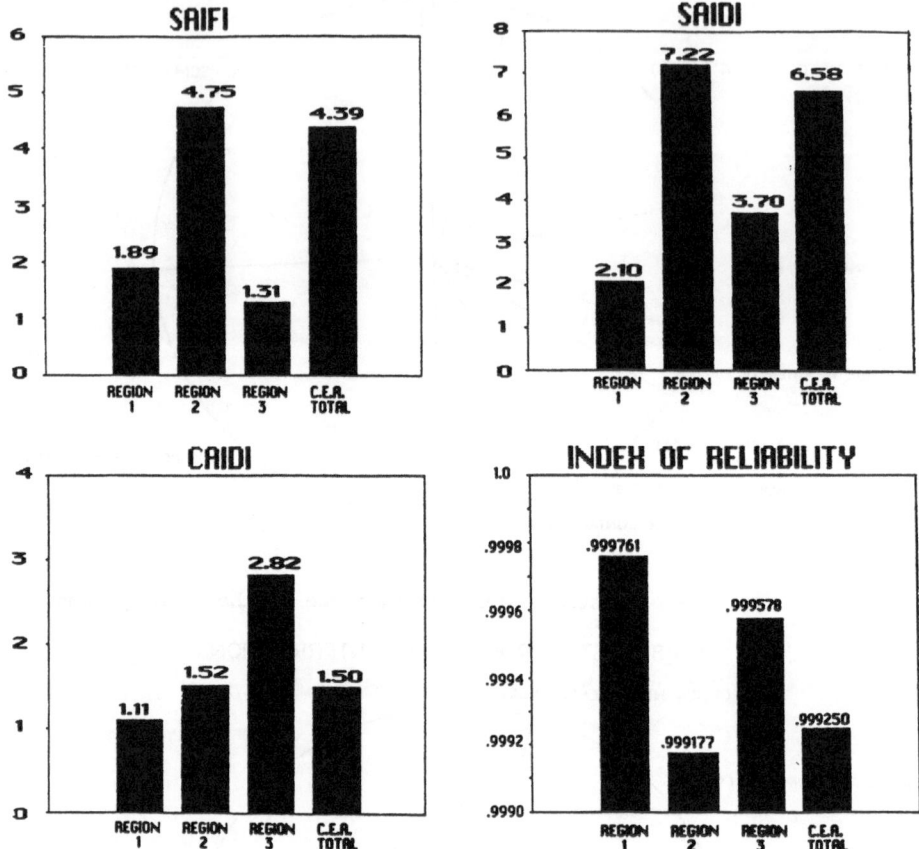

Figure 10. Basic system indices for each region and for the overall data.

The CEA Service Continuity Report contains detailed summaries of individual utility and aggregated system statistics which provide a valuable record of reliability performance at HLIII.

CONCLUSION

This paper has presented an overview of the CEA reporting procedures for assessing overall power system performance and providing generation and transmission outage statistics which can be used to predict future performance.

The Equipment Reliability Information System (ERIS) is presented with particular emphasis on the ability of the system to provide the necessary reliability parameters to support the conventional models used in HLI and HLII reliability evaluation. The Electric Power System Reliability Assessment (EPSRA) procedure is briefly described. This procedure when fully implemented should provide an extremely important source of comparative utility performance data at the HLII and HLIII levels.

ACKNOWLEDGMENT

The authors would like to acknowledge the contribution of all the participating utilities in the development of the CEA reporting procedures and particularly the contributions made by members and participants in the CEA Consultative Committee on Outage Statistics.

REFERENCES

1. Canadian Electrical Association Equipment Reliability Information System, 1988 Annual Report Generation Equipment Status, Canadian Electrical Association, Montreal, 1989.

2. Canadian Electrical Association Equipment Reliability Information System, Forced Outage Performance of Transmission Equipment 1982-86, Canadian Electrical Association, Montreal, 1987.

3. Billinton, R., Bibliography on the application of probability methods in power system reliability evaluation. IEEE Transactions on Power Apparatus and Systems, Vol. PAS-91, 1972, pp. 649-660.

4. IEEE Subcommittee on the Application of Probability Methods, Power System Engineering Committee, Bibliography on the application of probability methods in power system reliability evaluation 1971-1977. IEEE Transactions on Power Apparatus and Systems, Vol. PAS-97, 1978, pp. 2235-2242.

5. Allan, R.N., Billinton, R. and Lee, S.H., Bibliography on the application of probability methods in power system reliability evaluation 1977-1982. IEEE Transactions on Power Apparatus and Systems, Vol. PAS-103, No. 2, February 1984, pp. 275-282.

6. Allan, R.N., Billinton, R., Singh, C. and Shahidapour, S.M., Bibliography on the application of probability methods in power system reliability evaluation 1982-1987. IEEE Winter Power Meeting, New York, 1988.

7. Billinton, R. and Allan, R.N., Power system reliability in perspective. IEE Electronics and Power, March 1984, pp. 231-135.

8. Billinton, R. and Allan, R.N., Reliability Evaluation of Power Systems, Plenum Publishers, New York, 1984.

9. The Consultative Committee on Outage Statistics, Canadian Electrical Association, Significant power interruptions on bulk electricity systems 1985-88 report.

10. The Consultative Committee on Outage Statistics Canadian Electrical Association, 1988 annual service continuity report on distribution system performance in Canadian electrical utilities.

A DESIGN APPROACH TO FATIGUE RELIABILITY

DAOLI CHEN
DEPARTMENT OF MECHANICAL ENGINEERING
WUHAN IRON AND STEEL UNIVERSITY
WUHAN, PEOPLES REPUBLIC OF CHINA

P. MARTIN
DEPARTMENT OF MECHANICAL ENGINEERING
THE UNIVERSITY OF LIVERPOOL

ABSTRACT

It is usually vital to investigate the implications of fatigue failure
during the design of mechanical components. Conventional methods used
for analysis during the design of mechanical components are based on
the use of Miner's rule for the estimation of cumulative fatigue
damage.

The advent of low cost micro computers suitable for use as an aid to
design analysis means that it is now feasible to base the estimation
of fatigue reliability   during design   on procedures   involving
simulation. It is indicated that the procedures identified represent
a potentially general method for the estimation of fatigue reliability
from a knowledge of only two distributions of cycles to failure at
different loading levels.

The use of the method for the design prediction of fatigue reliability
is illustrated with examples and it is shown that the life reliability
of particular ball bearing and roller bearing types may be predicted.

The use of the method for the fatigue reliability analysis of a gear
drive shaft subjected to combined bending and torsion is also
considered with the aid of a numerical example.

INTRODUCTION

During the design of mechanical components a number of problems are
encountered in the estimation of fatigue strength reliability.
One problem relates to the estimation of cumulative fatigue
reliability under the application of several levels of load. The
estimation of simple fatigue reliability at a single load level
without a known distribution of the fatigue strength for a specific
cycle number represents another problem since the distribution of
fatigue strength cannot be obtained directly from tests. It can only
be estimated from tests using approximate methods such as the
staircase method as described by Frost et al [1].

For example, the estimation of the fatigue reliability for a specific
cycle number of more than 10^6 in the case of a steel component
represents a problem.

Methods for estimating fatigue reliability at the initial design stage
could provide valuable guidance in the initial proportioning of
mechanical components.

A number of attempts have been made to overcome the first of the above
problems. For example, Gardner [2] has proposed a method based on the
concept of equivalent cycles to represent the sequential loading
effect but it has been considered by Carter [3] that this method,
although applicable to initially flawed structures, may not reflect
real cumulative damage when crack nucleation is involved.

Although it cannot reflect the sequential effect Miner's rule has been
used as a criterion for cumulative fatigue in conventional design
methods for a long time. In practice this may not be a disadvantage
for initial design estimates, since in many fatigue loading cases
there is no obvious loading sequence.

Furthermore, if a load spectrum is used for estimating cumulative
fatigue, the consideration of the sequential effect is meaningless,
because a load spectrum has already lost the sequential information.

The method of analysis now considered therefore aims to provide an approach to fatigue reliability estimation for the purposes of initial design. It is restricted to the use of a load spectrum approach in conjunction with the application of Miner's rule and sequential effects are not considered.

However, it is very difficult mathematically to apply an exact analytical approach to determine cumulative fatigue reliability with Miner's rule when considering scatter of both cycles to failure and applied loads. The application of Miner's rule under these conditions is now considered.

APPLICATION OF MINER'S RULE

When the loading sequence is random some experimental results have been reported [2] which show that $\Sigma\, n_i/N_i \simeq 1$ where $i = 1, 2, \ldots\ldots$ etc., where n_i is the applied number of cycles and N_i is the number of cycles to failure at the ith loading level and k is the number of loading levels.

This provides some further evidence for the use of Miner's law when the loading sequence is random in the form

$$\sum_{i=1}^{k} n_i/N_i = 1 \tag{1}$$

However, Kececioglu [2] has observed that this does not usually hold when a definite loading sequence is involved but the following treatment is restricted to random loading and it assumes that Miner's rule is valid.

In considering applications involving potential fatigue failure in which there is scatter in both the applied loads and the cycles to failure in fatigue, it is not feasible to base the estimation of cumulative fatigue reliability using Miner's rule on the use of deterministic methods. A probabilistic approach has therefore been adopted.

If N is the total number of cycles to failure for a particular application and load spectrum, and if p_i is the proportion of the cycles at the i th loading level to the total cycles, then if Miner's Law is applied in the form given in equation (1) it can be shown that

$$N = 1 / \sum_{i=1}^{k} p_i/N_i \qquad (2)$$

Equation (2) may be regarded as a revised expression of Miner's rule in which p_i may be regarded as the probability that any load selected at random from a load spectrum will be at the level i.

In this form equation (2) may be used as the basis of a Monte Carlo method for the estimation of fatigue life for a required reliability when used in conjunction with a distribution model
of fatigue strength - cycles to failure.

DISTRIBUTION MODEL - FATIGUE STRENGTH/CYCLES TO FAILURE

It is assumed that the relation between cycles to failure N with a corresponding strength S conforms with the classical model usually expressed as The Basquin Equation in the form

$$NS^m = C \qquad (3)$$

where, S is the stress amplitude S_a, or stress range S_R and m and C are empirical constants. This is generally considered to be valid for the high cycle fatigue range [4].

Equation (3) may be expressed in the form

$$m \log S + \log N = C^1 \qquad (4)$$

where S denotes fatigue strength and N is the corresponding number of cycles to failure. Any dispersion of S or N may be regarded as due to the dispersion of the exponent m and the constant C or C^1.

According to the results of life testing at several loading levels, lifetimes at each loading level are typically distributed as demonstrated by the representative results given in [1] and [2]. At several loading levels it can be assumed that there exists a distributive band in a log S - log N space as show in Fig. 1 and this band consists of numerous log S - log N lines each of which corresponds to a specific probability of occurrence. These lines are equivalent to so called P - S - N curves.

Based on these arguments, if the distributions of cycles to failure are known at only two specific loading levels for a particular mechanical component or material type, then in principle the distribution of cycles to failure at any loading level and the distribution of fatigue strength for any number of cycles to failure can be obtained by Monte Carlo simulation.

As shown in Fig. 1, distributions of cycles to failure 1 and 2 correspond the loading levels S_1 and S_2 respectively.

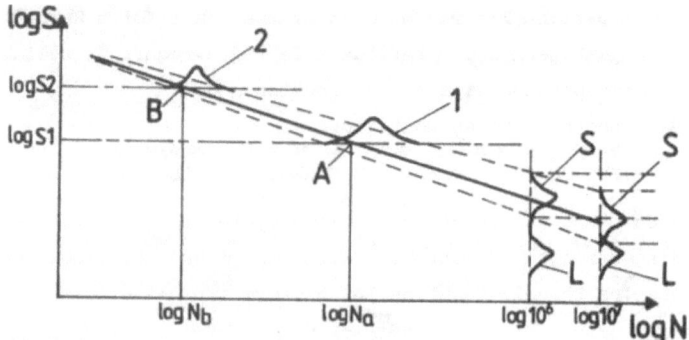

Figure 1. Distributed model of LOG S - LOG N

If the point A in the distribution 1 corresponds to a specific number of cycles N_a, for which the probability

$$P_a = P (N \leq N_a)$$

then another point B in the distribution 2 corresponds to another specific number of cycles N_b, for which the probability

$$P_b = P (N \leq N_b) = P_a$$

can always be established. By connecting these two points, a log S - log N line corresponding to the probability $P = P_a = P_b$ can be obtained and all points denoting cycles to failure along this line corresponding to all loading levels in a particular range can be established using interpolation or extrapolation methods.

As an example, let A correspond to P, log S_1, log N_a, and let B correspond to P, log S_2, log N_b, then the slope of the line is - 1/m.

This can be established using equation (4) to give

$$m = (\log N_a - \log N_b) / (\log S_2 - \log S_1) \qquad (5)$$

For any point ($\log N_i$, $\log S_i$) on the line corresponding to
$P - \log S_i - \log N_i$, it can be established from equation (4) that
$$\log N_i = \log N_b - m (\log S_i - \log S_2) \qquad (6)$$
and
$$\log S_i = \log S_2 - (\log N_i - \log N_b) / m \qquad (7)$$
where m is established from equation (5).

If the parameters of the two distributions of cycles to failure are known, then simulation can be used to generate pairs of random values from the two known distributions of cycles to failure corresponding to the same probability.

If a loading level S is fixed, the distribution of the corresponding N can be obtained from equation (6).

If a specific number of cycles to failure N is fixed, then the distribution of the corresponding fatigue strength S can be obtained from equation (7).

LOG-NORMAL DISTRIBUTION OF CYCLES-TO-FAILURE

If the distribution of the cycles-to-failure follows the log-normal model, then generally
$$P = \Phi (Z) \qquad (8)$$
and
$$Z = (\log N - \mu) / \sigma \qquad (9)$$
When μ_a, σ_a and μ_b, σ_b are the means and standard deviations of the distributions of the logarithms of cycles to failure at loading levels S_1 and S_2 respectively and then, if $Z_a = Z_b$. The condition that $P = P_a = P_b$ will then have been satisfied.

WEIBULL DISTRIBUTION OF CYCLES-TO-FAILURE

If the cycles to failure follows a WEIBULL DISTRIBUTION, then generally
$$P = 1 - r \qquad (10)$$
and
$$r = \exp [- (N/0)^\beta] \qquad (11)$$
When N_a, 0_a, β_a and N_b, 0_b, β_b are the variables, scale parameters and shape parameters of the Weibull distributions of cycles to failure at loading levels S_1 and S_2 respectively, then $r_a = r_b$.

The condition $P = P_a = P_b$ will then have been satisfied.

The application of these concepts depends on the ability to generate
random values of the normal and Weibull variate.

<h2 style="text-align:center">MONTE CARLO METHOD</h2>

Generation of random values of the Normal variate

A random value of the standard normal variate Z can be obtained by
using Teichroew's approximate method following the procedure described
by Naylor [5]. The normal variate X with mean μ and standard
deviation σ can then be obtained from

$$X = \mu + \sigma z \qquad (12)$$

From equation (14) it follows that a random value of the normal
variate is related to the standard normal variate in terms of equation
(13)

$$Z = (x - \mu)/\sigma \qquad (13)$$

By comparing equations (13) and (8) to (9), it follows that by
generating a single random value of the standard normal variate, two
points on the two distributions of cycles to failure at two loading
levels with the same probability of occurrence can be obtained by
letting $\mu = \mu_a$; $\sigma = \sigma_a$ and $\mu = \mu_b$; $\sigma = \sigma_b$ respectively.

Generation of random values of a Weibull variate.

A direct transformation can be used for generating a random value of
N. Generally from (10) and (11) , $P = 1 - r$

It follows that

$$N = \theta \, (-\log \, (r))^{1/\beta} \qquad (14)$$

where θ is the scale parameter, β is the shape parameter of the
Weibull distribution and r is a random number of the uniform variate
which can be generated by the computer directly. Comparing equation
(14) with (10) and (11), generating a single random number r of the
uniform variate, two points of the two distributions of cycles to
failure at two loading levels with the same probability of occurrence
$P = P_a = P_b$ can be obtained by letting $\theta=\theta_a$, $\beta = \beta_a$ and $\theta = \theta_b$, $\beta = \beta_b$
respectively.

APPLICATION OF MODEL

The general simulation procedure was developed in the form of a number of computer programs written in BASIC and implemented initially on an IBM 3083 mainframe and later on an IBM PC.

The following computer programs have been evolved :-

Procedure 1 - Determination of cumulative fatigue reliability

This procedure depends on the use of Miner's law as given in equation (1) and the chance of survival or the reliability R is given by the governing relationship

$$R = Pr. \ (\Sigma n_i/N_i < 1) \qquad (15)$$

Procedure 2 - Determination of the cycles to failure for a given reliability

This procedure depends on equations (5), (6) and (2) to determine the cumulative distribution of N, i.e. F(N). Then N corresponding to a given reliability R(N) = 1 - F(N).

Procedure 3 - Determination of simple fatigue reliability at specific cycle life

This is based on the interference theory and considers the dispersion of both fatigue strength and load. The basic formula is R = Pr.(S > L), where the variable of fatigue strength S is calculated from equation (7) and L is the variable of load.

These basic procedures were evaluated and then evolved by considering a number of specific applications. These are now described.

EXAMPLES OF APPLICATION

The accuracy of the simulation procedure was first examined by comparing simulated with published experimental results. The procedure was then used to determine the reliability of standard bearing types. A further determination established the endurance fatigue reliability of a typical mechanical component subjected to combined bending and torsion.

These examples of application are now considered :-

1. Accuracy of the simulation

Experimental data published by Kececioglu et. al. [2] has been used to examine the accuracy of the simulation. The original data is shown in Table 1.

TABLE 1. THE DISTRIBUTION OF PARAMETERS OF THE LOGARITHM OF THE CYCLES TO FAILURE AT THREE STRESS LEVELS.

Stress levels (psi)	Means of logeN	Standard deviations of logeN
70,000	11.3872	0.1974
80,000	10.6389	0.1972
100,000	9.3886	0.1969

From the normal distribution parameters of logN corresponding to the two stress levels, 70,000 psi and 100,000 psi, the distribution of logN corresponding to the stress level 80,000 psi has been determined by simulation. If tested with the CHI-square formula, the frequency distribution obtained from 4,000 samplings by the computer divided into 13 classes cannot be rejected as the normal distribution at the significant level of 0.05 with the mean 10.63178 and the standard deviation 0.1970627. The calculated value of the CHI-square test is 10.313 which is much less than the value of $\chi^2 0.05$ for 11 degrees of freedom, 19.675. Comparing the experimental data 10.6389 for the mean and 0.1972 for the standard deviation, the error of the parameter estimation from the simulation is within 0.0667 percent for the mean and 0.069 percent for the standard deviation respectively.

2. Determination of cumulative fatigue reliability

The original experimental data from [2] is again used as the basis in conjunction with the cycles applied. at different stress levels in [2] as shown in Table 2.

TABLE 2.   THE CYCLES APPLIED AT THREE STRESS LEVELS.

Stress levels (psi)	Cycles applied
70,000	10,000
80,000	6,000
100,000	3,500

The reliability calculated from 10,000 Monte Carlo simulative cycles is 0.9993. It is a little higher than the result 0.99878 obtained from the approach of Gardner [2].

If the failure probability is very small or if the requirement involves the saving of computer running time, the frequency distribution of

$$SM = \sum_{i=1}^{k} \frac{n_i}{N_i}$$

can be obtained by using a smaller number of simulative cycles, to which a standard form of distribution can be fitted.

Using the equation for a particular standard distribution and the parameters estimated, the reliability, i.e. the probability of SM<1 can be obtained. In the above case from 1,000 simulation cycles, a lognormal distribution can be fitted to the frequency distribution of SM according to the CHI-square test. The mean of the logarithm of SM equals -0.58732, the deviation equals 0.200409. So that

$$Z = \frac{0 + 0.58732}{0.200409} = 2.9306$$

$$R = \Phi(z) = 0.998305.$$

Comparing the value of the reliability obtained from 10,000 cycles of simulation with the value of R above, the error is within 0.0996 percent. The agreement between the values obtained from the full simulation and this approximation is thus shown to be very close.

3.   Determination of cycles corresponding to a given reliability in a cumulative fatigue process

According to Table 2, the proportion of applied cycles at the stress levels 100,000 psi, 80,000 psi and 70,000 psi are 0.5128, 0.3077 and 0.1795.

The calculated result shows that the cycles N corresponding to the reliability 0.99878 established by simulation is 19907 which is higher than 19500, which is the sum of 10,000, 6,000 and 3,500 given in table 2. These values have been quoted in [2]. This indicates that this method does not necessarily apply for the sequential loading situation. In order to reduce the running time of the computer, or if the digit number of the given reliability is too large, a standard distribution form can again be fitted to obtain cycles N from less simulation cycles. From the parameters estimated, the N corresponding to the given reliability can then be obtained.

4. **The determination of the reliability of simple fatigue for a specific cycle considering the dispersion of the applied load.**

Taking a typical commercial bearing as an example, the mean and the standard deviation of the applied load, which is assumed to be from a normal distribution are 480 and 32 lb. respectively, calculate the reliability at the required cycles of 10^4, 10^5 or 10^6.

From [6] the lifetime of rolling bearings can be approximated by a Weibull distribution of two parameters. The shape parameter β is taken as 1.17 and the scale parameter

$$\theta = 6.84 L_{10} \tag{16}$$

where L_{10} is the rating lifetime, i.e. the lifetime of the bearing corresponding to the failure probability of 10 percent, which is specified as 1000,000 cycles for all bearings.

The relation of the lifetime with the load applied is

$$L_{D10} = (C_r / F)^a L_{10} \tag{17}$$

where C_r is the rating load, F is the applied load, L_{D10} is the lifetime corresponding to F with a failure probability of 10 percent, and

$$a = 3 \qquad \text{for ball bearings}$$
$$a = 10/3 \qquad \text{for roller bearings.}$$

For the ball bearing FAG 635, $C_r = 480$ lb. Let $S_1 = C_r = 480$ lb., $S_2 = F = 600$ lb., according to equations (16) - (17), the parameters of the Weibull distributions of the bearing cycle life at loading levels S_1 and S_2, are listed in Table 3.

**TABLE 3. THE PARAMETERS OF THE WEIBULL DISTRIBUTION OF THE
CYCLES TO FAILURE OF THE BEARING FAG 635 AT TWO LOADING LEVELS.**

Loading levels (lb.)	Shape parameters of N	Scale parameters of N
480	1.17	6.84×10^6
600	1.17	3.50208×10^6

Using the simulation procedure, the calculated values of the
reliability of the bearing under the applied load L - N(480,32) at the
required lifetimes are shown in Table 4.

**TABLE 4. RELIABILITY OF THE BEARING FAG 635 UNDER THE LOAD L - N
(480lb., 32lb.) at different cycle lives required.**

Cycles required	Reliability
10,000	0.997
100,000	0.991
100,000	0.893

If it is assumed that σ_s = 0, the calculated reliability using the
programme is just 0.9 when the cycles equal 10^6, which conforms to the
definition of the L_{10} life. Taking another type of bearing FAG 6240,
the rating load of which is C_r = 46,500 lb. Let L - N(46,500 lb.,0),
he calculated result is also 0.9 which again conforms to the L_{10} life.

These values indicate that the model adopted here is reasonable.   The
fundamental reason for these results is the monotonicity of the
function S expressed in the form of equation (7) relative to N_b or N_a.
That is to say that the variates N and S are dependent on each other
in a fatigue situation.

This procedure therefore represents a potentially general method for
the determination of fatigue reliability from a knowledge of only two
distributions at different load levels.

5. The determination of endurance fatigue reliability with the consideration of the difference of stress ratio between the applied stress and the life testing.

It is assumed that the distribution of the endurance limit is the same as the fatigue strength at cycles to failure of 10^6 or 10^7. Taking 10^7 as the critical lifecycle is safer than taking 10^6, as shown in Fig.1.

Sometimes the stress ratio caused by the applied load in the component is different from that of life testing. When using the data from life testing, the fatigue strength derived from life testing in equation (7) must be converted into the equivalent fatigue strength corresponding to the stress ratio of the component. Let r_t be the stress ratio for the life testing, S_t is the fatigue strength derived from equation (7), S_u as the ultimate strength, r as the stress ratio of the applied stress, S_r as the equivalent fatigue strength and S_{al} as the fatigue strength corresponding to reverse load. According to the Gerber parabola as shown in Fig. 2,

$$S_t^2 + \frac{r_t \sqrt{(1 + r_t^2)} S_u^2}{S_{al}} S_t - (1 + r_t^2) S_u^2 = 0 \qquad (18)$$

$$S_r^2 + \frac{r \sqrt{(1 + r^2)} S_u^2}{S_{al}} S_r - (1 + r^2) S_u^2 = 0 \qquad (19)$$

from equations (18) - (19), the equivalent fatigue strength S_r can be determined.

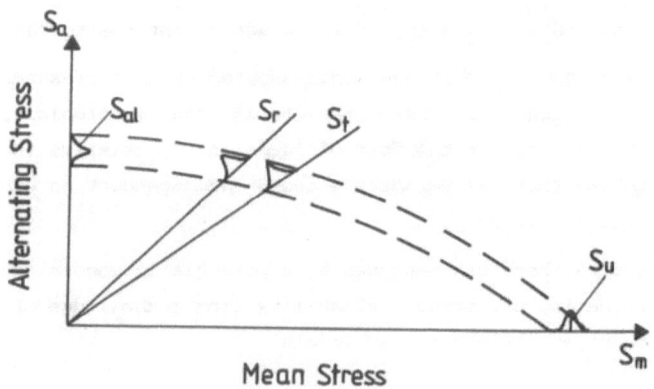

Figure 2. Distributional Gerber Parabola

As an example, calculate the endurance fatigue reliability of the
bevel gear shaft shown in Fig. 3, which is a driving shaft in a gear
box with a rated capacity of 125HP and an input rating rotational
speed of 1500 rpm.

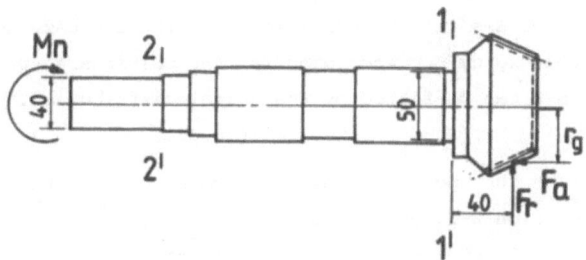

Figure 3. Bevel gear shaft and applied loads

The input torque Mn = 59.68 kg-cm and if the coefficient of variation
of Mn is 1/15, then the standard deviation of Mn is σ_{Mn} = 398 (kg-cm).

From the analysis of the force applied on the shaft, the axial force
F_a is resisted by the shoulder of the shaft, the tangential force of
the gear F_t is reacted by Mn, and the radial force of the gear F_r may
be determined in terms of rg, the pressure angle of the teeth α and
the taper angle of the bevel gear ß.

For one of the possibly most dangerous points, the outside point of
the cross section 1-1, the shear stress τ_1 and the bending stress S_b
can be determined allowing for an appropriate stress concentration
factor. The maximum distortion energy criterion of failure was used
to determine the mean failure governing stress S_m and the amplitude of
the stress $S_a = S_b$. From these values the resultant stress and the
stress ratio $r = S_a/S_m$ were determined.

The material of the shaft was taken as AISI 4340, applied in Ref.1.
The stress ratio of life testing of the material r_1 as stated in table
2 of Ref.1 is 0.83. The other data of the material are the same as
listed in table 1, Let $N = 10^7$.

The simulation program was modified to accommodate the determination
of S_r and S_t based on the Gerber parabola.

In this program, the ultimate strength S_u is considered as the normal random variable, with a mean of 116400 psi and a standard deviation of 1300 psi according to [6]. The Gerber parabola with a certain probability of occurrence is determined from S_t and S_u.

After 1000 simulative cycles, the calculated results were obtained giving : the mean of S-s is 3105.472 kg/cm^2 and the standard deviation of S-s as 194.811 kg/cm^2. According to the CHI-square test, the distribution of S-s can be fitted by the normal distribution. So

$$Pr.(S -s > 0) = 1 - \Phi(15.941)$$

For another one of the possibly most dangerous points, the outside point of cross section 2-2, using a similar procedure gave the result

$$Pr.(S - s > 0) = 1 - \Phi(11.865)$$

Section 2-2 is potentially more dangerous than 1-1 but the above figures suggest that the endurance reliability of the shaft can be considered to be intrinsically reliable [3].

COMPUTER RUNNING TIMES

As an indication of the computer running times involves in representative calculations, the CPU running times on the IBM 3083 mainframe and an IBM PC are given in each case for 1000 simulation samples.

TABLE 5.   COMPUTER RUNNING TIMES

	IBM 3083	IBM PC
1. CUMULATIVE FATIGUE (LOGNORMAL DISTRIBUTION)	5.28 SECONDS	2.5 MINUTES
2. CUMULATIVE FATIGUE (WEIBULL DISTRIBUTION)	4.68 "	2.0 "
3. SIMPLE FATIGUE	4.12 "	1.5 "
4. CYCLES FOR REQUIRED RELIABILITY	20.4 "	8.5 "

CONCLUSIONS

1. It has been shown that the use of a simulation method in conjunction with a knowledge of the form of the distribution of cycles to failure at two representative stress levels may be used as a basis for the determination of fatigue reliability for a given life and the life for a given reliability.

2. The computer running times may be low enough to potentially allow the method to be used for general design estimation.

3. The method is not intended to be applied in the sequential loading situation.

4. The method has been tested on a limited number of examples but further investigation is required to improve confidence in the approach.

REFERENCES

1   Frost, N.E., Marsh, K.J., Pook, L.P.
    Metal Fatigue, Oxford University Press, 1974.

2   Kececioglu, D.G., Chester, L.B. and Gardner, E.O.,
    Sequential cumulative fatigue reliability, Annals of the
    Reliability and Maintainability Symposium, Los Angeles, 1974, pp.
    533-9

3   Carter, A.D.S.
    Mechanical Reliability, 2nd. ed., Macmillan, London, 1986.

4   **Fatigue Reliability : Introduction,**
    The Committee on Fatigue and Fracture Reliability of the
    Committee on Structural Safety and Reliability of the Structural
    Division of ASCE., Journal of the Structural Division Proceedings
    of ASCE Vol. 108, No. ST1, Jan. 1982.

5   Naylor, T.H., Balintfy, J.L., Burdick, D.S., Chu, K.
    Computer Simulation Techniques, John Wiley & Sons Inc. 1968

6   Shigley, J.E.
    Mechanical Engineering Design, 3rd ed. McGraw-Hill Book Company,
    New York, 1977

7   **FAG Ball Bearings**, Fag Roller Bearings Catalogue 41000 E 1966
    ed., FAG Bearing Company Limited, Wolverhampton, 1966

8   Kececioglu, D.B. and Chester, L.B.
    **Combined Axial Stress Fatigue Reliability for AISI 4130 and 4340
    Steels**, Journal of Engineering for Industry, Transactions of
    ASME, Feb. 1976.

228

SOME NEW ASPECTS ON COMPONENT IMPORTANCE MEASURES

MIN XIE
Division of Quality Technology
Linköping University, S-581 83 Linköping, Sweden

KECHENG SHEN
Department of Machine Design
Lund Institute of Technology, Box 118, S-221 00 Lund, Sweden

ABSTRACT

Component importance measures are useful in deciding where to allocate further efforts to increase system performance, particularly system reliability and availability. Which of the components should be judged as important, and hence to be improved at the first hand, depends on the component structural position in the system, the improvement potential of the component and also the improvement action to be taken. In this paper we review some recent advances in measuring the importance of system components. A general methodology that incorporates these features is presented. We will also introduce some concepts concerning availability importance useful for repairable systems.

INTRODUCTION

Recently there have been both practical and theoretical concerns with the measures of the importance of system components. For a complex system consisting of a number of components, it is generally believed that some of them affect the system reliability more than others. To improve the system performance, these components should be studied and improved at the first hand.

In this paper we will discuss some of the recent advances in studying component importance. From a development's point of view, a component is important if an improvement of it increases the system performance more than others. However the ranking of the component importance should depend on the system structure, component reliability and also available improvement actions. We may have different rankings for different situations [1].

How different improvement actions can be incorporated in ranking components will be studied. This is of practical interest since at the design stage we have only limited resources and few choices of methodology of improving the components. It is essential that the important components are improved and the greatest gain of the system reliability performance increase is resulted.

In this paper we review some recent advances in measuring the importance of system components. A general methodology that incorporates these features is presented. We will also introduce some concepts concerning availability importance useful for repairable systems.

INCREASE OF SYSTEM RELIABILITY
BY IMPROVING A COMPONENT

Suppose that a system consists of n components and denote by p_i the reliability of component i. The reliability of the system is a function of the component reliabilities and we denote this by $h(\mathbf{p})$ where $\mathbf{p}$ is the vector of component reliabilities.

If we change the reliability of component i from p_i to p'_i, the system reliability is changed from $h(\mathbf{p})$ to $h(p'_i, \mathbf{p})$. It is shown in Xie and Shen [1] that this latter change is equal to

$$I_\Delta^{(i)} = (p'_i - p_i)(h(1_i, \mathbf{p}) - h(0_i, \mathbf{p})). \tag{1}$$

It is also noted that

$$I_B^{(i)} = h(1_i, \mathbf{p}) - h(0_i, \mathbf{p}) \tag{2}$$

is the Birnbaum [2] measure of the importance of component i. Further measures of component importance have been studied by Barlow and Proschan [3], Fussel [4], Natvig [5] and Xie [6] among others.

INCORPORATING IMPROVEMENT ACTIONS
IN MEASURING COMPONENT IMPORTANCE

Measures of component importance are useful in deciding where to allocate further development resources. Hence if for a component, an improvement of it results in a larger increase of system reliability, then it should be considered as important. For a general definition of the component importance the expression (1) may be used and it depends on the structural position of the component and the improvement action that will be taken. It is also noted that ranking of component importance may differ for different improvement actions. As a conclusion, the Birnbaum measure of component importance should not be used directly. However, it is central in studying component importance. Below we present some concrete importance measures incorporating different improvement actions of practical relevance. For other results and numerical values showing their difference and similarity, we refer to Xie and Shen [1] and Shen and Xie [7].

Parallel Redundancy Importance

A common improvement action of a component is to provide parallel redundancy. Especially for electrical and electronic components this is a widely used technique. In nuclear power plants a lot of such redundancies are used in, for example, radiation measurements to assure high safety.

Considering the case when we add an identical and statistically independent component to the existing one in the system. This will certainly increase the system reliability since the component position i will now have a higher reliability. In this sense we have improved component position i. In Xie and Shen [1] the increase of system reliability by such an improvement is derived as

$$I_P^{(i)} = p_i q_i (h(1_i, \mathbf{p}) - h(0_i, \mathbf{p})). \qquad (3)$$

It is seen by this expression that under the assumption of provision of parallel redundancy to the system components, that component which possesses the largest value computed from above should be chosen to be first given the redundancy. From this expression we can also observe that those features mentioned in Section 1 have been successfully incorporated into this measure. The Birnbaum measure expresses the structural importance of the component; the quantity of $p_i q_i$ is the increase of component position i by the redundancy since

$$p_i' - p_i = 2p_i - p_i^2 - p_i = p_i q_i$$

where $q_i = 1 - p_i$ is the failure probability.

An interesting result is obtained in Xie and Shen [8] where it is noted that the quantity defined in (3) is in fact the covariance of the component state and system state, i.e. if we let

$$X_i = \begin{cases} 1, & \text{if component } i \text{ is functioning} \\ 0, & \text{if component } i \text{ is failed} \end{cases}$$

and

$$\phi(\mathbf{X}) = \begin{cases} 1, & \text{if the system is functioning} \\ 0, & \text{if the system is failed} \end{cases}$$

then we have

$$\text{cov}(X_i, \phi(\mathbf{X})) = p_i q_i (h(1_i, \mathbf{p}) - h(0_i, \mathbf{p})). \tag{4}$$

Therefore this measure may be interpreted as the dependence between component i and the system. The stronger this dependence is, the more important the component is if parallel redundancy is a feasible improvement action. For details and examples concerning this measure, see Xie and Shen [8]. This measure has also recently been considered by Boland et al.[9], Iyer [10] and Råde [11]. Several similar results are also derived independently by these authors.

Standby Redundancy

One of the drawbacks of parallel redundancy is that it is difficult to implement except in electronic systems. In, for example, mechanical designs such a redundancy would be either impossible to, by the nature of the design, or not feasible because of constraints such as weight, volume and other practical considerations, implement.

This drawback may be overcome by use of standby redundancy by which an identical and statistically independent component in standby would be switched on whenever the on-line component fails. This is equivalent to addition of life times of the two components.

Let $F_i(\bar{F}_i)$ and $F(\bar{F})$ be the life distribution(reliability) of component i and the system respectively. By having a standby redundancy on component i, the reliability of this component position is changed to, see Xie and Shen [12],

$$\bar{G}_i(t) = \bar{F}_i(t) + \bar{F}_i * F_i(t) \qquad (5)$$

where $\bar{F}_i * F_i(t)$ is the convolution of $\bar{F}_i(t)$ and $F_i(t)$ defined by

$$\int_0^t \bar{F}_i(t-s) \, dF_i(s)$$

The increase is

$$\bar{G}_i(t) - \bar{F}_i(t) = \bar{F}_i * F_i(t). \qquad (6)$$

Substituting this into (1) we obtain the increase of system reliability

$$I_S^{(i)} = \bar{F}_i * F_i(t)(h(1_i, \bar{\mathbf{F}}(t)) - h(0_i, \bar{\mathbf{F}}(t)). \qquad (7)$$

Similar to the case of parallel redundancy the component which should be provided with a standby component in the first place is the one which gives the largest value of the above quantity. Note the time dependence of both the Birnbaum measure and the increase of reliability of component position i which was made implicit in the previous discussions.

Generally the evaluation of (7) is much more complicated than in the case of parallel redundancy and a ranking would be much more difficult even for some simple component life distributions.

Minimal Repair

The idea of minimal repair may be another alternative to improving reliability of components. It means that whenever a component fails it will be repaired and returned to its state just before the failure, i.e. the failure is simply erased. This has been studied by many authors, see Natvig [13] and Nakagawa [14]. Practically it may be interpreted as such a maintenance policy that a failed component is immediately repaired to be "just able" to function; no more effort is made. In practice, examples may be found that for the same failure a complete or total repair would perhaps take weeks while a quickly fixed repair may only take days or hours.

It is shown in Natvig [13] that the increase of reliability of component position i is given by

$$\bar{G}_i(t) - \bar{F}_i(t) = -\bar{F}_i(t) \ln \bar{F}_i(t) \tag{8}$$

so that the system reliability improvement is

$$I_M^{(i)} = -\bar{F}_i(t) \ln \bar{F}_i(t)(h(1_i, \bar{\mathbf{F}}(t)) - h(0_i, \bar{\mathbf{F}}(t)). \tag{9}$$

Apparently ranking of components in this case is also rather complicated and similar comments as in the case of standby redundancy are also valid.

Effect of Perfect Improvement of a Component — Improvement Potential

Theoretically no component can be perfect i.e. it will never fail. In practice, however, there may be cases when a component is considered very much more reliable than another, hence it is "perfect" i.e. it has reliability one.

When we consider to use such a perfect component to replace a non-perfect one in a system, the increase of the system reliability is maximized by improving this component. Assume $p_i' = 1$ in (1) we obtain the increase of system reliability

$$I_1^{(i)} = q_i(h(1_i, \mathbf{p}) - h(0_i, \mathbf{p})). \tag{10}$$

This increase is bounded only by the reliability of the replaced component. The more reliable the component is the less the system reliability increase will be. Therefore this effect actually expresses the improvement potential of component i i.e. how much more is left for it to be improved. This has also been considered by Aven [15] and Shen and Xie [7].

This improvement potential has another practical implication, namely, a component already having high reliability will not contribute much upon improvement and at the same time it will usually cost much more to improve it than the less reliable ones.

AVAILABILITY IMPORTANCE
OF SYSTEM COMPONENTS

The availability of a device is defined as the probability that it functions at time t. Denote by $A_i(t)$ the availability of component i at time t and $A(t)$ the system availability at time t. Then $A(t)$ may usually be expressed as a function of the availability of the components, see e.g. Barlow and Proschan [16]. We define the *availability importance* of component i at time t by the rate of the system availability increase due to the component availability increase, that is

$$I_A^{(i)}(t) = \frac{\partial A}{\partial A_i}\bigg|_{A_j=A_j(t)}, j = 1, 2, \ldots, n \qquad (11)$$

In the following of this paper we assume that we have a coherent system(for a general discussion, see e.g. Barlow and Proschan [16]). We also assume that the components are independent and have independent functioning and repair times.

Denote by $h(\mathbf{p})$ the system reliability where $\mathbf{p}$ is a vector of component reliabilities. From Barlow and Proschan [16] it follows that

$$A(t) = h(A_1(t), \ldots, A_n(t)).$$

Hence

$$I_A^{(i)}(t) \;=\; \left.\frac{\partial A}{\partial A_i}\right|_{A_j=A_j(t),j=1,2,\ldots,n}$$

$$=\; \left.\frac{\partial h(\mathbf{p})}{\partial p_i}\right|_{p_j=A_j(t),j=1,2,\ldots,n}$$

$$=\; I_B^{(i)}(\mathbf{A}(t)) \tag{12}$$

where $I_B^{(i)}$ is the so-called Birnbaum measure of component importance. Mathematically the Birnbaum importance measure and the availability importance measure have the same form, just like the relation between the system reliability and system availability. However we will concentrate here on the availability of the system and this is essential in studying repairable systems that make up most of the systems.

It is useful to introduce the following availability importance concepts defined on component i.

The limiting availability importance $I_A^{(i)}$ is defined by

$$I_A^{(i)} = \lim_{t\to\infty} I_A^{(i)}(t) \tag{13}$$

when the limit exists.

The average availability importance in $[0,T]$ is defined by

$$I_A^{(i)}(T) = \int_0^T I_B^{(i)}(t)\ dt/T. \tag{14}$$

The limiting average availability importance may be similarly defined by

$$I_{A_{av}}^{(i)}(T) = \lim_{T\to\infty} \int_0^T I_A^{(i)}(t)\ dt/T. \tag{15}$$

Since it is easily seen that limiting average availability importance is the same as the limiting availability importance when it exists, the most interesting concepts are the availability importance(AI) at time t, the limiting availability importance(LAI) and the average availability importance(AAI).

It should be pointed out here that a serious critique on the Birnbaum measure is its time-dependence. It can only give a ranking of components at fixed time points. The modification of the Birnbaum measure to time-independent cases is quite complex. One has to consider weighted averages of the Birnbaum measure [13]. For the availability importance measure, the problem is less essential since the most interesting situation is the limiting case in which the limiting availability importance is time-independent.

A component which is structurally important does not necessarily have high availability importance. The availability importance of a component may become lower if repair on it takes very short time compared with the others.

For coherent systems it follows that

$$I_A^{(i)}(t) = h(1_i, \mathbf{A}(t)) - h(0_i, \mathbf{A}(t)).$$

The following notations will be used in this paper,

$$h(\cdot_i, \mathbf{x}) = h(x_1, x_2, \ldots, \cdot_i, \ldots, x_n).$$

$$\bar{A}(t) = 1 - A(t).$$

For series or parallel systems, the following results that are valid for the Birnbaum measure hold for the availability importance measure.

For a series system,

$$I_A^{(i)}(t) = \prod_{j \neq i} A_j(t) = A(t)/A_i(t). \tag{16}$$

It is easily seen that in this case the component which has the smallest availability is the most important one.

For a parallel system,

$$I_A^{(i)}(t) = \prod_{j \neq i} \bar{A}_j(t) = \bar{A}(t)/\bar{A}_i(t). \tag{17}$$

Accordingly in this case the most available component is the most important one.

Other similar results associated with the Birnbaum importance measures are also valid here.

Acknowledgement: Part of this research is supported by CENIIT at Linköping University under a project titled "System Reliability". The authors are also grateful to Prof. Bo Bergman for many discussions.

REFERENCES

1. Xie, M. and Shen, K., On ranking of system components with respect to different improvement actions. *Microelectron. Reliab.*, 1989, **29**(2), 159-64.

2. Birnbaum, Z. W., On the importance of different components in multicomponent system. In *Multivariate analysis-II*, ed. P. R. Krishnaiah, Academic Press, New York, 1969, pp. 581-92.

3. Barlow, E. E. and Proschan, F., Importance of system components and fault tree events. *Stochastic Processes Appl.*, 1975, **3**, 153-73.

4. Fussel, J. B., How to hand-calculate system reliability characteristics. *IEEE Trans. Reliab.*, 1975, **R-24**, 169-74.

5. Natvig, B., A suggestion of a new measure of importance of system components. *Stochastic Processes Appl.*, 1979, **9**, 319-30.

6. Xie, M., On some importance measures of system components. *Stochastic Processes Appl.*, 1987, **25**, 273-80.

7. Shen, K. and Xie, M., The increase of Reliability of k-out-of-n systems Through Improving a Component. *Reliability Engineering and System Safety*, 1989, **26**, 189-95.

8. Xie, M. and Shen, K., The parallel redundancy importance of system components. *Technical Report, No. LiTH-IKP-R-566*, Linköping University, Sweden, May 1989.

9. Boland, P. J., El-Neweihi, E. and Proschan, F., Active redundancy allocation in coherent systems. *Probability in the Engineering and Informational Sciences*, 1988, **2**(3), 343-53.

10. Iyer, S., The increase in reliability and mean time to failure of a system due to component redundancy. *OPSEARCH*, 1989, **26**(3), 145-50.

11. Råde, L., Expected time to failure of reliability systems. *Math. Scientist*, 1989, **14**, 24-37.

12. Xie, M. and Shen, K., On the increase of the expected system yield due to component improvement, *Reliability Engineering and System Safety*, 1989, to appear.

13. Natvig, B., New light on measures of importance of system components. *Scand. J. Statist.*, 1985, **12**, 43-54.

14. Nakagawa, T. and Kowada, M., Analysis of a system with minimal repair and its applications to replacement policy, *European J. Oper. Res.*, 1983, **12**, 176-82.

15. Aven, T., Some Consideration on Reliability Theory and its Applications. *Reliability Engineering and System Safety*, 1988, **21**, 215-23.

16. Barlow, R. E. and Proschan, F., *Statistical theory of reliability and life testing, probability models*. To Begin With, Silver Spring, MD, 1981.

THE RECENT DEVELOPMENT ON
THE NON-DESTRUCTIVE INSPECTION RELIABILITY AND
ITS APPLICATIONS IN OFFSHORE STRUCTURAL INTEGRITY ANALYSIS

J C P Kam

Department of Mechanical Engineering
University College London
Torrington Place, London WC1E 7JE

ABSTRACT

Due to the many design uncertainties involved in the long term fatigue loading and performance of offshore structural components, in-service non-destructive inspections (NDI) are usually incorporated as part of the maintenance programme to ensure the actual performance abides by the design. However, the (un)reliability of NDI introduces another type of uncertainty into the assessment process. The aerospace and other industries, facing similar problems, carried out extensive studies in NDI reliability and statistical trials.

The differences in the structural configurations, and in the operational features mean that data exchange between industries are difficult. However, the exchange of modelling and analytical techniques is possible.

This paper briefly reviews the development of the fitness for purpose philosophy in structural maintenance and the associated research in NDI reliability. Emphasis, however, will be put on the scope of applications in offshore structures. Moreover, the paper will discuss some problems which are unique to the offshore industry and these require special considerations in obtaining and using the NDI reliability data. Some possible solutions to the above problems will also be discussed in this paper and illustrated with examples.

NOMENCLATURE

A, B	Weibull distribution parameters
$\hat{B}$	Bias factor for σ
C_j	Paris crack growth constants for segment j
$\hat{K}_t$	Non-dimensionalised stress concentration factor
P (X)	Structural reliability (probability of failure) before updating
P (X|θ)	Updated structural reliability (probability of failure)
P (θ)	NDI reliability (probability of sizing /detecting defect present)
a, a'	An arbitrary crack size
a_i, a_f	Initial and final crack growth sizes
a_{cr}	Critical crack size (Maximum size prior to failure)
a_{NDI}	Critical non-destructive inspection size
$\hat{a}$	A measured crack size

da/dN	Crack growth rate
$f_Y(X)$	Probability density function of Y when Y = X
g	State function
m_j	Paris crack growth exponents for segment j
p (a)	Probability density function of crack size a
$p(X|\theta)$	Probability density function of X given an event θ
ΔK	Stress intensity range
$\Phi(X)$	Cumulative standard Normal distribution at X
β	Reliability index
δa	An arbitrary, infinitesimally small crack size
μ_x	Mean of X
σ	Long term stress time history root mean square (RMS) value
σ_x	Standard deviation of X

INTRODUCTION

As the exploration and production activities progress into the deeper and more hostile regions of the North Sea, the maintenance of offshore structural integrity becomes more and more important to the safety and operational reliability of the installations. Fatigue was found to be one of the major factors affecting the long term integrity of fixed steel platforms. A typical welded tubular joint could experience up to 200 million wave and vibration induced stress cycle during the operational life [1]. As a consequence, fatigue cracks could appear in the joints which in turn could fail in the forms of leakage, loss of member stiffness and even fracture [2]. In order to maintain the integrity of these structures, the Offshore Industry has adopted a practice of regular inspections. If defects are detected in-service, they are usually removed by grinding or in some cases by large scale repair (re-welding, replacement or grouting). These types of underwater operations are very costly. The recent fatigue research in tubular structures, however, has shown that large fatigue cracks could develop in joints even well within the safe design life [2]. It may therefore be unnecessary to repair all the defects if the components in question can be shown to be "fit for purpose". These findings have opened new possibilities in the approach of offshore structural maintenance.

Furthermore, if the relative level of "fitness" - as classified by reliability analysis - can be established [such as 3, 4], the remedial operations can be more rationally planned and targeted to the more critical joints. If repair is preferred, the analysis should also show whether the operations could be delayed until a better weather window is available. This will have the added benefits of improved workmanship for works carried out under less hostile weather conditions. Another possibility is to re-schedule the inspection programme to maintain the required level of reliability and thus eliminating the necessity for repair (at least for the short term). These new options could provide the Industry with more flexibility in rationalising the expensive maintenance operations.

From the above discussion, it can be seen that non-destructive inspection (NDI) is an important part of the maintenance strategy. Because NDI is used to measure the current status of a structural member, the reliability of the NDI measurements will therefore affect the expected reliability of the member. However, depending on the strategy being used, the requirements in NDI reliability information can differ. These are discussed in the following section.

NDI RELIABILITY IN FITNESS FOR PURPOSE ASSESSMENTS

The term 'fitness' could be referred to a state of any object which are properly designed and maintained to serve its intended purposes. This can be translated into reliability terms as having a failure probability below a prescribed level. The aerospace industry has pioneered some of the early research. One of such work is the definition of the critical inspection size (a_{NDI}) with the use of fracture mechanics and crack growth analysis [such as 5]. The idea is shown in Figure 1 and it is predominantly a deterministic methodology. A "worst" crack development curve is obtained from analyses using subjective choices of bounding values for the input

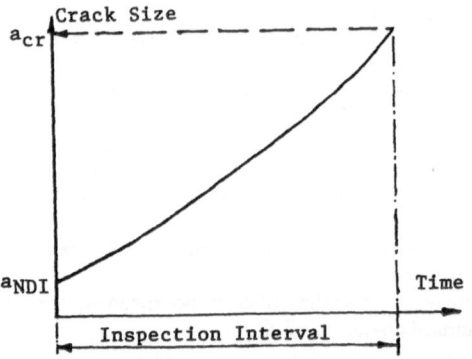

FIGURE 1   SCHEMATIC DIAGRAM SHOWING a_{NDI} APPROACH

parameters. By fulfilling the requirement that no crack will grow beyond the critical (failure or fracture) size (a_{cr}) within a single inspection interval, a starting size could be determined. This starting size is the a_{NDI} and is specified as the minimum defect size that the NDI must be able to detect with a high probability of success. This probability of success is also known as the probability of detection (POD).

In this strategy, a positive detection could be easily defined. For example in the case of ultrasonic reflection inspection, a positive detection could be classified as the characteristic response signal being larger than a "threshold" level. However, whether this signal is just above the threshold or many times larger, is not a major concern to the maintenance strategy. This is because there are usually only two courses of actions to choose after inspection, namely, to retire the component for the cause of having a defect larger than the maximum acceptable size, or to return the component to service because there is still enough remaining strength in the component to tolerate a defect of any size upto the maximum acceptable value.

The major disadvantage of this approach is that the "worst" crack growth is determined subjectively. The predicted crack growth could be made excessively conservative if only bounding values are used (upper or lower bounds wherever relevant). Furthermore, the lack of a uniform "fitness" measurement makes it difficult to compare the reliability of different components.

A similar approach could also be applied in the offshore industry. However, many of the welded structural components have a long fatigue life and during this life, relatively large cracks can be developed. Moreover, offshore structural components cannot be easily replaced (or even repaired) as in the case of aerospace components. Therefore, the above approach could not be applied as efficiently as in the aerospace industry. Instead, the fitness for purpose concept could be applied in a flexible way to accept defects of measurable sizes. There is no "maximum acceptable size" and the defect will be sized as well as detected so that fracture mechanics can be used to update the current reliability level, to schedule inspection intervals and even to optimize the maintenance strategy. However, this new approach requires reliability information on the NDI accuracy (for sizing), while previously only data on sensitivity (for detection) were needed. The basis of the reliability based fracture mechanics analysis is outlined below.

STRUCTURAL RELIABILITY METHODS

The idea of structural reliability analysis can be demonstrated by a simple situation where loading interacts with strength (Figure 2). Considering a state function of the form,

$$g = Strength - Loading \qquad (1)$$

If both the loading and the strength are random variables with defined probability distributions, g will also be a random variable with some probability distribution. When $g > 0$, the structural component is said to be in the "safe state". Failure occurs when $g < 0$ (when loading is larger

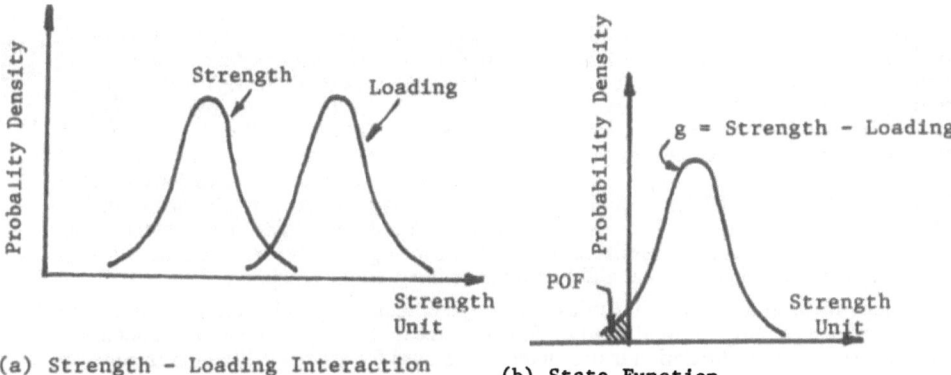

(a) Strength - Loading Interaction   (b) State Function

FIGURE 2   THE CONCEPT OF STRUCTURAL RELIABILITY

than the strength). The limiting case where $g = 0$ is called the limit state. If g is Normally distributed, $N(\mu_g, \sigma_g)$ (Figure 2b), then the ratio between $(g - \mu_g)$ and σ_g is standard Normal $N(0, 1)$. A probability of failure (POF) can be evaluated as,

$$POF = \int_{g<0} f_g(g)\, dg = 1 - \Phi(\mu_g/\sigma_g) \qquad (2a)$$

$$POF = 1 - \Phi(\beta) \qquad (2b)$$

where f_g is the probability density function and Φ is the cumulative standard Normal distribution function. The ratio (μ_g / σ_g) is called the reliability index (β) and is used as a convenient indicator for the comparison of the reliability of structures in similar conditions. The same concept could be extended to "g" with several controlling variables,

$$g = g(x_1, x_2, \ldots, x_n) \qquad (3)$$

Various methods have been developed to calculate the POF for the above general state function, and these can be classified (loosely) into 3 levels according to the uncertainties information used in the analyses and the corresponding expected accuracies [such as 6]. Level I includes the deterministic techniques which are suitable for design calculation. The characteristic values (prescribed percentiles in the distributions of the parameters) are used in these methods and the state function is evaluated only once to check whether failure has occurred. In modern (limit state) design codes these characteristic values are usually determined from the 'probabilistic' level II or level III analyses.

The "exact" POF is calculated in the Level III methods and this is obtained by numerically integrating the joint distribution of all the controlling parameters. In practice, however, the complex joint distribution is difficult to determine except for very simple distributions and the integration procedure is also time consuming.

The POF is only approximated in the Level II methods, although the accuracy of the approximation varies with approaches. The methods require the first two moments (mean and variance or covariance) and the probability distribution for each parameter. Several implementations using Rosenblatt transform [6] or orthogonal transform [such as 7] are available and these are usually referred to as the advanced Level II methods. The methods calculated the shortest distance from the origin to the failure surface (limit state surface) and this distance is found to be a good approximation of β [8]. In fact, for linear state functions comprised of uncorrelated and Normally distributed parameters, these methods give the exact POF. The point at which the distance is calculated is traditionally called the design point. The accuracy of the reported implementations could be further improved by a higher order hyper-plane approximation [6] of the state function. The effect, however, is relatively small unless the state function has a small curvature at the design point.

Another popular method is the Monte Carlo simulation method. The technique is meant to calculate the "exact" POF. However, because the method is basically a numerical sampling technique, the results are susceptible to statistical errors. Large sample size is usually required and only an approximation to the level III result can be obtained. For example, 39600 times of state function evaluation are required [9] to obtain a POF of 1% (β = 2.33) with 95% confidence that the true value lies within the range 0.9% to 1.1%. As a comparison, the advanced Level II algorithms require about 200 times of evaluating the state function consisted of ten parameters. The efficiency of the Level II methods is achieved through transformation and optimization techniques. The advantages in using the Monte Carlo method include its simplicity and the fact that the results nearly always converge. Moreover, the state function is used explicitly and no numerical differentiation or integration is required.

There are also techniques (such as the antithetic sampling) which could reduce the required number of samples significantly. However, due to the large amount of computations involved, Monte Carlo simulation is only likely to be used for special calibration rather than routine analyses.

Reliability Fracture Mechanics Analysis

Fracture mechanics crack growth analyses can be expressed as a multi-variate state function as in equation (3). Investigations using advanced level II methods have been reported by several researchers [such as 3, 4, 10, 11] on fatigue state functions of various complexities. The features include random loading, realistic crack growth models and materials response in corrosive environment. The details will not be repeated here.

An example taken from [11] is shown below to illustrate the type of input requirements and results in a typical analysis. Table 1 contains the input data for a corrosion fatigue crack growth analysis of a tubular T joint under axial loading. The load range distribution is characterised by a two parameter (A, B) Weibull distribution. The uncertainties in the hot spot stress concentration factor and the long term stress history RMS are represented by the non-dimensionalised bias factors, $\hat{K}_t$ and $\hat{B}$, respectively. Using a characteristic value of A = 1.2, the fatigue life estimated by the Department of Energy Guidance [12] is 24 years. The choices of the values for the parameters in the reliability analysis have been discussed in further detail in [11] and will not be repeated here.

The standard result for the reliability crack growth analysis is the β history (Curve O in Figure 3). This curve shows the decay of reliability with time (as the reserved life decreases during crack growth).

From the above example, it is clear that in order to facilitate the reliability analysis, crack size distributions need to be used as opposed to a deterministic a_{NDI}. This will require the information on sizing reliability. The following reviews the current understanding in the study of NDI reliability for offshore applications and outlines the future development in this area.

NDI RELIABILITY TRIALS

Usually a large number of known sample defects are inspected with an NDI technique in a trial and the results are analysed statistically. One of the early findings was that the defects need to be realistic for the applications in question. This was also confirmed by the NDI reliability studies carried out in the nuclear industry [14]. Another finding was that the operational environment will have to be simulated as closely as possible because a ten fold increase in non-detectable crack size was found as the NDI operations were moved from the laboratory to the site facility [15].

Moreover, defects of the same nominal size may have characteristics which affect the relative POD performance. It is therefore recommended that as many defects (though of the same nominal size) should be used in the trials instead of using repeated inspections on a small number of samples. It has also been suggested that repeated inspections may alter the "expectation" of the NDI personnel and may create bias in the data. This change in expectation could be represented by a correlation parameter for the POD but the value of such a parameter is difficult to quantify. Zero correlation has been assumed in the past to justify repeated

245

Table 1 The Input Parameters for the Example

Effective chord length = 4000. mm
Brace diameter = 767. mm   Brace wall thickness = 45 mm
Chord diameter = 1080. mm   Chord wall thickness = 45mm = a_f
Cyclic rate = 7.5 Mcyc / yr   Target life = 20 years
$\sigma = 0.7\ MPa$ (see $\hat{B}$ for uncertainties)
Hot spot stress concentration factor = 10 (see $\hat{K}_t$ for uncertainties)

Variable	Mean	Standard Deviation	Type of Distribution
A	1.53	0.322	Normal
$\hat{B}$	0.7	0.35	Log Normal
$\hat{K}_t$	0.85	0.20	Normal
F*	0.4696	0.014	Log Normal
p*	0.0555	0.011	Log Normal
a_i	1.4mm	0.67mm	Normal
C_1 +	3.6337E-12	3.1237E-12	Log-Normal
C_2 +	1.3829E-11	1.0220E-11	Log-Normal

* = parameters used in the fracture mechanics model [13]
+ = for da/dN in (m / cycle), ΔK in (MPa $\sqrt{m}$),
and $m_1 = 3.77$, $m_2 = 2.99$

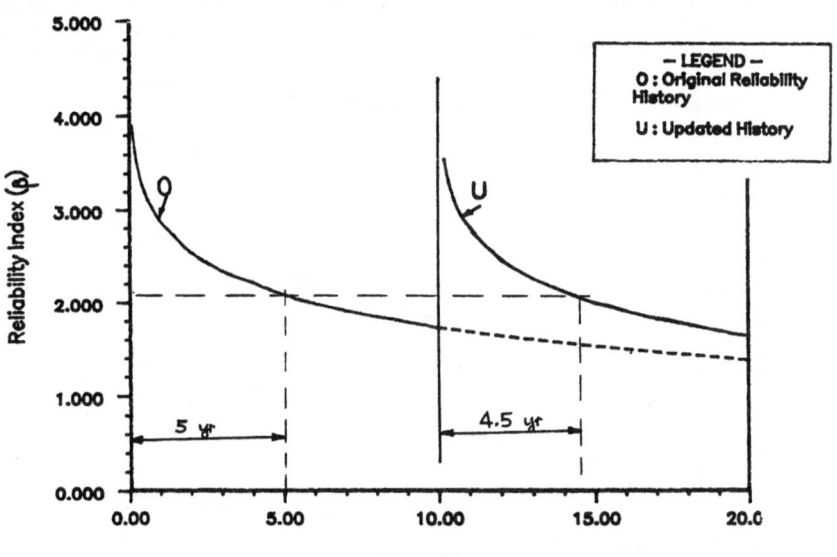

FIGURE 3   TIME DEVELOPMENT HISTORY OF RELIABILITY INDEX

inspections in a trial. As a consequence of these early findings, the current large scale trials in the Offshore Industry involve large sample size, realistic structures and defects, and underwater operations.

Another problem discovered was the difficulty in defining a success in detection. For a defect of size "a", it is likely that only a fraction of "a" is detected (measured) by the inspection. Sometimes, spurious indication is also obtained near the true defect, the measured size will then be larger than "a". The difficulties lies in choosing the best level of accuracy to define a correct (or successful) detection.

To illustrate the problems involved, the mean POD curves for an inspection technique are shown in Figure 4. These curves are derived from a single batch of trial data but constructed under different accuracy level and so different "success criteria". For an inspected size ranging between -90% to +90% of the true size (ie 10% accuracy), the POD performance is good. If the accuracy requirement is 90%, that is when the inspected size lies between -10% to +10% of the true value, the POD performance for the same NDI technique is much poorer. This dependence of POD on accuracy requirement also shows that the POD curves alone do not necessary provide all the information concerning the NDI reliability. In some cases the accuracy criterion is only applied to the lower bound of sizes. In other words, a 10% accuracy means accepting a measured size from -90% to +∞% of the true size as a successful detection. This will ignore all the errors in overestimating the true size and result in a general improvement of the POD curves. This, however, does not alter the relative POD performances under different accuracy levels and so the same problems remain.

It could be argued that if the required level of accuracy is kept at a minimum, then the POD performance will always be high. However, this will cause problems in practice because noise and weld irregularities could create spurious signals. If the accuracy level is set too low, many of this spurious signal will be "detected" as defects. Usually further works such as re-inspection or simple grinding may be carried out on a detected defect and consequently unnecessary (and expensive) works will be carried out on non-existing defects. Statistically the positive detection of a non-existing crack can be called a type II error and the negative detection (missing) of an existing crack can be termed as a type I error. There exist statistical techniques related to the operating characteristic (OC) curve method, which can assist in

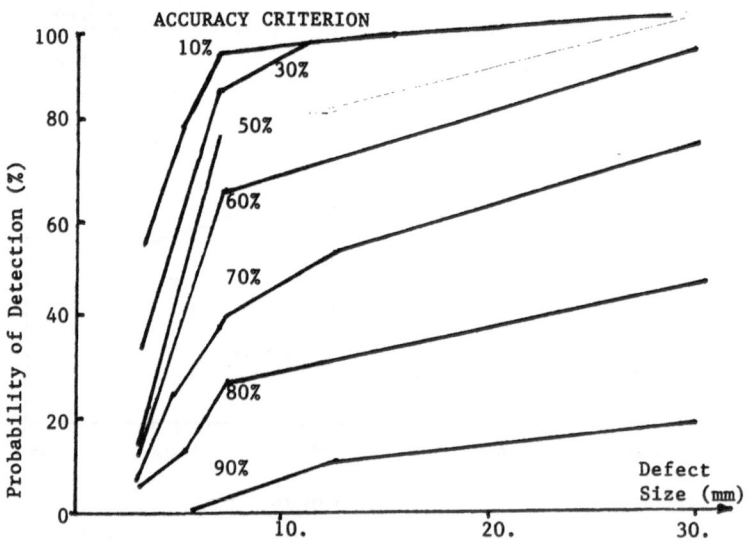

FIGURE 4 EFFECT OF ACCURACY CRITERION ON PROBABILITY OF DETECTION

choosing the accuracy level in order to minimize the two types of errors. The applications of the OC curve technique in POD data have been reported in some recent literature [such as 16]. One requirement in using the technique is that the relative "trade offs" of type I and type II errors must be quantified. These trade-offs could be expressed in reliability terms or economic terms which in turns could be determined by using reliability analyses.

The relationship between detection and sizing is further complicated as it was observed that the sizing reliability exhibits a transition in the characteristic distribution with size (Figure 5). The amount of results is not enough to justify a final conclusion but the implication is that POD data cannot be directly applied as sizing reliability information. In other words, the successful detection does not always indicate an accurate sizing of a defect. These recent observations also affect the applications of NDI data in reliability assessment. One of these areas is the updating of reliability level due to inspection.

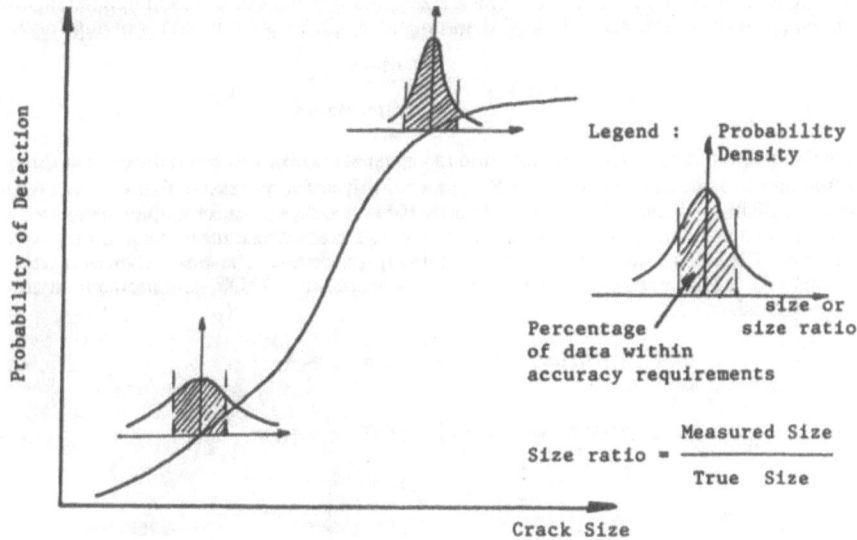

FIGURE 5   SCHEMATIC DIAGRAM SHOWING THE BEHAVIOUR OF INSPECTION ACCURACY WITH RESPECT TO CRACK SIZE

UPDATING STRATEGY

The concept of updating is exemplified by the development of the Baysian statistics [such as 17]. This branch of statistics relates data uncertainties with the degree of belief and as new information becomes available, the degree of belief can be updated. The basic application of Bayes' theory in the reliability assessment can be expressed as,

$$P(X \mid \theta) = \frac{P(X \cap \theta)}{P(\theta)} \tag{4}$$

where P(X) is the structural reliability (probability of failure) before updating, P(θ) is the NDI reliability (probability of sizing / detecting defect present). P($X \cap \theta$) is the probability of detection given the current reliability. P ($X \mid \theta$) is the updated structural reliability given that some results of inspection have been obtained. The situation can therefore be analysed as a parallel system and the resultant reliability is calculated by system reliability techniques [6].

This approach has two main difficulties. Firstly, the structural reliability level is updated

248

as one single process although many parameters are involved in the evaluation. The uncertainties in each of these parameters are therefore updated implicitly according to their relative importance in the reliability assessment. This procedure will imply that a better estimate has been achieved for the differences between the "design" and the "true" values of the parameters. However, there are uncertainties in the parameters which fall outside of this category. One example is the materials response data. Experience has shown that there are "intra-data" uncertainties during crack growth and the level of variations remain the same throughout the fatigue life of any specimen. Secondly, the computation in equation (4) assumes that a random size is directly measured when detected and this size is related to the POD information. As it has been discussed above that sizing and detection are two separate issues for NDI reliability assessment, the assumption is therefore not completely acceptable.

An alternative procedure is to update the crack size only. NDI can be considered as providing new information on current crack sizes, and it is therefore reasonable to constraint the updating to the crack sizes. Bayes' theorem can again be used in the following form,

$$p(a \mid \hat{a}) = \left\{ \frac{p(\hat{a} \mid a)}{\int_a p(\hat{a} \mid a) \, p(a) \, da} \right\} p(a) \tag{5}$$

where the bracketed { } term in equation (5) is usually known as the standardised likelihood function and is a measure of the NDI sizing reliability for a measured size of $\hat{a}$. Figure 6 shows a probability density function of the possible real defect sizes for a given measured size (curve 2). The standardised version (as defined in the bracketed term in equation (5)) of this function will therefore be used in the updating procedure. The prior (before updating) probability function can be evaluated by a series of reliability (POF) calculation with different ending crack sizes,

$$P(a) = P\left(a' - \frac{\delta a'}{2} < a < a' + \frac{\delta a'}{2}\right)$$

$$= POF\left\{a_f = \left(a + \frac{\delta a}{2}\right)\right\} - POF\left\{a_f = \left(a - \frac{\delta a}{2}\right)\right\} \tag{6}$$

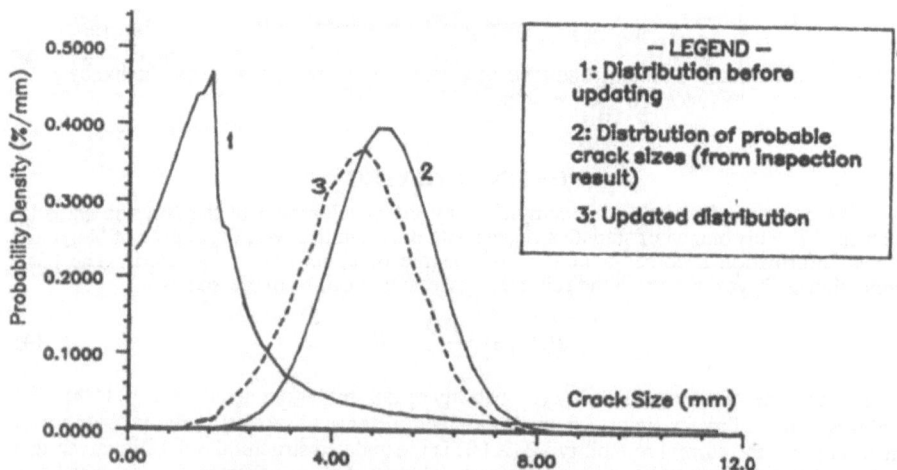

FIGURE 6. DISTRIBUTION OF CRACK SIZES AFTER 10 YEARS AND UPDATED CRACK SIZE AFTER INSPECTION

The density function could then be evaluated by numerically differentiating P(a). Using this approach, all the other input statistics will not be updated unless there are new data available (such as new wave loading information through in-situ monitoring). The procedure is applied to the sample case to update the reliability at the 10th year. The measured crack size is 5mm and the corresponding true size distribution is taken as Normally distributed with a standard deviation of 20%. The updated crack size distribution has been shown in Figure 6 (Curve 3) and the updated reliability history is shown as curve U in Figure 3.

APPLICATIONS

As it has been shown above, the β history is updated due to a measured crack size in service. If the original specification requires one inspection every 5 years, the next inspection will need to be carried out within 4.5 years to maintain (without repair) the same level of reliability as before. If repair is preferred, the β curve shows when the work must be carried out and thus gives some flexibility in maintenance planning.

If several joints are involved, the updated β histories could be used in setting priorities for the operations [18]. Moreover, underwater repair usually does not restore the joints to their original "perfect" conditions. The relative economy and benefits of repair can then be compared with other options (such as more frequent inspections) by using similar reliability analysis.

It has also been demonstrated [3] that a worse acceptable growth curve could be obtained from the reliability analysis. The idea was to construct a curve on which all crack sizes (including the final size) are reached with the same reliability from the starting size. This curve is therefore the "worst acceptable" case to reach the final size with the originally designed reliability. If a crack is found at any time in service to be larger than the corresponding size given by this curve (having taken into account the uncertainties of the NDI technique), the in-service crack growth is considered to be too fast and the original level of reliability cannot be maintained. Therefore this type of "worst acceptable" curves could be used as a tool for fast assessment of crack acceptance before more detailed analyses are carried out.

Sensitivity analyses could also be carried out and the results for the example are shown in Figure 7. The figures show the relative effect on reliability (and POF) for small changes in the input values of the means and standard deviations. The sharper the gradient, the more influence the parameter has on reliability. This provides the ranking of the importance of various input uncertainties and could be used to assess the priorities in quality assurance programmes. However, the ranking of sensitivity depends very much on the input statistics and the fracture mechanics model used in the analysis. Therefore, the confidence in the input statistics must be taken into account when conclusions are drawn from the sensitivity data.

CONCLUDING REMARKS

The foregoing has reviewed the applications of reliability techniques in the fatigue crack growth analysis of offshore welded joints. Now it appears to be the right time to consolidate the experience and understanding of the applications and formulate effective strategies for the maintenance of structures.

According to the experience and the recent findings of NDI reliability, it is recognized that NDI reliability data must be compiled in formats appropriate to the maintenance strategies. POD data must be used with care. In structural reliability based assessment, information on NDI sizing accuracy will also be needed.

The discussion on the approaches of updating reliability has not yet ended. The performances of the various approaches need to be compared in practical situations. Factors of finance, economics, cost and benefits will also need to be taken into consideration (the so called level IV design).

The sensitivity analysis shows that improved input data quality could result in higher reliability of the structure. Although extensive data analyses have been carried out in the past

250

[such as 19, 20, 21, and others], the results have not all been presented in formats readily applicable in reliability analysis. Therefore, one of the more important tasks should be the pooling and presenting data statistics for the use in reliability analyses.

As a conclusion, although more development will still be made, major advances have been achieved in the reliability / integrity assessment of offshore structures subject to fatigue cracking. With the new data in NDI reliability, the Industry is being equipped with a set of tools to rationalise the costly but necessary effort in maintaining offshore structural integrity.

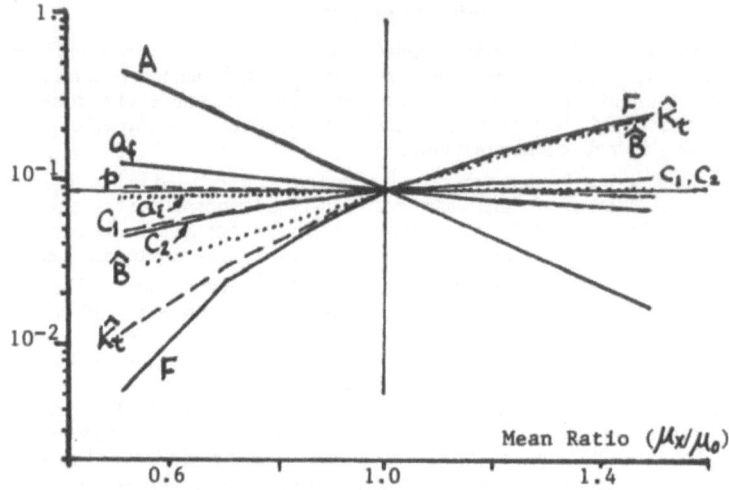

(a) Effect of Mean Values on Reliability (All Variables)

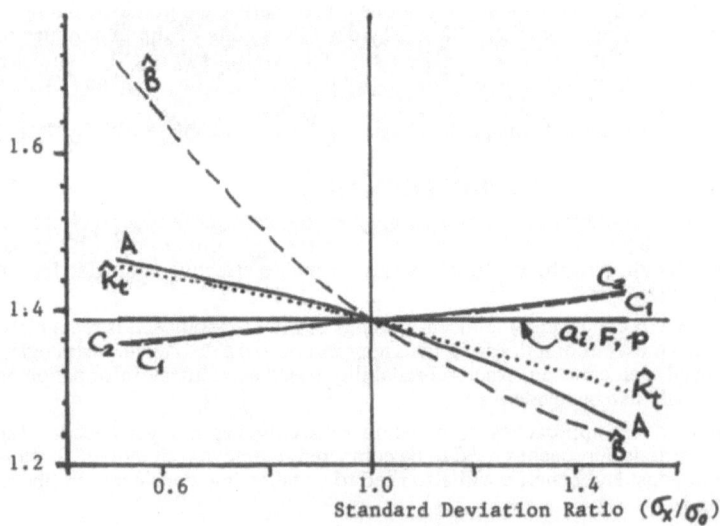

(b) Effect of Standard Deviation Values on Reliability (All Variables)

FIGURE 7   SENSITIVITY ANALYSES

REFERENCES

1. Marshall, P.W. : "Problems in Long Life Fatigue Assessment for Fixed Offshore Structures", Preprint 2638, ASCE, National Water Resources and Ocean Engineering Convention, San Diego, California, Apr., 1976.

2. Cohesive Programme of Research and Development into the Fatigue of Offshore Structures. July 1985 - June 1987. Final Report, ed. by Dover W.D., Dharmavasan, S., UCL, 1989.

3. Kam, J.C.P. 1988 : "Fitness for Purpose Assessment of Tubular Welded Joints in Offshore Structures", Proc. 7th Int. Offshore Mechanics and Arctic Eng. Sym., ASME, Houston, Feb., 1988.

4. Kirkemo, F. : "Applications of Probabilistic Fracture Mechanics to Offshore Structures", Applied Mechanics Reviews, Vol 41, No. 2, Feb., 1988.

5. Rummel W.D., et al, 1988. "Methodology for Analysis and Characterization of Non-Destructive Inspection Capability Data", Review of Progress in Quantitative NDE, Vol 7, 1988.

6. Madsen, H.O., Krenk, S., Lind, N.C. : Methods of Structural Safety. Published by Prentice - Hall Inc., 1986.

7. Thoft-Christensen, P., Baker, M.J., : Structural Reliability Theory and its Applications, Published by Spring Verlag, 1982

8. Hasofer, A.M., Lind, N.C. : "An Exact and Invariant First Order Reliability Format", Proc. ASCE. Jnl. Eng. Mechanics Div. 1974, pp 111-121.

9. Shooman, M.L. : Probabilistic Reliability : An Engineering Approach. McGraw - Hill Book Co., New York, 1968.

10. Baker, M.J., Private communication on fatigue reliability of offshore structures.

11. Kam, J.C.P. : "A Study on the NDI and Corrosion Fatigue Crack Growth in Tubular Welded Joints Using Reliability Methodologies", Proc. 9th Int. Conf. Offshore Mechanics and Arctic Engineering, ASME, Houston, 1990.

12. Department of Energy : Background Notes to the New Fatigue Design Guidance for Steel Welded Joints in Offshore Structures HMSO, London, 1983.

13. Kam, J.C.P., Dover, W.D. : "Corrosion Fatigue of Welded Tubular Joints: Fracture Mechanics Modelling and Data Interpretation", Proc. 8th Int. Offshore Mechanics and Arctic Eng. Sym., ASME, The Hague, Apr, 1989.

14. Nichols, R.W. : Non-Destructive Examination In Relation to Structural Integrity (PISC-I). Applied Science Publishers, 1984.

15. Rummel, W.D., et al, 1974. Detection of Fatigue Cracks by Non-Destructive Testing Methods, NASA Contractor Report, NASA-CR-2369, Feb 1974.

16. Forli, O., et al : A Comparison of Radiographic and Ultrasonic NDE, The Danish Welding Institute, 1983.

17. Ledermann, W., et al, ed., 1984. Hand-book of Applicable Mathematics. Vol VI: Statistics. Part B, Published by John Wiley & Sons, 1984.

18. Kam, J.C.P. : "The Efficient Maintenance of Offshore Structural Integrity Using Reliability Analysis", Quality and Reliability Eng.. Int. Jnl. Vol 5, No. 3, 1989.

19. Wirsching, P.H., Chen, Y.N. : "Considerations of Probability Based Fatigue Design for Marine Structures", Int. Jnl. Marine Structures. Design. Construction & Safety. Elsevier Applied Science, Vol 1, No. 1.

20. Bertini, L. : Private Communication on the Corrosion Fatigue Crack Growth Data Base University of Pisa, Italy, 1988.

21. Buchan, A.W. : A Comparison Between the Parametric Equations in Calculating Hot Spot Stresses, Working Group Report for Fatigue Crack Growth Software (FACTS), Sept, 1989, Cohesive Programme of Research and Development into Fatigue Crack Growth of Offshore Structure, 1987-89.

PROPORTIONAL HAZARDS ANALYSIS OF ELECTRONIC COMPONENT RELIABILITY DATA

J M MARSHALL, D W WIGHTMAN AND S J CHESTER
Department of Mathematics, Statistics and Operational Research
Nottingham Polytechnic
Burton Street, Nottingham

ABSTRACT

This paper addresses the challenge of analysing electronic component reliability data arising in diverse environments and applications. Most reliability data has such potential associated explanatory variables, the electronic component reliability data analysed in this paper is no exception. MIL HDBK 217E accounts for these explanatory variables by incorporating π factors. These factors are determined a-priori and give the magnitude and effect of the explanatory variables.

The data analysed in this paper originates from the field failure of electronic components database held at Loughborough University of Technology (LUT). This database is part of a joint project funded by the British MOD, with participants from both electronics companies and academic institutions, namely, STC, GEC, Plessey, LUT, Nottingham Polytechnic and the Danish Engineering Academy (DIA).

The proportional Hazards Model (PHM) has been employed in an exploratory approach and includes such covariates as system operational environment and data source. This paper presents the results on the application of PHM to field data on Bipolar Transistors (T1A) and provides an insight into those factors which significantly influence the failure of these types of devices.

INTRODUCTION

The reliability of electronic components has been a source of study for about 40 years. The majority of such studies have been concerned with the determination of failure rates under 'artificial' conditions such as simulation or accelerated life tests. Failure rates determined from such tests are extrapolated to what might be expected under field conditions. However, these extrapolations are often subject to doubt if only because multiple failure mechanisms may be involved.

Due to some of the problems associated with accelerated life test data and the lack of field data, a study centred at LUT sponsored by the British MOD was initiated to investigate the failure behaviour of electronic components used in equipments in the field [1].

After discussion with many systems manufacturers, three major British Electronic companies became the main data providers ie STC, GEC, Plessey, together with two Danish Electronic companies subcontracted through the DIA. In addition a contract was placed with Nottingham Polytechnic to provide statistical input to the project.

The project has progressed and now has an established database on the field failure of electronic components [2]. This database was designed specifically for the purpose of exploring the data in order to identify those characteristics which may contribute to the failure of electronic components, for example, screening level, environment, encapsulation, mounting. In addition, both non-failed information and failure data is collected in order to establish the lifetimes of those electronic components contained in the database. Data has been coming into the database from sources for approximately 3 years and thus the database is expanding both in terms of volume of data and the number of sources involved.

THE PROPORTIONAL HAZARDS MODEL (PHM)

Previous analyses of the data contained in the database have been concerned with the calculation of constant failure rates for different component types and the comparison of such failure rates for components in different environments and using different screening methods. These analyses have proved both interesting and useful for electronic reliability engineers. However, they may not provide a true reflection of the failure behaviour of electronic components.

Further analysis using hazard plotting techniques for bipolar transistors have shown that these components do not in fact fail exponentially but fit a Weibull distribution with a decreasing hazard [3]. In addition this analysis highlighted the differences in the failure pattern for T1As in ground mobile and ground benign environments. Investigation of other component types such as Rectangular Connectors and Coil Activated Relays have also exhibited a similar decreasing hazard. Further details on these analyses can be found in [4].

Due to the many possible factors which may affect the reliability of electronic components and the uncertainty over the distributional form of failure times (exponential or not), the Proportional Hazards Model which allows for an unspecified distributional form and explanatory factors was employed.

The proportional hazards model is a particularly flexible model, which can be used to isolate the effects of many factors on the hazard function of an electronic component [5]. As previously mentioned there are many factors which have potential impact on electronic component reliability, PHM can incorporate and quantify some of these factors as explanatory variables.

The proportional hazards model is structured upon the hazard. It is assumed for each component type that the associated hazard function can be decomposed into the product of a base-line hazard function, common to all components of a certain type, and an exponential term incorporating the effect of the values of the explanatory factors specific to each individual component, ie

$$h(t, z_1, \ldots, z_k) = h_o(t) \exp (\beta_1 z_1 + \beta_2 z_2 + \ldots \beta_k z_k) \qquad (1)$$

where z_i's are the values of the explanatory factors, $h_o(t)$ is the base-line hazard function and t is the time metric of performance for components, ie operating hours.

The z values can be either measured values, for example percentage time an equipment is switched on, or indicator variables representing, for example, encapsulation methods, environments, etc. The β's are unknown parameters of the model and represent the effect on the hazard of the explanatory factors. These β's are required to be estimated and tested to see whether each explanatory variable has an effect on the variation in observed times to failure. These explanatory variables are assumed to have a multiplicative effect on the base-line hazard function, thus the hazard functions for components, say, with different encapsulations are proportional to each other.

The base-line hazard function h_o represents the hazard a component would experience if the explanatory factors all take the value of zero.

It is fairly obvious then that the application of PHM to the reliability of electronic components would be of great value.

PHM owes its origins to a seminal paper given by Professor D R Cox [6]. At first PHM was mainly applied in the biomedical field, but in recent years the number of applications to reliability data has been increasing [7].

However, in the field of electronic component reliability, there have been very few applications.

The models in MIL HDBK 217E for electronic components are of a similar form to PHM. However the distribution is assumed exponential and the covariates are estimated from accelerated life tests data [8]. Landers and Kolarik, in 1986, published a paper comparing PHM and MIL HDBK 217E [9]. However, this approach is limited because of the assumption of a constant hazard rate and the validity of the accelerated life test data used.

Similarly, Ascher published a paper on the application of PHM to transistors [10]. However, this application was based on a hypothetical transistor and was intended to illustrate the potential use of PHM in the reliability of transistors but real data would obviously have been of greatest interest.

To date, the two papers discussed, [9] and [10] are the only published applications of PHM to non-repairable electronic components.

THE ELECTRONIC COMPONENT RELIABILITY DATA

The database contains information on 1.8 million components and 98 billion component hours. However, there are in fact only about 4,500 failures and about 50% of those are due to 'no fault found'. Thus the amount of censorings is vast compared to the number of failures. Due to the volume of information available for analysis the proportional hazards modelling computer routines employed at Nottingham Polytechnic had to be substantially changed to cope. See Appendix 1. There is information on about 90 out of 125 different component generic descriptions. Obviously, a decision as to which component types should be analysed first must be made [3].

Due to the sparseness of the data, many component types with data cannot be disaggregated by source, environment, screening etc. Apart from a few component types information across data sources is sparse. The data set chosen for the application of PHM contained data on bipolar transistors. This was chosen since exploratory data analysis had already been carried out because there was information across different data sources and environments.

The time metric of the data is operating time (in hours), t in equation (1). The censoring times are calculated from the difference between the up and running date and the end of the observation period. This figure is then multiplied by an estimate of the percentage of time the system was operational (usage), thus giving an operating time for a censored equipment. The operating time to failure of component is collected by the companies. The accuracy of the estimation of usage is currently under investigation and is obviously dependent on the data source and their methods of data collection [11].

In order to extract the censored information for a component, we find for each different equipment type the total number of bipolar transistors. Subsequently, for each equipment serial number of a particular type which did not fail, we can extract the operating time and therefore finally arrive at the censoring times for all those components in equipments which did not fail. Extracting the operating time to failure of a bipolar transistor is straightforward since the database contains such data. The uniqueness of this database is in the fact that all circuit boards are tracked and therefore the previous use and operating time to failure this time around is known.

POTENTIAL EXPLANATORY VARIABLES FOR ELECTRONIC COMPONENTS

The variables which are of most interest and therefore should be investigated for each component include:

> System Operational Environment
> Encapsulation Method
> Data Source
> Mounting Technology
> Screened Level
> Usage

Most of these variables are of the type which can be found in MIL 217E models. Table 1 shows the variables for T1As.

TABLE 1
Bipolar Transistor Variables

Encapsulation	Mounting	Environment
Hermetically sealed, glass ceramic	Radial wired	Ground Benign
Hermetically sealed, not known	Axial wired	Ground Mobile
Hermetically sealed, welded	Dual in line	
Plastic Transfer Moulded	Surface Mount	
Hermetically sealed, solder	Flat pack	

From Table 1 there are two different environments, 5 different encapsulation methods and 5 different mounting technologies. In addition to the information in Table 1, there is data from 6 different data sources (company).

With regard to the screening level of a bipolar transistor, because there is only one screening level for each data source it is not possible to include screening level separately. For example all data from companies 1, 3 4, 5 are screened to BS9000 level and data from companies 2 and 6 are company screened. In addition, the system operational environment is also linked to screening level and the data source i e companies 1, 3, 4 and 5 operate in ground mobile environments with BS9000 screened components and companies 2 and 6 are operating in ground benign environments. Table 2 summarises this information.

TABLE 2
Screening level/Environment for Company

Company	Screening Level	Environment
1	BS900	Ground Mobile
2	Other	Ground Benign
3	BS9000	Ground Mobile
4	BS9000	Ground Mobile
5	BS9000	Ground Mobile
6	Other	Ground Benign

Due to the information contained in Table 2 it was decided that we could not use covariates for the company and the environment in the same model because they are linearly dependent.

The information of interest concerning encapsulation is the comparison between the reliability of components which are hermetic or plastic encapsulated. This is because in general, hermetically sealed components are more expensive than plastic. Also plastic encapsulated components are not suitable in high humidity environments or for high temperature operation.

The inclusion of usage ie the percentage of the time the equipment is operational, is useful as an indication as to whether low use equipments which are most probably switched frequently off and on are less reliable than those systems which are continuously operational. This may not be conclusive since there may be errors in the calculation of usage. However, it may give an indication as to whether the differences in figures for usage contribute to the failure of bipolar transistors.

All the information mentioned above was thought to be of great interest and would therefore be incorporated in a proportional hazards model.

ANALYSIS OF BIPOLAR TRANSISTOR DATA

The analysis reported in this paper concentrates on bipolar transistors(T1A), with data provided by each of the data suppliers involved in the project. The investigation of this particular bipolar transistor data was not only to see if proportional hazards modelling could be utilised in the analysis procedure for the database, but if successfully applied, to establish a structural approach for the analysis of other component types.

The population of bipolar transistors in the field is in excess of 586,000. The failure data available broken down by company, encapsulation, mounting, usage and environment for this data set is given in Table 3.

TABLE 3
Failure Information for Bipolar Transistors

Company	Encapsulation	Mounting	Usage	Environment	No. of failed components
1	0	0	27	1	2
1	0	4	27	1	1
1	2	0	27	1	39
2	1	0	90	0	1
3	2	0	7	1	1
4	2	0	5	1	20
4	2	0	7	1	1
4	2	0	8	1	1
5	0	0	5	1	30
5	3	0	5	1	1
6	0	0	23	0	1
6	0	1	23	0	2
6	2	0	23	0	17
6	2	1	23	0	1
6	4	0	23	0	3

Total = 121

Note that the failure data matrix is sparse and the relationship between different cells, for example, failures on ground benign are almost totally associated with company six (code 0 in environment column).

With 117 of 121 failures occurring with mounting type coded 0 there is insufficient information to make comparisons of different mounting types. As stated in Table 3 the usage is highly associated with company so that the inclusion of covariates for company and usage would result in the problem of multicolinearity in the covariates. For this reason covariates for usage were omitted from the analysis of this component type. Since encapsulation is almost entirely hermetic comparisons with plastic cannot be made. For other component types and future analyses the problem of relationships between the covariates should occur to a lesser extent.

Since company almost solely defines the other information available the initial analysis concentrated on using only covariates for company. In Figure 1 a breakdown is shown of the times to failure for the component for each company.

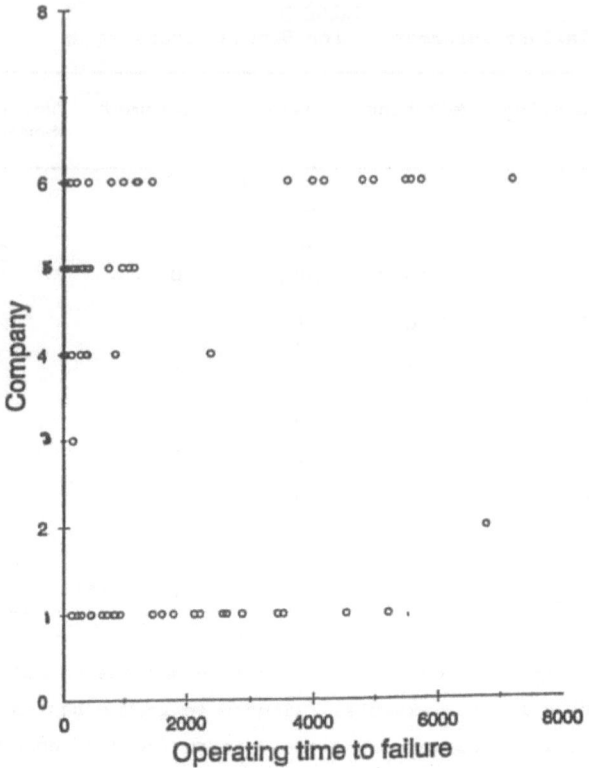

Figure 1. Operating Time to Failure/Company.

Figure 1 shows the pattern of failures through time for each company. From
this it can be observed that there is only one failure for companies 2 and
3. Note also that for company 4 there are 8 failures at 0 hours. The
pattern revealed in Figure 1 proved to be extremely useful in interpreting
the results from the proportional hazards analysis.

For the initial proportional hazards modelling analysis, company 6 was
taken as the base since it was associated with ground benign. Due to the
small numbers of failures for companies 2 and 3, these were also included
as the base. We therefore had covariates for company 1, 4, 5 to compare
against the base. This analysis highlighted the problem of near monotonicity [12] for the covariate for company 4. The problem of near monoton-

icity occurs when there is little overlap of failure times between the covariates causing the problem. The most immediate solution is to omit the covariate from the analysis.

Looking again at Figure 1 it can be seen that the vast majority of failures for company 4 occur before the first failure for company 1. In fact, over half the failures for this company occurred during the first hour of observation resulting in the problem with analyses. Apart from one failure for company 5 at time 0, it is seen that company 4 is the only one to encounter switch on problems.

Omitting the covariate for company 4 and running the proportional hazards model with a backward stepwise regression procedure, whereby in each run the most non-significant covariate is removed from the model and the model rerun with the remaining covariates, we found only company 1 to be statistically significant and different from the rest. As an alternative to dropping company 4 from the analysis we changed the base to companies 2, 3 and 5. The results were consistent and are shown in Table 4.

TABLE 4
Comparisons between different companies

Company	β Coefficient	Z-score	P value one-sided
1	-0.74349	-0.38724	0.0001
4 *	-0.11368	-0.3475	0.3475
6 *	-0.37286	-1.5035	0.0664

* Values when covariate omitted from model.

From Table 4 the β coefficient for company 1 (-0.74) indicates that the hazard for this company is smaller than for the rest and thus on average the time to failure for bipolar transistors is longer. A Weibull hazard plot for the base-line hazard for this analysis is shown in Figure 2. The straight line on the plot indicates the appropriateness of the Weibull distribution for the base-line. The value of the shape parameter (0.32) taken from the plot is in agreement with previous analysis on the individual cells in the data matrix [3].

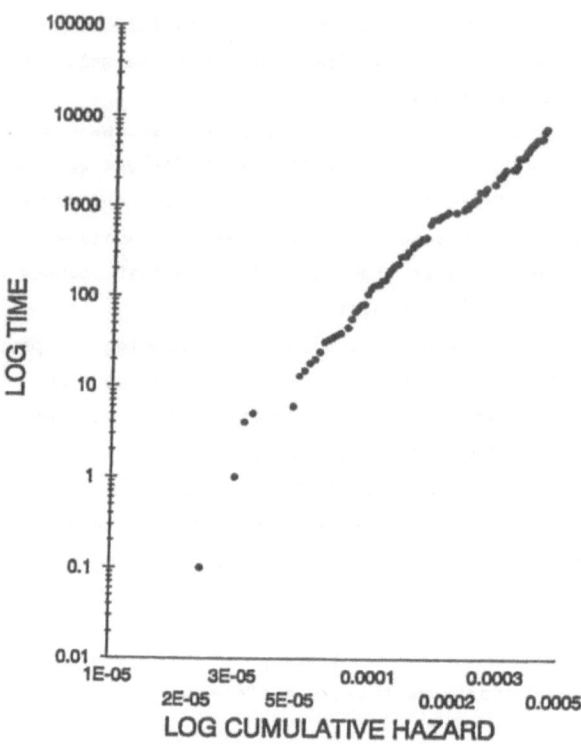

Figure 2.　Base-line hazard plot.

In Figure 3 a proportionality plot is shown for company 1.　This plot
investigates whether the covariate does indeed, as assumed, have a proportional effect on the hazard.　The plot is obtained by splitting the data
on the covariate (in the case of a binary covariate into two groups) and
running the analysis procedure on each group.　If the assumption is valid,
then the plot on appropriate axes should show constant vertical separation
between the groups [7].　Figure 3 clearly shows the assumption to be valid.

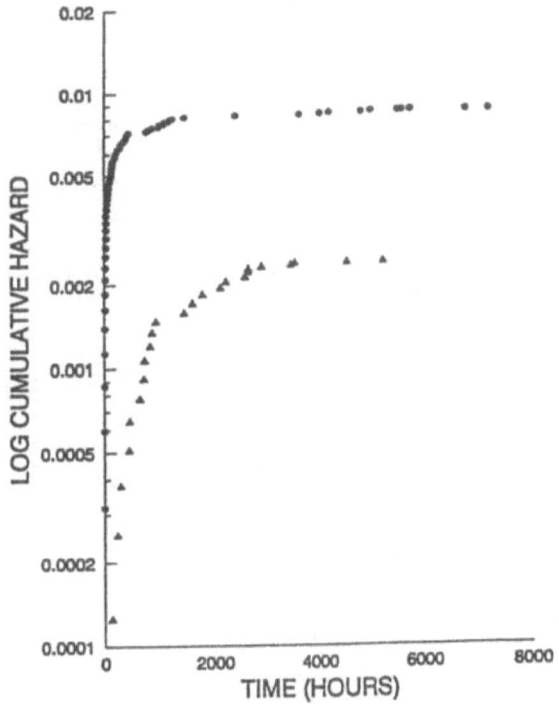

Figure 3. Proportionality Plot.

The Cox and Snell (1968) variance stabilised residual plot [13], [14], [15] is shown in Figure 4. If the model is appropriate then the residuals from the model should be similar to a sample of observations drawn from the unit exponential distribution. Plotting on appropriate scales, the observed residuals against expected residuals should give a 45 degree line (through the origin), this is the case for Figure 4.

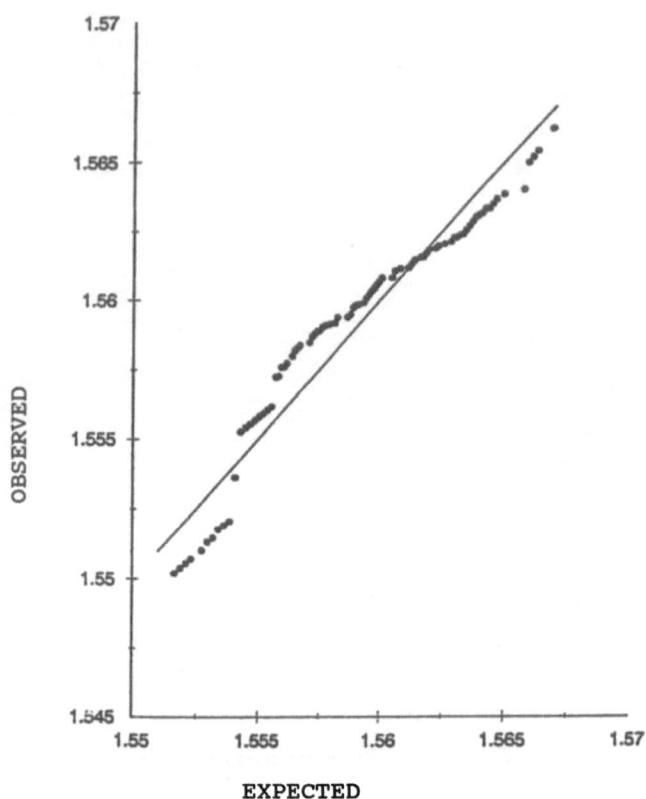

Figure 4. Variance Stabilised Cox and Snell Residuals.

Notwithstanding the diagnostics suggesting the good fit of the model
in the initial analysis, consideration was given to the appropriateness of
including information from companies 2, 3 and 4. With the majority of
company 5's failures occurring before the first failure on company 1, the
information for company 5 was therefore omitted from the next analysis
discussed here.

We were thus left with 66 failures and just over 410,000 censorings
for the comparison of companies 1 and 6. With company 6 as the base and
running the model we obtained the results as shown in Table 5.

TABLE 5
Comparisons of companies 1 and 6

β	z	p value (one sided)
-0.27964	-1.0707	0.1421

Although the β coefficient indicates that company 1 has a lower hazard than company 6, the β is non-significant on a two-tailed 5% test. Figure 5 shows a Weibull hazard plot for the data (since there are no significant covariates the plot is the result of combining the data from companies 1 and 6). The straightness of the plot shows the appropriatenesss of the Weibull, the scale parameter again reinforcing previous convictions. Due to the relationship between environment and company this meant that they could not be contained in the same proportional hazards model.

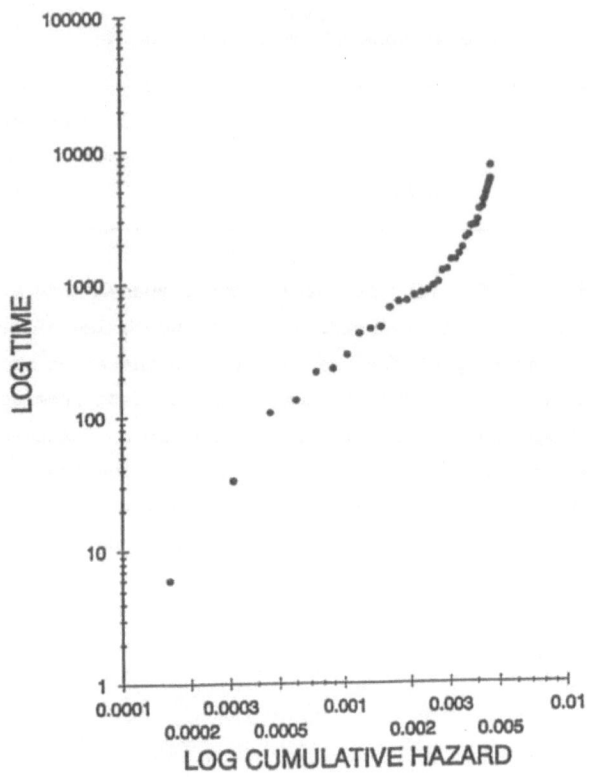

Figure 5. Weibull hazard plot.

Utilising the full data set analysis was focussed on the environmental information, with ground mobile as the base comparison for ground benign. The results obtained from this analysis showed the β coefficient for ground benign (-0.74141 with a p-value of 0.0007) was significant and that ground benign had a lower hazard compared to ground mobile.

We present in Figure 6 the Schoenfeld [22] partial residuals for this covariate. The partial residuals are essentially the observed value of the covariate minus the expected value for each failure point. From these plots we are looking for outlying values or gaps where no residuals are shown. The two lines in Figure 6 are typical for a binary covariate. However, in the upper band, corresponding to failures on ground benign, there is a period from 1600 to 3500 where no failures occurred, thus

267

separating the failures into two distinct groups. The reason for the apparent grouping is not obvious.

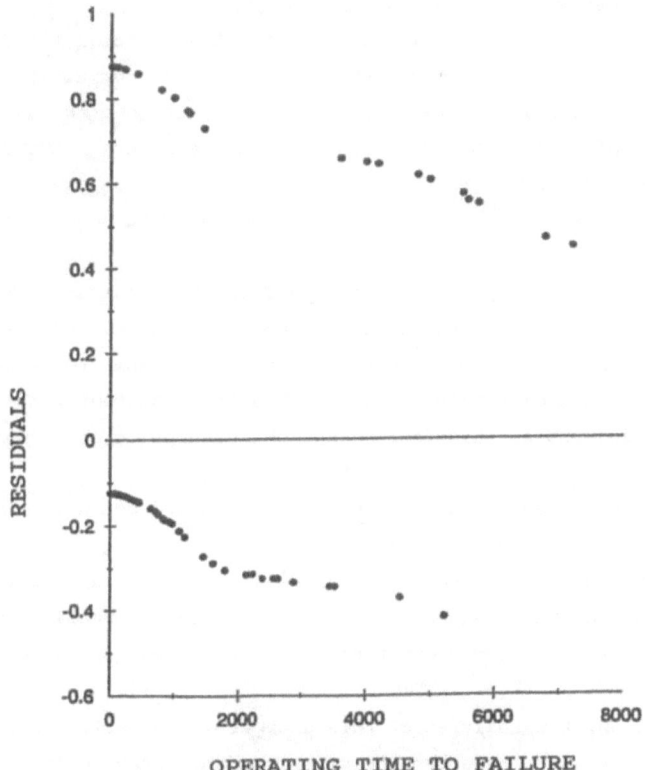

Figure 6. Schoenfeld Partial Residuals.

From Table 3 the ground benign information can be seen to have been supplied by companies 2 and 6. For company 2 over 98% of the censoring information occurred after the last failure observed on all of the data. This unusual data set together with company 3's information which contained only 1 failure, was omitted and subsequent analysis then showed the comparison of ground mobile and ground benign to be statistically non-significant.

SUMMARY AND CONCLUSIONS

Due to the sparseness of the data on bipolar transistors it was not feasible to include all the variables of interest. In addition, because

of the relationship between data source and environment it was not possible
to include both data source and environment as covariates in the same
model. Thus they were considered separately.

The results from this analysis indicated that bipolar transistors from
company 1 were more reliable than those in the other companies. However,
caution must be used when interpreting these results due to the particular
peculiarities of each of the different company data sets. Such peculiarities include the large number of 0 hour failures in company 4, the few
numbers of failures for companies 2 and 3 and the vast number of censoring
for companies 2 and 6.

Several positive aspects have arisen from this analysis. These
include the consistency of the shape of the base line hazard. In both
analyses the failures for bipolar transistors have exhibited a Weibull
distribution with a decreasing hazard. This is consistent with previous
analysis carried out on this data. Thus the assumption of a constant
hazard rate is not valid for this data.

The possible reasons for this decreasing hazard could be that the
burn-in procedures are not sufficient to eradicate weak components. Similarly, perhaps the methods employed for designing printed circuit boards
are done in an iterative manner whereby unless a design problem is encountered with a particular component type then the design is assumed to be
correct.

In addition, the analyses of this data provided useful experience in
the steps required to transform the data from the database into the
required format for PHM. This gained experience can be built upon as more
appropriate data becomes available on other component types.

Although this application has been limited it has shown that when
more data becomes available PHM will be a useful tool in the analysis of
electronic component field reliability data.

ACKNOWLEDGEMENTS

This work is being carried out with the support of the Procurement Executive, Ministry of Defence, and we are grateful to them for permission to
publish this paper.

We would also like to thank Mr J A Jones at Loughborough University
for his help in extracting the data from the database.

269

APPENDIX 1
ADAPTATION OF PROPORTIONAL HAZARDS MODELLING PROGRAMS

The application of proportional hazards modelling to electronic components
(non-repairable items) with the auxiliary information held in the database
causes no problem in itself. The problems arise in the analyses when the
number of each component, especially censored observations, in the field is
considered along with the auxiliary information that forms the covariates.
It is of course possible to consider subsets of the data or to concentrate
only on the failure information. However, this may well produce unsatisfactory analysis.

The adaptation of the proportional hazards modelling programs employed
at Nottingham Polytechnic to allow for the analysis of extremely large data
sets has two main themes:

1. Instead of holding failure, censoring, covariate values and other
    required information within the programs this information is contin-
    ually read from a data file. To allow for this change extra prepara-
    tion is required in the creation of the data file for proportional
    hazards modelling analysis. In particular the data in the file must
    be ranked from largest to smallest time, with a column to indicate
    the rank of the observation.

2. The grouping of observations with the same rank and covariate informa-
    tion. Again to allow for this change extra preparation must be
    investigated in the creation of the data file for analysis. Also to
    accommodate this change the proportional hazards modelling equations
    such as partial log-likelihood, first partial differntial of the log-
    likelihood and information matrix had to be changed.

The changes discussed in 1 and 2 have a major impact on how the diagnostics for the model are obtained. For the base-line hazard it is now
more convenient to use Breslow's [16], [17] formulation rather than in the
original Kalbfleisch and Prentice [18] routines the Kalbfleish and
Prentice [19] estimate. Breslow's estimate is a first order approximation
to that of Kalbfleisch and Prentice. The base-line hazard is used to
check appropriate distributional forms such as Weibull or Lognormal for
the model. The base line hazard is also used in the proportionality plots,
[20], [21].

It was found that the Schoenfeld [22] residuals require little
adaption, mainly because they are defined only at failure points. We have

placed a limit of 100 different possible combinations of covariates for the failures at each failure point. This of course is easily extended. The Schoenfeld residuals are used to compare observed covariate values against the expected covariate value.

Since empirical influence functions (see [24], [25] and [7]) are defined at both failure and censoring points and for each observation the value of the empirical function for each of the significant covariates depends on the number of risk sets the observation is a member of and whether a failure or not then the introduction of 1 and 2 require substantial programming changes. For each covariate in the model the set of empirical influence values are used in an informal manner to investigate the effect of each observation on determining the β coefficient.

The Cox and Snell [13] residuals for the model (see [20]) also requires a substantial amount of reprogramming. If the model is appropriate for the data then the Cox and Snell residuals should be somewhat similar to a sample from the unit exponential distribution.

REFERENCES

1. Campbell, D.S., Hayes, J.A. and Hetherington, D.R. The organisation of a study of the field failure of electronic components. Quality and Reliability Engineering International, 3, pp 251-258, 1987.

2. Marshall, J.M., Hayes, J.A., Campbell, D.S., Bendell, A. An Electronic Component Reliability Database. 10th Arts, Bradford, pp 40 53, 1988.

3. Marshall, J.M., Hayes, J.A., Campbell, D.S., Bendell, A. The Analysis of Electronic Component Reliability Data. 6th Euredata, Siena, pp 286 - 309, 1989.

4. Marshall, J.M. The Organisation and Statistical An alysis of an Electronic Component Field Failure Database. PHD Thesis, Loughborough University of Technology, 1990.

5. Wightman, D. and Bendell, A. The Practical Applications of Proportional Hazards Modelling. Reliability Engineering 15, 1986.

6. Cox, D.R. Regression models and life tables (with discussion). J.R. Stat. Soc, b34, 1974.

7. Wightman, D.W. The Application of Proportional Hazards Modelling to Reliability Problems. PhD Thesis, Trent Polytechnic, 1987.

8. Mil Handbook 217E. Reliability Prediction of Electronic Equipment. October 1986.

9. Landers, T.L. and Kolarik, W.J. Proportional Hazards Models and MIL HDBK 217. <u>Microelectronics and Reliability 26 No.4</u>.

10. Ascher, H.E. The use of regression techniques for matching relia-bility models to the real world. Software System Design Methods – <u>The Challenge of Advanced Computing Technology</u>. Springer – Verlag.

11. Chester, S.J. and Bendell, A. A discussion of Usage information within the LUT Electronic Component. Database. <u>Internal Report</u>, <u>Nottingham Polytechnic, 1989</u>.

12. Bryson, M.C. and Johnson, M.E. The incidence of monotone likelihood in the Cox model. <u>Technometrics</u>, Vol. 23, 381-383, 1981.

13. Cox, D.R. and Snell, E.J. A general definition of residuals (with discussion). <u>J.R. Statist.Soc., B</u>. 30, 248-275, 1968.

14. Cox, D.R. <u>Analysis of Binary Data</u>. Methuen, London.

15. Clayton, D and Curick, J. The EM algorithm for Cox's regression model using GLIM. <u>Appl. Statist, 34,</u> No.2 148-156, 1985.

16. Breslow's Discusion: Cox, D.R. Regression model and life-tables (with discussion). <u>J.R.Statist.Soc.B.34</u>, 187-220, 1972.

17. Breslow, N.E. Covariate analysis of censored survival data. <u>Biometrics, 30</u>, 89-99, 1974.

18. Kalbfleisch, J.D. and Prentice, R.L. <u>The Statistical Analysis of</u> <u>Failure Time Data</u>. John Wiley and Sons, Chichester, 1980.

19. Kalbfleisch, J.D. and Prentice, R.L. Marginal likelihoods based on Cox's regression and life model. <u>Biometrika, 60</u>. 267 – 278, 1973.

20. Kay,R. Proportional hazards regression models and the analysis of censored survival data. <u>Appl. Statist</u>. 26, 227, 237, 1977.

21. Walker, E.V.

22. Schoenfeld, D. Partial residuals for the proportional hazards regression model. <u>Biometrika, 69</u>. 239 – 241, 1982.

23. Cain, K.C. and Lange, N.T. Approximate case influence for the proportional hazards regression model with censored data. <u>Biometrics,</u> <u>40</u>, 493 – 499, 1984.

24. Reid, N and Chapman, H. Influence functions for proportional hazards regression. <u>Biometrika, 72</u>, 1 – 9, 1985.

Strategies for Reliability Data Analysis

JI ANSELL
Department of Management Systems and Sciences,
University of Hull

and

MJ PHILLIPS
Department of Mathematics
University of Leicester

ABSTRACT

Recent developments in the building of expert systems for statistical analysis stress the need for the development of strategies for analysis. Also good practice dictates the development of clear strategies.

Until recently statisticians have not concerned themselves with the production of such strategies for data analysis, leaving analysis as an art rather than a science. The paper examines the possibility of producing strategies for Reliability data Analysis. The Strategies evolved are based on practical experience. Unfortunately producing such strategies highlights a number of concerns in reliability data analysis. The most important of which being the effect of context and objective on data analysis.

The last section of the paper considers the possibility of the building of an expert system or interactive consultation system for reliability data analysis. This includes consideration of the audience for such systems and their likely successes.

INTRODUCTION

Science, and therefore engineering, employs systematic approaches to gain understanding of phenomena. This so far has not been the general approach taken in reliability data analyses. Generally ad-hoc approaches have been taken, excused on the basis that most analysis is exploratory with confirmation not through statistics but through further engineering investigation. Reliability data analysis has been used either to identify problems or quantify the size of the problem.

Into the future though there are already indications of change. Firstly within statistics there has been increasing attention paid to strategy. This is particularly true of elements within statistical computing and expert systems, but can be said to arisen out of Cox and Snell [1]. Secondly design as understood by statisticians is increasingly becoming important in engineering through the need for quality and the work of Taguchi, Deming and others. Even in the area of field trials recent work by Kalbfliesch and Lawless [2] has indicated the importance of design. Hence we are on the threshold of change, as reliability data analysis moves from an art to a science.

The change is also mirrored in other areas associated with reliability. In operational research there has been increased criticism of blindly applying mathematical techniques without due attention to objective and context. This suggests greater care has to be taken over the formulation and identification of a study.

This paper therefore attempts to address some of the questions that arise as reliability data analysis makes this transition. We also hope to point the direction which reliability data analysis could take including 'automation' via expert systems.

Before we describing possible strategies we need to consider the 'state of the art'. Unlike other papers we intend not to concentrate on techniques but on the approach. Hence techniques will only appear as examples.

'STATE OF THE ART'

Statistics, and hence data analysis, is only one of a range of tools which might be employed in a reliability study. Many reliability studies do not require statistical inputs. Whilst, though, statistics does not necessarily play a central role within reliability, statistics along with probability do provide a set of tools to cope with uncertainty. It would not be possible within the current paper to describe all the applications of statistical within reliability. Even within the limited area of the application of data analysis it would be too lengthy to delineate all the possible techniques. Hence we will limit ourselves to general comments about approaches which arise out of experience rather than giving technical detail. For details of techniques and their application see Ascher and Feingold [3], Ansell and Phillips [4] or Bendell and Walls[5].

A high percentage of data analysis carried out in reliability is standard, attracting little attention. It is possible to forget that the routine collection of data and its simple analysis is carried out in many industries on a daily, weekly

or monthly basis. Yet even at a basic level we should stress
the need to be critical of the analysis and ask whether a
technique is useful or appropriate. Most importantly the
analyst must ensure the information obtained from the analysis
allow the analyst, or the manager, to attain her/his
objective?

O'Connor [6] has pointed out that much reliability analysis
is often based on false premises such as independence of
components, constant failure rates and that a system's
failure rates can be assessed from the 'sum' of the component
failure rates. Each of these assumptions and many others
which are made are invalid for most systems. It is therefore
very likely that our predictions will themselves be very
unreliable, as is frequently found in field trials. O'Connor
also criticises the use of inappropriate historic data to
predict future events, especially when there is no sound
theoretical background to the extrapolation. We will return
to these specific points in a later section.

The lack of criticism and the use of inappropriate techniques
has arisen for many reasons. There are vested interests which
have to preserve themselves. For example many data bases can
only be defended on the grounds of habit and custom rather
than their useful value. This might have arisen because of
the lack of guidance from statisticians about data collection.
This has lead to the erroneous concept that all data is good.
Few authors have commented on what may be, or more importantly
may not be, achieved through data analysis.

In the past statisticians have over emphasised the
techniques without detailing broader issues surrounding the
techniques use. A new data analytic technique arrives on the
scene, for example Proportional Hazard Modelling (PHM) or
Discriminant Analysis, and it becomes the flavour of the month
being used almost indiscriminately. We must be willing to
challenge the use of these techniques and remain critical of
their use. Models produced are often not realistic, with an
emphasis on mathematical tractability or elegance rather than
practical use. Reliability objectives should define the
requirements of models. If this was the case the dialogue
between engineers and statisticians would hopefully increase
which would be positive for both sides.

When embarking on an analysis there is a need if a systematic
approach is to be taken to define a strategy. Few authors have
attempted to describe how statistical analysis should be
carried out even in general statistics. Cox and Snell [1] is
a rare counter example. Those that have entered into this
field have often left many problems in their wake. For
example Ascher and Feingold's flowchart,(AF-Flowchart), has
the main drawback that it does not lead the user to a unique
solution for model selection for repairable data. Other
authors have tended to give specific guidance about a given
technique.

It would not be wise to concentrate too long on the problems

of the past detailed above. There is a need to define a
positive role for statistics/data analysis within reliability.
In the discussion of Ansell and Phillips [4] several
contributors attempted to describe the role of statistics,
data analysis, within reliability. Professor Barlow [7]
suggested that 'prediction was of crucial importance to
engineers', a sentiment that we would agree with though we
would also add assessment. Other authors have suggested that
pattern recognition or identification is the key use of
statistics. This is not a major divide as some authors wish
to suggest, as to predict one must identify and assess
effects, and identification by necessity includes being able
to assess effect.

Statistics therefore offers the reliability engineer a wide
variety of tools for identification, prediction and
assessment. To be useful, though, these tools must enhance
the engineer's study and not detract from it. An analysis
should be problem-led rather than technique-led as suggested
in [4] and [8], in which Ansell and Phillips centred on data
rather than techniques. To achieve this clear objectives for
the study must be set at the outset of the analysis and
reviewed throughout the analysis. These objectives will be
rarely defined by the statistician. More usually they it will
arise from contributions from a team of specialists or from
management. This team will be serving some management
function. The aim may be to achieve some stated level of
performance or surpass a previous standard. The objective
may be achieved with or without the service of a statistician,
as stated above. We will restrict attention to the cases in
which a statistician is needed.

A FRAMEWORK FOR THE STRATEGY

In this section, based on the previous section, we start to
consider a strategy and will assume that the reliability
study requires data analytic techniques. A strategy is a
series of steps which can be taken towards some objective.
The problem, though, is that there are going to be a variety
of approaches defined by the differing possible objectives and
contexts. Differing forms of data and the range of models
will also play a role. In such circumstances there is a need
to produce a general approach rather than a specific approach.

As mentioned earlier approaches such as the AF-flowchart
where emphasis is placed on finding a statistical model is too
rigid. It does not necessarily relate back to the objective
of the study. Hence we need an approach which takes into
account not only data and models but also objectives.

Before exploring further this idea let us consider the
idealised form of a 'statistical analysis' which is often
repeated in many texts. It is a quasi-scientific approach
based on iteration. Some tentative hypotheses, or models,
start the process and data is collected about the context. An
analysis is then carried out in line with the hypotheses and

the data collected. The information gained updates the
knowledge about the models and allows for refinement of the
models, or hypotheses. New models may emerge which can then
be explored, with either new data or the old. The cycles
continue until some point is reached when the experimenter is
content with the model or models produced. It is generally
assumed that the experimenter will start with a very simple
model and other models will evolve in a parsimonious fashion.

This idealised description is rarely followed in practice.
The approach fails to take account of the fact that the
sponsor of the analysis may have a different aim than
producing the model. The objective of a reliability study
will usually have another end point than model selection
dependent on context. The objective will also affect the
analysis to be carried out. Hence to produce an adequate
framework for reliability data analysis there is a need to
account for the objective of the study.

Rather than having two poles, data and model, a third is
added: objective, see figure 1. Each of these poles will
represent possibly more than one object within an analysis.
An analysis becomes the attempt to related the three poles.
It may be relating the current data to models and objective,
the current objective to data and models or the current model
to data and objective.

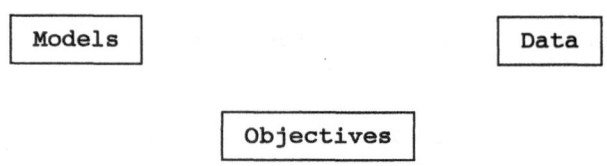

Figure 1. General Framework

For example suppose we wish to know whether to employ a
replacement strategy or not for a component. Our objective
initially might be to establish whether the component's
distribution has the property of IFR or DFR. A wide range of
possible models exist which may assist in deciding between the
two. Also a wide range of data sets would satisfy these
models.

Given a set of data then we would not wish necessarily attempt
to find a particular model but might be satisfied by a
graphical approach. Having established whether it is IFR or
DFR then the objective can then be reinterpreted and hence
further analysis carried out.

The general strategy will therefore consist of data, models
and objective. It is unfortunately likely that data will be
collected before the objective of a study is defined.
Therefore the strategy must allow for the possibility of the

data defining in the first place the study as much as the objective. It is to be hoped that models will not play a decisive role, since they must be perceived only as tools. In the next section we consider each of these areas and suggest plausible approaches for each.

DATA MODELS AND OBJECTIVES

Given the comments in previous section it is important to examine these separate areas in a little more detail before describing strategies.

As stated in the introduction one major change currently taking place is the emphasis which is increasingly been attached to design. This arise from the quality movement, Deming, Box, Taguchi and others, and is echoed by papers such as Kalbfliesch and Lawless [2] in field studies. This has two effects, firstly analysis may be proscribed by design but equally influential is the quality of data collected. In the past data was collected often as an addendum to other activities. This has lead some to suggest that 'reliability data are themselves typically less reliable than some other sorts of data', Bendell [9].

The improvement in data collection will also hopefully affect data definition. To be useful data must be clearly defined. Newton [10] gave three analyses of the same set of numeric values which suggested different aging behaviour based on reinterpretations of the data. To analyse any set of data the context must be clearly defined otherwise the analysis will have no use value. As earlier indicated lack of clarity in definition brings into doubt use of historic data in future prediction, [6]. We need to be clear that the historic data are relevant to the particular analysis before proceeding with their use. Relevance can only be assessed by knowing the context of the historic data.

To describe the range of possible likely data sets/forms which might be encountered in reliability would be a paper in itself. Briefly the types of data which are likely are:

1) Time to failure data for a single system/component.
2) Time between failures data for a repairable system, may include repair times, etc.
3) Multiple failure time data.
4) Failure data with covariates.

Any systematic approach to analysing reliability data must have a vocabulary for data that extends over at least this range. It must be adaptable to take in other forms easily.

The variety of statistical models which have been used within reliability in recent time cover almost the whole spectrum of applied statistics. In [3] for repairable systems the authors list 5 generic types of possible models: Homogeneous Poison Process (HPP), Non-Homogeneous Poison Process (NHPP), Renewal

Process (RP), Superimposition Renewal Process (SRP), Branching Poison Process (BPP). Kimber [11] suggests that beyond Lewis [12] there have been few applications of BPP within reliability, however Walls and Ansell [13] do suggest that it or its extensions might be of use to model 'common cause' failure. McDonald and Richards [14] describe two related sets of distributions which cover many of the standard distributions. They also suggest some methods for model selection.

The models described above cover only single time variable models. Most data when the context is included in the description will be multivariate. Hence the models required should include multivariate models. This, of course, covers regression models such as accelerated life models and proportional hazards models as well as regression models based on a specific distribution, Weibull regression, etc.

The most difficult area to define is the range of possible objectives. In a previous section we suggested some objectives for reliability data analysis, identification, prediction and assessment. These are obviously included but also valid are general objectives such as improvement in quality or decisions about replacement strategies. It would be possible to continue almost without limit in attempting to define all possible objectives.

GENERAL STRATEGIES

It is obvious from the comments made so far that the area under discussion is very large. It could include almost all areas within applied statistics and when objectives are also included then it subsumes aspects of management and engineering. Hence it is not plausible to talk about a single strategy which would be effective in all these diverse areas. Ansell and Phillips [8] also suggests that the strategy adopted should be flexible to cope with the variety of objectives and contexts.

We have already inherited a variety of approaches or outlines which may be followed, such as the general statistical approach or the AF-flowchart. Along with these are general strategies concepts such as starting with simple description and only revising if the model is inadequate. In [8] we have suggested that one starts with basic statistics, evolve through graphical tools to possibly more complex models such as PHM.

Given our comment above we feel it is more sensible to tackle the problem in a modular form. The broad areas are objective, data and model. Within each there is a need to have possibly more than one strategy. As indicated above it is not sensible to indicate which of these areas should be the starting point of the study, however we will take the order: objective, data and then models.

For the objective the need is to define/clarify the objective.
The defined objective may also have to be interpreted into a
statistical form. This might be regarded as diagnosis where
out of a set of possible objectives an objective or set of
objectives are selected. The main problem is the openness of
the objectives, any set might be regarded as subset of the
possible list. The objective will effect both the data
required or models to be considered. An objective may suggest
specific forms of data or limit to some forms the nature of
the data.

Other parts of the strategy may define the form of the data
required. There may be limits on the type of data available.
Given an objective the available data may not be able to
assist in achieving the objective. Assuming the data may
achieve the desire objective the initial stage is to obtain a
clear description of the data and hopefully its context.

There will be a number of stages. Firstly form of the data,
number of variables, number of cases and any structure within
the data. If a design has been used to collect the data it
will naturally give rise to a structure within the data. The
next stage is to validate the data and check for consistencies
and any abnormalities. After this stage there is the need to
describe the context. This is more similar to categorisation
as for the objective. Generally the problem as with the
objective is that all data forms have to be catered for. As
with the objective the available data will affect both the
objectives which can be studied as well as the models which
can be fitted.

Models unlike the other areas can be regarded as abstract.
It is a description which does not directly relate back to the
real world, though outcomes from it may impinge on the real
world. A model should be selected which as closely reflects
the situation as possible. The models again may be defined by
the other parts of the analysis. It is however the case that
models should play no major part in defining the other
availability of data or objectives. If models did then the
strategy employed would be technique-led rather than problem-
led.

Within statistical models fortunately usually a framework or
hierarchy has been discovered which links models. Hence for
many models structures exist which progress from the general
to the specific. The statistical literature is helpful to
differentiate between models. For example the A-F flowchart
help to differentiate to a degree between HPP, BPP and NHPP.
Whilst other models satisfy the selection criterion, they are
usually the simplest and hence by the argument of parsimony
they should be fitted first.

Decisions on whether a model fits are usually based on whether
they provide an adequate description and whether the
assumptions made are valid. In the example of regression-type
modelling the simplest model is the null model. If such a
model adequately predicts the data then there is no point in

fitting more complex models. If the model does not perform
well then there is a need to add variables. The process of
selection of further variables in most regression-type
analysis is problematic as generally no best strategy has
evolved, but there usually are a number of good strategies
which might be employed. Having found the best model for
prediction there is the need to validate the model. Do the
residuals from the model behave as they ought? This second
stage model checking is frequently overlooked. In many areas
it will be easier, than in regression, both to select an
appropriate model and to validate the model.

Summarising the strategy employed has a modular form. The
data and objective both need clarifying. The objective may
need to be translated into a statistical form. Given the data
and the objective the selection of model should be relatively
easy.

In the next section we examine how the strategy might be
employed to produce an aid to analysis.

EXPERT SYSTEMS

Expert systems are buzz-words of the eighties which whilst
promising advances have failed to convince. The concept has
however attracted considerable research effort and monies and
several expert systems have been developed. There exist a
number of such system in reliability, for example the fatigue
model of Stead et al, [15].

In statistics there have been a variety of expert systems
developed. KENS [16], GLIMPSE [17] and HUGIN [18] typify the
different forms. KENS is a knowledge enhancement system,
which can be regarded as an electronic textbook. Its domain
of application is non-parametric statistics. GLIMPSE which is
an aid to statistical analysis using generalised linear models
is more sophisticated. It can be regarded as a user friendly
front end for GLIM with expert knowledge incorporated to help
with analysis. HUGIN is a statistical expert system which
works over a probabilistic network. Hence HUGIN can be used
as a diagnostic tool. An idealised expert system for
reliability might include elements of each of these elements.

An expert system has to be designed with an audience in mind.
Generally the audience can be split into two: the
knowledgeable user and the naive user. The naive user
requires considerable support and assistance, and is thus very
absorbent of development time. For the knowledgeable user
development time will be much shorter. The systems mentioned
above were designed assuming a knowledgeable user. One may
question designing for such a group but in complex areas it
will be difficult to descend to the level of naive users in
the current epoch.

Assuming the knowledgeable user then the aim of the system
will be to aid with analysis. The first requirement of the

system will be to decide whether the case is data or goal driven. Is the data available or has the objective been defined? In some senses it does not matter the order of definition of data or objective, provided they are available during the analysis.

Let us assume again the objective is the starting point. The next stage is 'diagnosing' the objective. Most systems should be able to cope with at least a specified list of objectives. However the sheer range of possibility and the requirement to be open and flexible is possibly currently out side the capability of the technology. Whilst systems can feebly learn they still have difficulty in extend the range of their comprehension. It may be plausible if the extension sought is within the lexicography of the system but if it is beyond then the system cannot make the quantum jumps required.

A typical strategy for defining the objective would be to follow the approach of INTERNIST, see [19]. This would consist of grouping the objectives into families, then selecting from the families specific forms, narrowing from general objectives to a specific objective. INTERNIST allows for more than one objective arising out of the analysis.

For data definition and entry it may be easier to follow the pattern that has already emerged through GLIMPSE. The data forms would have to be more general than those allowed for in GLIMPSE. It would also be hoped that the form of evolved software would be easier to use than GLIMPSE, see Al-Doori[20]. Non-standard forms would have to be initially ignored if a system was to evolve in the near future.

Given the data and the objective have been initially specified then a logical structure may be used to produce a model area. The selection of model would require some scoring system over the range of possible models. This seems plausible in the current technology. The analysis would be partially defined also by the objective including an end point. The movement to that end point would be following either a hierarchical model structure and/or a specified form of analysis dependent on objective. If we are selecting some stochastic model, we might employ the AF-flowchart. If it is regression-type exercise then variable selection would always be followed by residual analysis.

The building of such a system is within the capability of the current technology, the limitation would be the number of people-years required for specific design. There will, of course, be a number of systems evolving within the area in the near future, hopefully they may come somewhere near this idealised model.

CONCLUSION

The paper has consider the current state of data analysis. It has stressed the need for critically review of any statistical technique used with an emphasis on the requirement to clearly state the objective of the study. Based on these comments the paper has defined a general strategy within a framework. Using this strategy it has considered the possibility of building an expert system to automate the analysis. Unfortunately such a system will take time to evolve, though some parts may develop in the near future.

REFERENCES

1. Cox, DR, and Snell, E.J., <u>Applied Statistics: Principles and Examples</u>, Chapman-Hall, London, 1981, pp. 20-27.

2. Kalbfliesch, J.D., and Lawless, J.F., Estimation of reliability in field-performance studies (with discussion) <u>Technometrics</u>, 30, 1988, pp. 365-388.

3. Ascher, H, and Feingold, H, <u>Repairable Systems Reliability</u>, 1984, Marcel Dekker, New York.

4. Ansell, JI, and Phillips, MJ, Practical problems in the statistical analysis of reliability data (with discussion), <u>Appl. Stat.</u>, 38, 1989, pp. 205-247.

5. Walls, L.A., and Bendell, A., Exploring reliability data, <u>Quality and Reliab Eng Int</u>, 1, pp. 37-51.

6. O'Connor, P.D.T., Statistics in quality and reliability: lessons from the past, and future opportunities, talk given to RSS Seminar on 'Reliability and Statistcs', 1989, to appear.

7. Barlow, RE, Discussion of [4], 1989.

8. Ansell, JI, and Phillips, MJ, Practical reliability data analysis, 1990, to appear.

9. Bendell, A, Discussion of [4], 1989.

10. Newton, D.W., Some pitfalls in reliability data anlysis, talk given to RSS Seminar on 'Reliability and Statistcs', 1989, to appear.

11. Kimber, A, Discussion of [4], 1989.

12. Lewis, P.A.W., A branching Poisson process model for the analysis of computer failure patterns, <u>J. R. Statist. Soc. B</u>, 26, pp. 347-368.

13. Walls, L.A., and Ansell, J.I., Dependency Modelling, 11th ARTS, to appear.

14. McDonald, J, and Richards, D.O., Model slection: some generalized distributions, <u>Proc Reliability '87</u>, 1987, 2B/1/1-15.

15. Stead, J.P., Strutt, J.E., and Billingham, J., Progress towards an intelligent computer aided risk evaluation system, <u>Proc 9th ARTS</u>, 1986, A2/4/1-9.

16. Hand, D.J., A statistical knowledge enhancement system, <u>J. R. Statist. Soc</u>, A, 150, 334-345.

17. GLIMPSE manual, 1989.

18. HUGIN manual, 1989.

19. Alty, JL, and Coombs, MJ, Expert Systems: Concepts and Examples, 1984, NCC Pub, Manchester.

20. Al-Doori, M, Review of GLIMPSE, <u>Prof Statist</u>, 1990, to appear.

FLEXIBILITY AND RELIABILITY OF PROCESSING SYSTEMS

E. N. Pistikopoulos and T. A. Mazzuchi
Department of Mathematics & Systems Engineering
Koninklijke/Shell-Laboratorium, Amsterdam
PO Box 3003, 1003 AA Amsterdam, The Netherlands

ABSTRACT

In design and operation of processing systems, two categories
of uncertainty play an important role:

- uncertainty with respect to the realization of
  continuous parameters (product demands, feedstock
  qualities, hest transfer coefficients, etc.)

- uncertainty with respect to discrete states (especially
  those related to equipment availability).

This paper presents novel analytical tools to simultaneously
account for both type of uncertainties in process design and
evaluation.

INTRODUCTION

Reliability assessment of production systems provides
quantitative information on the probability of system
failures based on the reliability characteristics of
individual elements and a systems model. Availability
assessment addresses the fraction of system uptime in a given
time interval by taking into account maintenance aspects.
Furthermore, systems effectiveness involves the assessment of
the expected production capability by including unit
production capacities [1]. In existing techniques, the
production capacity is typically defined for a single product
in direct relation with equipment availability and capacity
[2,3]. For complex multiproduct systems, however, the major
question is how to measure the overall system effectiveness
by explicitly accounting for potential interactions between
different subsystems in the additional presence of continuous
parameter uncertainty.

Due to the long design lives, chemical and refinery environments are typically faced with continuous parameter uncertainty corresponding to variations in either external parameters (such as feedstream quality, product demand, environmental conditions, prices) or internal process parameters (such as heat transfer coefficients, reaction constants, physical properties). These variations are usually specified through fixed ranges or continuous probability distribution functions. The ability of a system to accommodate continuous uncertainty is termed process flexibility [4, 5]. Process flexibility accounts for process interactions and can readily be used in complex multiproduct environments. However, it does not explicitly consider reliability and maintenance aspects.

The need for coupling flexibility with reliability has only recently been realized in different disciplines, such as production manufacturing [6], product quality management [7], and chemical engineering [8,9]. Currently, however, there is no available systematic framework for measuring flexibility, reliability, and availability in an integrated fashion. In order for such a measure to be meaningful and useful in practical applications, it must be based on:

- a thorough mathematical model of the process,
- a description of the uncertainty involved (both continuous and discrete),
- an explicit consideration of repair time and maintenance policies.

In the sequel, this paper will deal with the motivation and initial effort for developing a metric for the simultaneous flexibility and reliability assessment, without including maintenance aspects. The paper is organized as follows. First, a motivating example will introduce the need for such an integrated framework. Next, the concepts of process flexibility will be briefly reviewed and a new stochastic flexibility index will be presented. The development of the combined flexibility-reliability index will follow and its use will be illustrated with an example problem.

MOTIVATING EXAMPLE

Consider the following in-line blending system

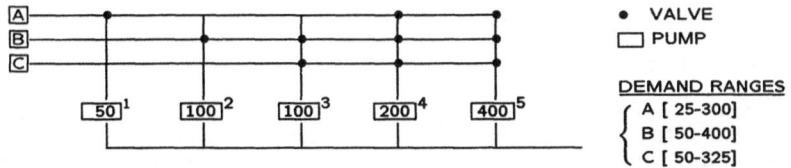

Figure 1. Example in-line blending system.

286

The basic pieces of equipment involved in a typical in-line blending system are pumps (which account for the transfer of materials), headers (where the materials are blended), and valves that determine which pump(s) transfer a specific material to a specific header. The role of an in-line blending system is to transport a number of intermediate components (materials) through the upper grid of valves via the pumps through the lower grid of valves to the headers. There, intermediate components are mixed into final products according to specified blend recipes. These recipes are usually specified by a range of expected demand (i.e. final product rates with different compositions from the same set of intermediate materials). The valves account for the transfer of the desired materials via the right combination of pumps. Each pump is characterized by an operating envelope defined by its turn-down ratio (the ratio of the lower bound of flow capacity over the upper bound). During operation, splitting of a material flowrate through a number of pumps is allowed, however, mixing of components through a single pump is prohibited. An in-line blending structure should meet the demand specifications while being able to deliver the final products within a specified time interval.

The in-line blending system of Figure 1 consists of three materials (A, B, C), and one header. The desired product recipe (demand) is a blend of all three materials and is assumed uniformly distributed within the range indicated. The structure involves five pumps with maximum capacities 50, 100, 100, 200, and 400 (m³/hr) respectively, and turn-down of 0.5. Valves are suitably located so as to enable the transfer of material A via the first, fourth, and fifth pump; the transfer of material B via the second, third, fourth, and fifth pump; and finally, the transfer of C via the second, fourth, and fifth pump. The stochastic nature of pump and valve failures (repairs) can be modeled using some well known distributional forms, such as the exponential or Weibull distribution.

In assessing the performance of the given blending system to satisfy uncertain (continuous) demand subject to equipment availability, current systems reliability and availability techniques usually proceed by:

(i)   focusing on a specific (average or maximum) demand requirement,
(ii) discretizing the demand range at prespecified levels, or
(iii) considering system units in isolation, apart from process interactions.

Considering each of the above approaches:

(i) The average of the three product ranges is the recipe (137.5, 175, 137.5) for which there exists a feasible pump-valve assignement. On this basis, one may conclude that the system is (on average) operable within the demand range, and that the system reliability (and availability) can be obtained directly from the reliability characteristics associated with the assigned pump-valve combination. However, it will be shown that demand subranges exist for which the system is not operable, even though all equipment are operational. In addition, the size of infeasible ranges is dependent on (but not directly proportional to) the availability of equipment; i.e. a failure of particular equipment may result in a different pump-valve assignement for a specific demand pattern (process interactions), and consequently its effect on the overall system performance may be limited.

(ii) The major disadvantages of the discretization approach are the explosive nature of possible combinations for which the reliability analysis must be applied (10 discrete points for each material result in 1000 recipes), and the arbitrary nature with which intermediate demand points are selected. The former relates to computational difficulties, whereas the latter leads to problems as discussed in (i).

Both cases, (i) and (ii) share the underlying limitation that general conclusions cannot be drawn for the entire continuous demand range, based solely on information for the discrete points considered.

(iii) To show how process interactions can play a governing role in operability assessment, consider the failure of pumps 4 and 5.. Considering their operation in isolation, one may draw the intuitive conclusion that the larger capacity pump (5) is the more critical pump in terms of system reliability (it covers a comparatively wider demand volume). It will be shown, however, that due to process interactions (structure complexity, material splitting, pump-valve assignements, and actual demand patterns), counter-intuitive phenomena may occur. In particular, it will be shown that the relative importance of pump 4 is in fact higher.

The major reason for the inadequacy of the above approaches to fully assess the inherent operability performance of the blending system is due to the absence of an explicit flexibility consideration. It is the purpose of this work to demonstrate that for reliability and availability assessment to be meaningful within a process context (in the presence of continuous uncertainty), they have to be considered in close relation to process flexibility.

FLEXIBILITY ANALYSIS

Review
The physical performance of a processing system can typically be described by a set of equations (i.e. heat and mass balances) and inequalities (i.e. quality specifications, bounds). The set of equalities, however, can easily be

eliminated, especially for linear models (viz. by using MACSYMA). This leads to a set of reduced linear inequality constraints of the following form:

$$f_j(d, z, \theta) = b_j^T \begin{bmatrix} d \\ z \\ \theta \end{bmatrix} + b_j^0 \leq 0 \qquad j \in J \qquad (1)$$

The vector of design variables d defines the structure of the process and equipment sizes. These variables are fixed at the design stage and remain constant during plant operation. z is the vector of control variables, which represent the degrees of freedom that are available during operation, and which can be adjusted for different realizations of the uncertain parameters; θ is the vector of uncertain parameters. The inequalities in (1), then, determine feasibility or infeasibility of operation for a given design d.

Grossmann and Floudas, [10], exploited the fact that sets of active constraints are responsible for feasible operation for a given design d, and thus potentially limit process flexibility. They showed that, under certain conditions, each set consists of n+1 active constraints (n is the number of control variables). For each active set J_A^k, a feasibility function $\psi^k(d, \theta)$ of design d at the parameter value θ qualitatively represents the adjustments of control variables z to minimize the maximum violation of the constraints, and is defined as follows:

$$\psi^k(d, \theta) = \min_{z, u} u$$
$$\text{s.t.} \quad f_j(d, z, \theta) \leq u \qquad j \in J_A^k \qquad (2)$$

where u is a constant. If $\psi^k(d, \theta) \leq 0$ for all values of θ and all n_{AS} active sets J_A^k, $k=1,\ldots,n_{AS}$, then the feasibility of operation can be ensured for design d. If $\psi^k(d, \theta) > 0$ for some values of θ, the design d is infeasible (i.e. incapable of handling these parameter values).

Pistikopoulos and Grossmann, [11], obtained an important analytical expression for the feasibility function $\psi^k(d, \theta)$ of a given active set J_A^k in terms of the design variables d and uncertain parameters θ, independent of the control variables z. Specifically, they showed that :

$$\psi^k(d, \theta) = \sum_{j \in J_A^k} \lambda_j^k f_j(d, z, \theta)$$

$$= \sum_{j \in J_A^k} \lambda_j^k \left(b_j^T \begin{bmatrix} d \\ z \\ \theta \end{bmatrix} + b_j^0 \right) \qquad k=1,\ldots,n_{AS} \qquad (3)$$

and

$$\frac{\partial \psi^k}{\partial z} = \sum_{j \varepsilon J_A^k} \lambda_j^k \frac{\partial f_j}{\partial z} = \sum_{j \varepsilon J_A^k} \lambda_j^k b_{j,z}^T = 0 \qquad (4)$$

(i.e. $\psi^k(d,\theta)$ is independent of z), where λ_j^k are Lagrange multipliers for constraint j in (2). Note that the Kuhn-Tucker conditions of (2) yield a linear square system of n+1 equations, whose solution uniquely determines the values of λ_j^k for each active set. Substituting (4) into (3) leads to:

$$\psi^k(d,\theta) = \sum_{j \varepsilon J_A^k} \lambda_j^k \left(\hat{b}_j^T \begin{bmatrix} d \\ \theta \end{bmatrix} + b_j^0 \right) \qquad k=1,\ldots,n_{AS} \qquad (5)$$

where $\hat{b}_j^T = \begin{bmatrix} b_{j,d} & b_{j,\theta} \end{bmatrix}$.

The importance of the expression in (5), for a fixed design d, is that it transforms the original constraint space f_j, $j\varepsilon J$, in (z,θ) to the space of explicit feasibility functions ψ^k in (θ). This then defines the feasible region in the space of uncertain parameters θ, by taking into account the fact that control variables are optimally adjusted during plant operation.

Stochastic Flexibility Index
As mentioned previously, for a design d to be flexible to operate under changing conditions θ, it suffices that the feasibility functions $\psi^k(d,\theta)$, k=1, ..., n_{AS}, are non-positive. A value of $\psi^k(d,\theta) = 0$ (for any k) defines the boundary of the constraint set projected in the θ-space. Qualitatively, then, the flexibility of a design d represents the fraction of the parameter space that lies within the region defined by the use of the feasibility functions. Since

$$\psi(d,\theta) = \max_k \left\{ \psi^k(d,\theta) \right\} \qquad k = 1, \ldots, n_{AS} \qquad (6)$$

and $\psi^k(d,\theta)$ is given by the expression in (5), an index for flexibility can be defined in the following way:

$$\begin{aligned} F &= \Pr\{\psi(d,\theta) \le 0\} \\ &= \int\cdots\int_{\{\theta:\ \psi(d,\theta)\ \le\ 0\}} P_D(\theta)\ d\theta \end{aligned} \qquad (7)$$

where $P_D(\theta)$ is the joint distribution function of the uncertain parameters θ. A graphical illustration for the definition of F for a design involving two uncertain parameters with joint probability distribution function is shown in Fig. 2. Note that from (7), F can be obtained as the integral of the probability distribution over the feasible region (i.e. the shaded area). Figure 2(a) corresponds to the case of a joint uniform distribution function, whereas Figure 2(b) corresponds to the case of a bivariate normal density function. Contours of the probability function are shown, with which bounds for F from inscribed and circumscribed contours can be obtained.

The importance of the expression in (7) is that it provides a general definition for the flexibility index for linear or nonlinear systems, where uncertainty is specified through any form of probability distribution function.

The evaluation of such an index will involve either the use of numerical integration techniques or the determination of the appropriate distributional properties for the function $\psi(d,\theta)$. For example, Pistikopoulos and Mazzuchi [12] have shown that by assuming a linear model describing the process system, a transformation to the space of a new set of stochastic variables can be achieved via the expressions in (5). By properly exploiting the probabilistic nature of the problem, the task of integration can be greatly reduced. For the special case of normal parameters, they showed that these transformations preserve the statistical properties of the original uncertain parameters:

$$F = \Pr\{\psi(d,\theta) \leq 0\}$$

$$= \Pr\left\{\psi^1(d^E,\theta) \leq 0, \ ..., \ \psi^{n_{AS}}(d^E,\theta) \leq 0\right\},$$

$$= \Pr\left\{\sum_{j \in J_A^k} \lambda_j^k \mathcal{B}_j^T \theta + \mathcal{B}_j^0 \leq 0, \ k=1, \ ..., n_{AS}\right\} \tag{8}$$

$$= \int_{-\infty}^{0} \cdots \int_{-\infty}^{0} \phi_{MVN}(\psi^1, \ ..., \ \psi^{n_{AS}}) \ d\psi^1 \cdots d\psi^{n_{AS}},$$

where ϕ_{MVN} is the multivariate normal probability density function of the correlated random variables ψ^k with mean vector and variance-covariance matrix defined through the linear transformation.

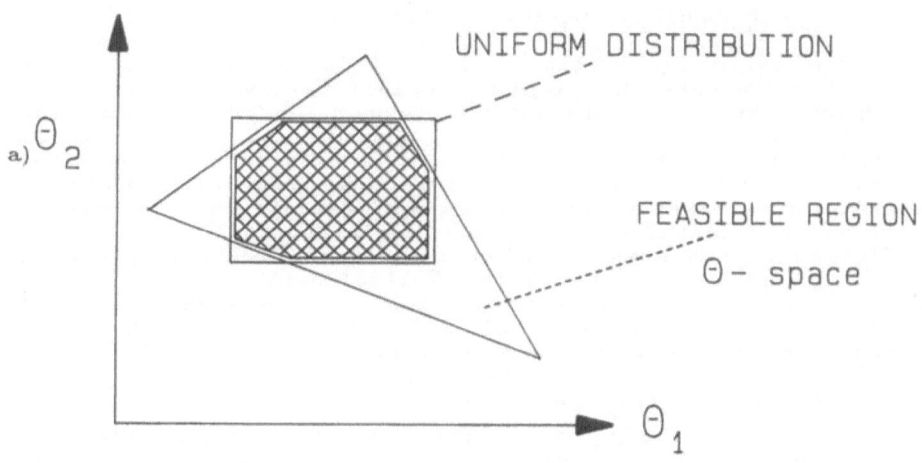

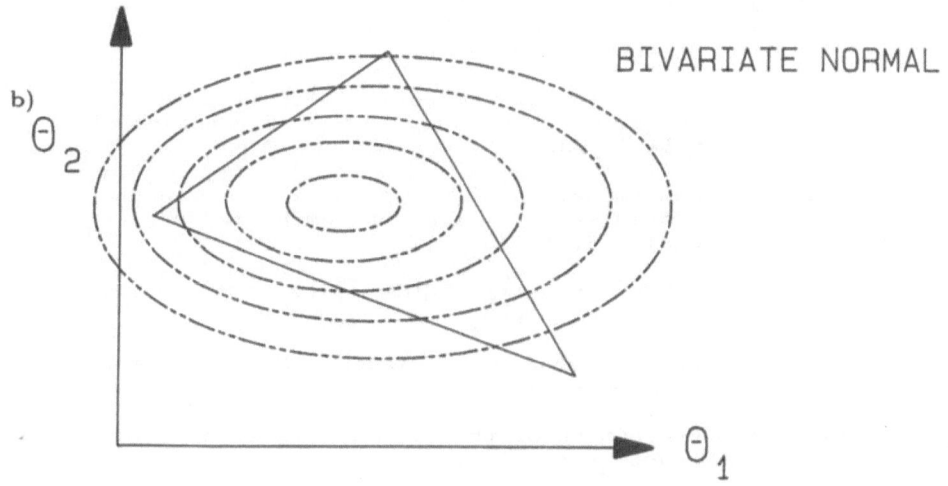

Figure 2. Graphical illustration of stochastic flexibility for
(a) uniform density, (b) bivariate normal density

A COMBINED FLEXIBILITY-RELIABILITY INDEX

The stochastic flexibility index presented in the previous section represents the <u>maximum</u> probability level for a design to operate in a feasible way (the proportion of time that the design is expected to meet demand and specification requirements). In this respect, this index also represents the reliability of the process system (capability of process to operate) at a fixed system (equipment) state in the presence of continuous uncertainty. Since any equipment failure defines a different system state with an associated flexibility index, then by modeling the random lifelength of equipment, flexibility as a function of time can be established. A combined flexibility-reliability index can then be introduced, which measures the capability of the design to simultaneously accommodate the inherent uncertainty in the equipment availability and the process parameters.

A systematic algorithmic procedure for determining such a combined flexibility-reliability index can be summarized in the following basic steps :

STEP 1. Identification of all operational states $i=1,...,$ N of the system with an associated non-zero flexibility index, F_i.

STEP 2. Evaluation of the reliability of each equipment k, k = 1 ,..., N as a function of time, $R_k(t)$.

STEP 3. Determination of the probability of each system state i, e.g., assuming independent equipment failures:

$$P_i(t) = \prod_{k \varepsilon S_i} R_k(t) * \prod_{k \varepsilon \bar{S}_i} \{1 - R_k(t)\}$$

where S_i ($\bar{S}_i$) is the index set for operational (failed) equipment.

STEP 4. Evaluation of the flexibility index F as a function of time:

$$F(t) = \sum_{i=1}^{N} P_i(t) * F_i$$

STEP 5. Determination of the combined flexibility - reliability index for a given interval [0,T]:

$$\bar{F}(T) = \frac{1}{T} \int_0^T F(t) \, dt$$

Qualitatively, this metric represents the average flexibility index over all possible system states within the specified time interval [0, T]. Therefore, it assesses the average system performance under both continuous uncertainty and random equipmment failures. The closer this index is to the flexibility index of the initial (no equipment failure) state, the more reliable the system is.

Note that from the current expression for $\bar{F}$, as T goes to infinity, the combined flexibility-reliability index goes to zero. This is consistent with the assumption of no repair mechanism for the equipment. In this respect, the index corresponds to a "worst-case" situation. Incorporation of maintenance and repair time distributions, however, is a straightforward task within this framework, and it will be reported elsewhere.

The potential advantages of the proposed framework to assess process flexibility, system reliability (and availability) in an integrated fashion, will be demonstrated by revisiting the motivating example.

EXAMPLE

Consider again the in-line blending system of Figure 1. For convenience in the presentation, only pump failures will be considered. The first step of the combined flexibility-reliability framework yields the stochastic flexibility index for each state (assuming uniform demand) as given in Table 1 (see [13] for details). It should be noted that the flexibility index of the original blending structure (all equipment operational) is F = 0.37; i.e. the structure is capable of accommodating demand requirements (in the specified range) approximately 37% of the time. Note also, that since this blending system involves mixing of three materials via five pumps, this implies that any simultaneous failure of three pumps (or more) will result in the inability of the system to meet quality requirements (thus a zero flexibility index). The significant implication of the latter is that the flexibility index evaluation needs to be performed for only sixteen (out of thirty-two) states.

Step 2 of the proposed framework requires a characterization of component (pump) reliability. For illustrative purposes, a stochastic model is assumed based on a Weibull distributional representation of the failure behavior of the pumps as follows:

$$f(t) = \exp(-\alpha t^{\beta}) \ , \qquad \beta , \alpha > 0$$

An attempt has been made to illustrate the relative importance of each pump by varying its failure behavior. A relatively large (small) value of β results in a rapidly (slowly) increasing failure rate (for $\beta > 1$). Figure 3 presents the time-dependent flexibility index for two

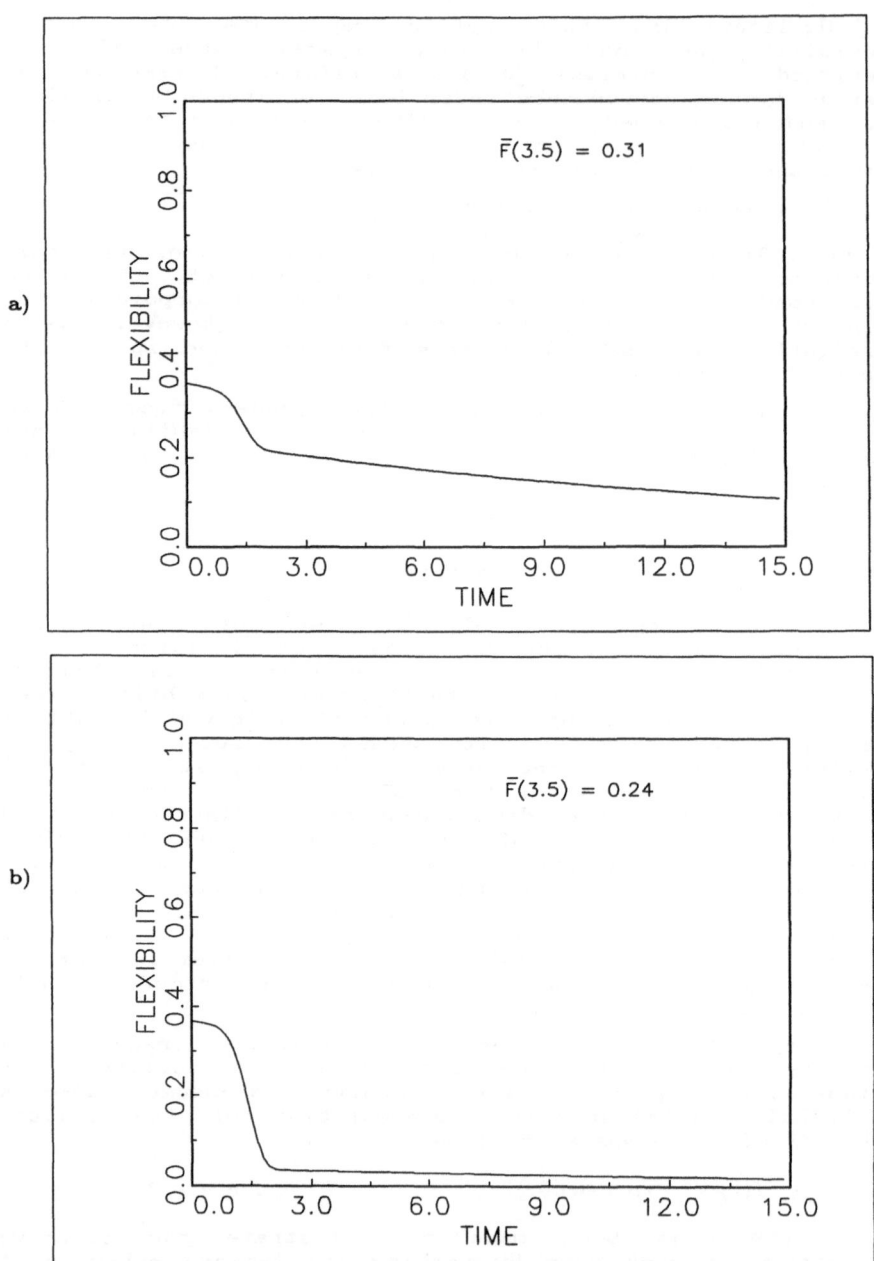

Figure 3. Time-dependent flexibility index, with $\underset{\sim}{\beta}$ values
(a) (5 1 1 1 1),   (b) (1 1 1 5 1)

TABLE 1
Equipment states and corresponding flexibility

FAILED PUMPS	STATE	FLEXIBILITY
none	11111	0.3672
1	01111	0.2409
2	10111	0.2066
3	11011	0.1074
4	11101	0.0390
5	11110	0.0449
1, 2	00111	0.0425
1, 3	01011	0.0992
1, 4	01101	0.0094
1, 5	01110	0.0094
2, 3	10011	0.0307
2, 4	10101	0.0059
2, 5	10110	0.0047
3, 4	11001	0.0153
3, 5	11010	0.0094
4, 5	11100	0.0024
(Other combination)		0

different choices of β values (α parameter fixed at 0.01). Comparing the two flexibility curves, it can be seen that the failure behavior of pump 4 (case b) has a more profound effect on the system flexibility than that of pump 1 (case a). This can be used to identify possible reliability bottlenecks (or key equipment) within an integrated process context. Table 2 summarizes the results obtained for different β values and a fixed time interval [0, 3.5]. Typically, in design, these parameters must be estimated from actual pump failure data or engineering principles.

TABLE 2
Combined flexibility-reliability index for different
Weibull shape parameters

PUMP 1	PUMP 2	PUMP 3	PUMP 4	PUMP 5	$\bar{F}(3.5)$
(no	failures)				0.3672
1	1	1	1	1	0.3470
5	1	1	1	1	0.3072
1	5	1	1	1	0.2967
1	1	5	1	1	0.2656
1	1	1	5	1	0.2442
1	1	1	1	5	0.2459
5	5	1	1	1	0.2410
1	1	1	5	5	0.2126
5	5	5	1	1	0.2189
5	5	5	5	5	0.1905

296

In general, the results follow the intuitive rule that a larger decrease of the combined flexibility-reliability index is associated with higher pump failure rates. However, due to the complexity of the structure, the following unexpected and counter-intuitive feature is revealed: the effect of the 200 capacity pump failure on the flexibility and the combined flexibility-reliability index is larger than the effect of the 400 capacity pump failure, although both of them transfer all three materials. This result clearly points out that caution should be exercised when trying to adapt conventional rules to complex systems. Material splitting, pump interactions due to valve allocations, and demand patterns, all play a role in the performance of an in-line blending structure to simultaneously account for demand uncertainty and pump failures.

CONCLUSIONS

The aim of this paper has been to illustrate the potential advantage of including process flexibility as an explicit objective in the operability assessment of complex multiproduct processign systems. In such systems, process interactions play a dominant role and existing tools for reliability and availability analysis seem to be, very often, insufficient.

A stochastic flexibility index was introduced, which accurately measures the inherent resilience of a design to withstand continuous parameter uncertainty. Based on this development and quantitative information regarding equipment reliability characteristics, a combined flexibility-reliability index has been proposed. This index allows for the simultaneous assessment of system performance with respect to both flexibility and reliability (and availability). The significance of such an index is that it provides, at early design stages, quantitative means for comparing different structural alternatives as well as for identifying potential bottlenecks of critically unreliable and/or uneffective equipment. Insights provided by the proposed index may also have profound implications for optimal maintenance and repair policies. Extension of this work to include more involved reliability and maintenance models is under investigation.

REFERENCES

1. Matrz, H. F. and Waller, R. A., Bayesian Reliability Analysis, John Wiley and Sons., New York, 1982.

2. Mazumdar, M., Approximate computation of power generating system reliability indexes, in Handbook of Statistics: Vol .7: Reliability and Quality Control, Eds. P. R. Krishnaiah and C. R. Rao, Elsevier Publishers, Amsterdam, 1988, pp. 55-72.

3. Dhillon, B. S. and Rayapati, S. N., Chemical-System reliability: A review, IEEE Trans. Rel., 1988, 37, 199-208.

4. Grossmann, I.E., Halemane, K.P. and Swaney, R.E., Optimization strategies for flexible chemical processes, Comput. chem. Engng., 1983, 7, pp. 439-62.

5. Swaney, R.E. and Grossmann, I.E., An index for operational flexibility in chemical process design, AIChE Jl, 1985, 31, pp. 621-30.

6. Cheng, T.C.E., An economic production quantity model with flexibility and reliability considerations, European Journal of Operational Research, 1989, 39, pp. 174-9.

7. Phadke, M.S. and Dehnad, K., Optimization of product and process design for quality and cost, Quality and Reliability Engineering International, 1988, 4, pp. 105-12.

8. Straub, D.A. and Grossmann, I.E., Integrated statistical metric of flexibility for systems with discrete state and continuous parameter uncertainties, Paper 165c presented at the Annual AIChE Meeting, San Francisco, November 1989.

9. Odi, T.O. and Karimi, I.A., A general stochastic model for intermediate storage in noncontinuous processes, Paper presented at the Annual AIChE Meeting, San Francisco, November 1989.

10. Grossmann, I.E. and Floudas, C.A., Active constraint strategy for flexibility analysis in chemical processes, Comput. Chem. Engng., 1987, 11, pp. 675-93.

11. Pistikopoulos, E.N. and Grossmann, I.E., Optimal retrofit design for improving process flexibility in linear systems, Comput. Chem. Engng., 1988, 12, pp. 719-31.

12. Pistikopoulos, E.N. and Mazzuchi, T.A., A novel flexibility analysis approach for processes with stochastic parameters, submitted for publication, Comput. Chem. Engng., 1989.

13. Pistikopoulos, E.N., Mazzuchi, T.A., Maranas, K. D. and Thomaidis, T.V., Simultaneous assessment of flexibility, reliability, and availability for in-line refinery blending systems: a unified framework for analysis and retrofit design, Technical Report, Department of Chemical Engineering, University of Thessaloniki, 1990.

A COMPOSITE MODEL FOR EVALUATING THE OPTIMUM DESIGN, MAINTENANCE, CONDITION MONITORING AND DISPOSAL FOR ANY EQUIPMENT

JOHN WOODHOUSE & COLIN LABOUCHERE
Directors, Optimis Ltd.

INTRODUCTION

As industrial management strategy becomes more and more analytical, both the technical and human problems are increasingly highlighted. In human terms, this involves fostering a greater business awareness, launching quality programs, teaching the basics of reliability theory etc. On the technical side, there are the problems of insufficient or highly-variable data, the complex interactions of risk and financial mathematics and the constant pressure on analytical timescales. The following paper describes a combination of analytical disciplines and a flexible, risk modelling calculator developed by Optimis over the last twelve years.

THE DISCIPLINES

The demand for analytical efforts is determined by the management decisions that need to be made. A large measure of improvement can be achieved, therefore, by applying basic disciplines to the decisions themselves - principally the cost/benefit trade-off that is often involved. Before I introduce the supportive calculation methods, I would like to summarise these disciplines: the value of a 'front-line' analysis tool is almost totally dependant upon the procedures or context for its use.

The establishment of systematic procedures can be a major programme in itself - training programmes, methods for identifying problems, active management support etc. In summary form, however, the disciplines involve the following four steps:

ASK THE RIGHT QUESTIONS i.e. develop systematic checks and procedures to ensure that the right questions are asked and that the decision-maker is not too close to the problem.

OF THE RIGHT PEOPLE Human judgement or experience will always be a major source of relevant information and most management decisions involve conflicting interests in several departments: personal involvement and objectives need to be catered-for.

IN THE RIGHT WAY In the absence of hard data, range estimates are always possible, whereas challenging for a 'spot estimate' generates concerns over accountability: more detailed investigation is only needed if the range estimates are found not to be sufficient.

AND SHOW THEM HOW THE ANSWERS ARE USED Visibility of the analysis process not only lends credibility to the conclusions but also encourages support from those who provide the basis for future analyses - those whose knowledge is most needed.

The existence of a decision to be made reflects either a problem to be solved or an opportunity to be considered. In either case, the stages of systematic analysis involve

a) Describing the current situation (problem)

b) Quantifying the potential opportunity (or solution)

c) Balancing the degree of improvement against the costs involved

Of course, in a wider context, several possible solutions need to be considered: a reliability 'problem' such as excessive failures of a pump might be solved by design change, by a preventive maintenance schedule or by living with the problem but reducing the impact (installed spare capacity etc.). Nevertheless the above sequence applies in each case - the only additional step occurs at the end: the potential solutions are ranked and the best is selected.

CREATING A MODEL OF THE PROBLEM

The picture of the current situation must be as comprehensive as possible - it is not enough to describe only the obvious elements or the most easily measured ones. In this way, even if a problem is perceived as one of unacceptable failure rates, the questions about operating efficiency, statutory requirements and even employee morale need to be asked. The full picture incorporates models of the following variables:

RELIABILITY: Failure <u>probability patterns</u> and failure <u>consequences</u>

EFFICIENCY: Production rates, fuel consumption, conversion performance etc.

LIFE EXPECTANCY: The usable life and capital value of the equipment

ABSOLUTE LIMITS: Statutory requirements, project timescales etc.

"SHINE" FACTORS: Public image, morale, environmental pressures etc.

THE RELIABILITY MODEL

In the case of reliability data, the descriptive stage is particularly difficult. There are considerations of possible failure modes, statistical limitations and probability interactions. A systematic approach can be developed, however, along the following lines:

Step 1: Equipment and Timescale selection.
a) Specification of component, assembly, equipment or process boundary for the model.

b) Choice of timescale to reflect any variations in failure probability etc. (e.g. Life Cycle, Time since Overhaul, Time since last Inspection)

Step 2: Failure Responses.
Classification of each failure type into one of two groups: Restore failures that cause the chosen 'Time since' clock to be reset, and patch-and-Continue failures.

Step 3: Quantified Failure Modes.
a) Continue Failures: Estimates of failure rates made at different times since new/last maintenance etc. Range estimates are entered into the computer and curves fitted to them as necessary.

b) Restore Failures: Separate consideration of Burn-in, Random and Wearout modes. Described in terms of survival characteristics, the computer can be used to fit curves to example points and, from the survival curves, calculate the corresponding failure density function and hazard rate characteristics.

This is an area of substantial technical challenge. In the usual situation of insufficient or only partially relevant failure data, it is necessary to describe the problem via either the observed or the anticipated <u>symptoms</u>. MAINOPT accepts information in the following format:

Burn-in, introduced risk, infant mortality (depending on and determined by the timescale that has been selected).
e.g. 0-10% fail in the first month.

Random failures (<u>with</u> <u>respect</u> <u>to</u> <u>the</u> <u>selected</u> <u>timescale</u>) are entered as either a probability level or the resulting survival percentage.
e.g. 80% reach (or would reach) 12 months.
or risk = 0.0143 per month

Wearout, deterioration or failures whose probability increases with the selected timescale (note that this may include "random" failures whose likelihood of <u>detection</u> is related to inspection frequency).
e.g. Earliest possible failure = 12 months
80-100% would have failed if left to 5 years

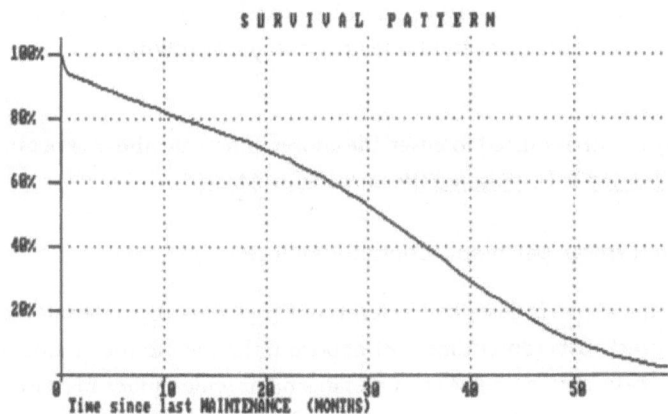

Step 4: Hazard Rates and Survival Curves.

The quantified Hazard Rate h(t), can be be described through the Survival Curve R(t), by the introducing the Failure Density Distribution f(t), as follows:

Since
$$f(x) = -d\frac{R(x)}{dx} \tag{1}$$

and
$$f(x) = R(x) \times h(x) \tag{2}$$

therefore
$$h(x)dx = -\frac{1}{R(x)}dR(x) \tag{3}$$

integrate both sides
$$-\int_0^t h(x)dx = [\ln R(x)]_0^t \tag{4}$$

so
$$R(t) = \exp\left[-\int_0^t h(x)dx\right] \tag{5}$$

See also Barlow & Proschan (1) or Cox (2).

Step 6: Cost Information

Based upon the reliability pattern (described by range-estimated example points), a financial optimisation can be made. This requires cost estimates, of course, and these are arranged as follows:

    a) Action required to 'reset' the chosen timescale: the cost of both Planned & Unplanned (Restore Failure) tasks

    b) Typical 'patch-and-Continue' Failure

Each of the above is described in terms of Direct (labour, materials etc.) and Indirect costs (downtime, lost opportunities etc.) to the company. In many cases, these will be estimates in similar wide ranges to those of failure probabilities. In addition, it is valuable to be able to describe a schedule of possible costs, weighted by a further distribution of probabilities:

Percentage of Occasions	Cost per Occasion
80%	£0
15%	£10,000
5%	£40,000

In a similar fashion, costs may vary by other mechanisms. Particularly in the case of optimised inspection frequencies, the cost of the performing the work is dependant upon the condition found at the time - the necessary corrective work may increase with inspection interval as there is more chance of finding accumulated deterioration if the equipment is left for longer periods. MAINOPT also allows a schedule of such trends, again by fitting curves to example points. Thus:

Inspection Interval	Cost per Occasion	Based on £50 inspection
1 month	£100	+ 5% risk of £1,000 repair
3 months	£10,050	+ typically £10,000 repairs
6 months	£80,050	+ typically £80,000 repairs

Step 7: Economic Optimisation
The final stage is the combined risk exposure and total cost calculation. This is achieved by MAINOPT once a range of planned task intervals or life cycles is specified. The economic optimisation of the reliability elements is as follows:

For a specific cycle length and an age-based inspection, maintenance or replacement policy (Barlow (3)):

Failure Cost
Rate
(£/month)

$$FCost = \frac{P(failed) \times Cost\ Per\ Failure(F\,££)}{Ave.\ Achieved\ Life} \qquad (6)$$

Ave.
Achieved
Life

$$\int_0^t R(x)dx \qquad (7)$$

from (5)

$$R(x) = exp\left[-\int_0^t h(x)dx\right]$$

so, failure
cost rate

$$FCost(\tau) = \frac{(1-R(\tau)) \times F££}{\int_0^\tau R(t)dt} \qquad (8)$$

and planned
cost rate

$$PCost(\tau) = \frac{(R(\tau)) \times P££}{\int_0^\tau R(t)dt} \qquad (9)$$

(8) plus (9)
(Total Cost)

$$TotalCost(\tau) = \frac{(1-R(\tau)) \times F££ + R(\tau) \times P££}{\int_0^\tau R(t)dt} \qquad (10)$$

For a straightforward optimisation of planned and unplanned costs, this formula can be used directly - simply differentiating or computing successive points to find the minimum. In larger, multi-variable models of industrial problems, the pattern of survival must also be used to factor the impact of, say, falling performance or operating costs before they can themselves be converted to a total cost rate.

Also, the above mathematics relates to an Age-based schedule of planned tasks, where the planned task is performed after a selected time if there has been no prior failure. If a Block maintenance policy (the planned task is performed periodically, irrespective of intervening failures) is applied, the relationships take on a further level of treatment.

EXAMPLE RELIABILITY PROBLEM

The following details just some of the reliability analysis options available in the MAINOPT analysis tool:

BURN-IN (Infancy)	BURN-IN PERCENTAGE	5	%
	BURN-IN PERIOD	.1	MONTHS
RANDOM	PERCENTAGE SURVIVING TO START OF WEAROUT —OR—	80	%
	RANDOM FAILURE RATE	.0143	/MONTH
WEAROUT (Old Age)	START OF WEAROUT	12	MONTHS
	PERCENTAGE FAILING	95	%
	BY TIME	60	MONTHS

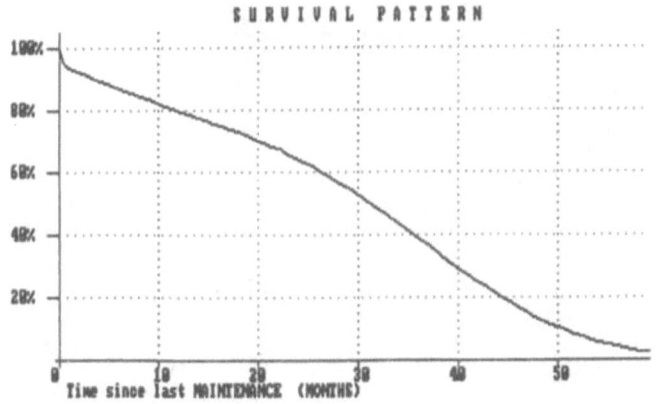

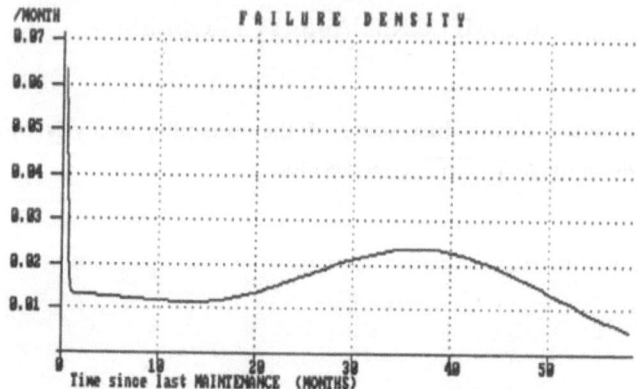

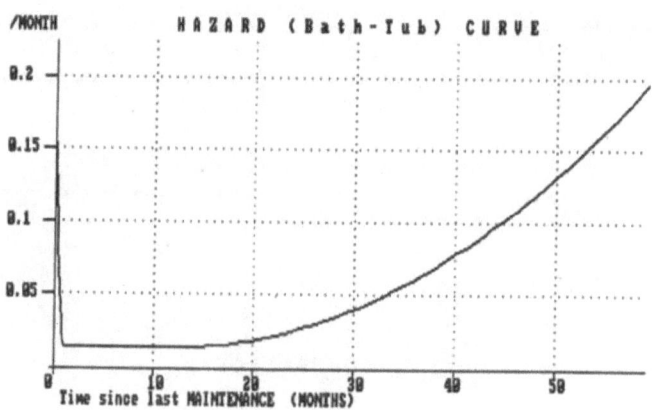

COST INFORMATION		
	DIRECT COST	PENALTY COST
PLANNED MAINTENANCE TASK	£ 5000	£ 0
RESTORE FAILURE	£ 8000	VARIES

PERCENTAGE OF OCCASIONS	PENALTY COST OF RESTORE FAILURE
80.00 %	£ 0
15.00 %	£ 10000
5.00 %	£ 40000

RESULTS

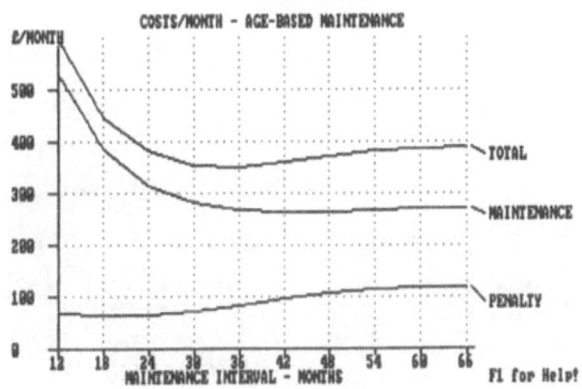

COSTS/MONTH - AGE-BASED MAINTENANCE

SENSITIVITY TEST

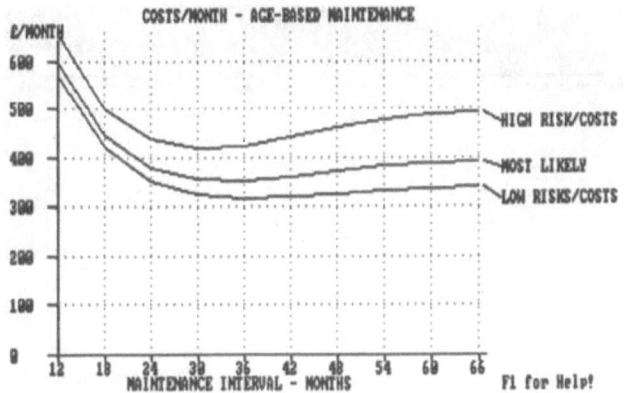

COSTS/MONTH - AGE-BASED MAINTENANCE

310

```
SCHEDULE DETAILS for  AGE-BASED MAINTENANCE at 36 MONTHS            Page 1


RELIABILITY                          AVERAGE COSTS

Mean Time Between Failures (MTBF)    Direct Costs

   Restore Failures    41.9 MONTHS     Planned Maintenance £ 76.2/MONTH
   Continue Failures   None
   Overall MTBF        41.9 MONTHS     Restore Failures    £ 190/MONTH
                                       Continue Failures   None
Mean Achieved Life     25.5 MONTHS
                                     Penalty Costs
% Surviving without
      Restore Failure  39 %            Planned Maintenance £ 0/MONTH
                                       Restore Failures    £ 83.4/MONTH
                                       Continue Failures   None
```

```
SCHEDULE DETAILS for  AGE-BASED MAINTENANCE at 24 MONTHS            Page 1


RELIABILITY                          AVERAGE COSTS

Mean Time Between Failures (MTBF)    Direct Costs

   Restore Failures    53.9 MONTHS     Planned Maintenance £ 166/MONTH
   Continue Failures   None
   Overall MTBF        53.9 MONTHS     Restore Failures    £ 148/MONTH
                                       Continue Failures   None
Mean Achieved Life     19.2 MONTHS
                                     Penalty Costs
% Surviving without
      Restore Failure  64.2 %          Planned Maintenance £ 0/MONTH
                                       Restore Failures    £ 64.9/MONTH
                                       Continue Failures   None
```

CONCLUSIONS

The MAINOPT Cost Optimisation model is much more flexible than can possibly be illustrated here - optimisation opportunities apply to

> Preventive Maintenance Justification & Timing
> Component or Equipment Replacement Decisions
> Inspection Strategies
> Evaluation of Condition-monitoring Techniques
> Design Improvement Evaluations
> Degree of Spare Capacity
> Plant Shutdown Decisions

With a combination of simple disciplines to establish appropriate range estimates for the data required in each case, and the software tools to convert estimates into Total Cost measurements, industrial management decisions are set to become much more analytical. The business considerations and access to full reliability analysis techniques can be brought forward to the front-line decision-makers.

Over the last twelve years, it has been noticeable that the larger process industries (such as Shell, Esso, CEGB, the Mars Group etc.) have been quickest to exploit this opportunity. In these and other industries, the limiting factor has often been the human one - the culture changes required to introduce greater business awareness and reliability concepts. With both a comprehensive education programme and the availability of suitable tools, the constraints are finally diminishing.

REFERENCES

(1) Barlow, R.E. & Proschan, F., The Mathematical Theory of Reliability, Wiley, New York, 1965.

(2) Cox, D.R., Renewal Theory, Chapman & Hall, London, 1962

(3) Barlow, R.E., in Statistical Theory of Reliability, ed. Marvin Zelen, Wisconsin Press, 1963

HUMAN FACTORS IN MAINTENANCE WORK IN THE

IN THE BRITISH COAL INDUSTRY

JONATHAN WINTERTON

Lecturer in Industrial Relations
University of Bradford Management Centre
Emm Lane, Bradford, West Yorkshire BD9 4JL, UK

ABSTRACT

New coal mining technologies involving computerized monitoring and control have
been designed to reduce human involvement, remove skilled work and improve machine
reliability. Tension between the production objectives of the new systems and human
factors is revealed in relation to health and safety and work organization. Craft work
is being reorganized to create more flexible, less autonomous maintenance workers
serving the requirements of an inflexible technology. Production delays caused by
random breakdowns show that the full potential of the computerized systems is not
being realized. Teams of analytical trouble shooters have been established to harness
the skills and knowledge of the workforce which the systems have been designed to
circumvent. A human-centred design strategy could have produced a superior
technical system which enhances workers' skills and improves the working
environment.

INTRODUCTION

The first oil shock in 1973 caused a wordwide re-evaluation of coal, which in Britain
led to the 1974 Plan for Coal. This tripartite agreement to expand coal production
involved a programme of investment in new capacity and new technologies. Through
adopting a systems engineering approach, the Operational Research Executive of the
National Coal Board (now British Coal) established a framework for the automation of
production, personnel and materials activities throughout the industry [1].

The automation programme was to raise productivity by increasing production,
reducing manpower and improving reliability. Work measurement had established that
as much as two-thirds of the available shift time was lost in operational delays due to
machine breakdown etc., and man-made delays arising from natural breaks taken by

face teams [2]. Systems were to be developed to improve machine reliability and human reliability [3]. A key objective of automation was to increase management control over operations and to minimize human intervention, thereby raising the productivity of labour and capital [4].

The design strategy, essentially Taylorist or Fordist in character, was profoundly influenced by the historical problems of supervision at the coal face [5]. These traditional problems were inflamed by the political context of Plan for Coal. The miners had been unified under national pay bargaining since 1966 and even before the oil crisis restored some of their former bargaining power, large-scale strikes had returned to the coalfields after an absence of almost fifty years [6]. Moreover, as the 1974 strike led to the demise of the Heath Government, it was clear that the leadership of the National Union of Mineworkers was moving to the left [7]. In addition to the key objective of expanding coal production, the automation programme also therefore adopted strategies designed to limit the miners' power at the point of production [8].

MINOS

By the late 1970s the systems engineering programme had enabled the Mining Research and Development Establishment to design MINOS, a computer system for centralized monitoring and control of colliery activities [9]. MINOS comprises various sub-systems devoted to particular functions such as coal clearance, face delays and coal preparation, all of which are overseen by mini-computers in a surface control room. The VDUs of the control console present mimic displays of the coal face, conveyor systems and other operations, indicating the status of items of plant. When the conveyors are not running, when the shearer is stopped or when ventilation fans have failed, the system alerts control-room operators. Much of the information that is passed to the control room is analysed by a secondary computer accessible only to management, and at the top of the system hierarchy another computer collates the information obtained from individual pits to compare performance. Figure 1 gives a schematic view of the system.

The effects of MINOS on mineworkers, both in terms of the labour market and the labour process, have been extensively analysed elsewhere [10]. Without repeating these arguments in the present paper, it is necessary to examine particular subsystems to assess their implications for reliability, maintainability, productivity and health and safety.

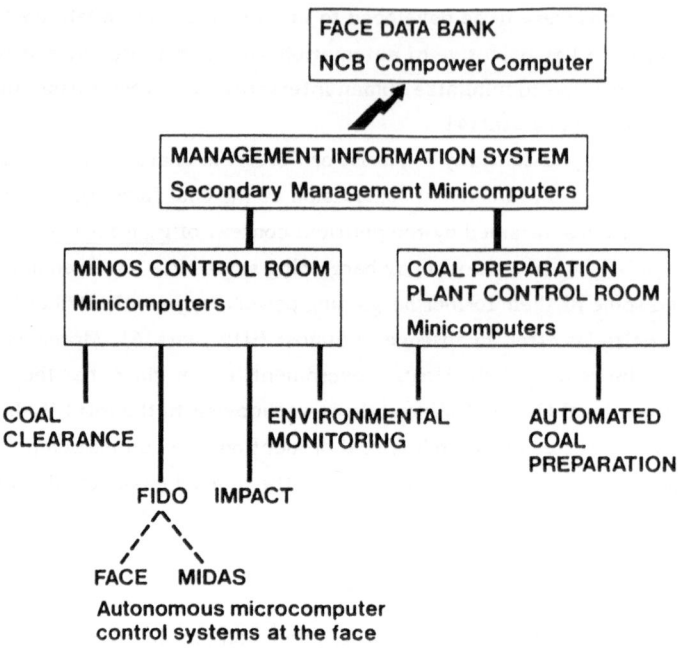

Figure 1. Schematic representation of MINOS.

At the coal face several subsystems were developed to monitor and control face activities. FIDO (Face Information Digested On-line) monitors the 'man-made' delays. General purpose monitoring outstations at the coal-face send digital information to the control room computer which indicates on the console VDU the position of the shearer along the face, as well as the cause and duration of any delays. Delays of less than twenty minutes are attributed to the men and adversely affect their bonus pay. Delays of twenty minutes or more that are found to have arisen out of operational difficulties beyond the control of the face-team are allowed as contingencies under the pay scheme. FACE (Face Advance Control Equipment) controls the advance of the conveyor and roof supports. The automation of the shearer involves two further subsystems : horizontal guidance [11] and MIDAS (Machine Information Display and Automation System), which controls the vertical guidance of the shearer and monitors the condition of the machine [12]. The monitoring of the condition of the shearer via MIDAS is aimed at reducing operational delays at the coal face. A similar facility is offered for all plant and machinery throughout the pit with IMPACT (In-built Machine Performance and Condition Testing). IMPACT reports on breakdowns and, through the monitoring and analysis of information such as oil pressure, bearing temperature and motor torque, can predict failures before they occur. The system specifies the nature of a fault, corrective action to be undertaken and even the personnel requirement [13].

Before innovations which would increase face output could be introduced, the systems approach dictated that MINOS coal clearance was installed to increase the capacity of conveyors and bunkers. From the MINOS control room conveyors can be started or stopped and bunkers loaded or discharged, these activities being monitored by the system and displayed in mimic form on the control console VDU. Automatic sequence starting of conveyors prevents overruns and chute blockages at transfer points, while conveyors are automatically stopped if overheating or a torn belt is detected. In conjunction with coal clearance, the first phase of MINOS included facilities to control fixed plant - ventilation fans, pumps and other equipment - and to monitor the working environment. The environmental monitoring subsystem continuously measures atmospheric pressure and methane concentration, and analyses mine air. The surface control room is alerted of any abnormal conditions, while continuous monitoring provides early warning of methane concentrations building up. The separate MINOS subsystems are linked together in the colliery control room where changes in the operations of one subsystem are made automatically in response to conditions monitored by another subsystem. Beyond the level of day to day operations, the secondary computer collates and analyses information of interest to management. The MIS (Management Information System) comprises software to analyse summary data relating to shift performance, machine availability, maintenance and reliability.

OPERATING RESULTS

There can be little doubt that MINOS has surpassed even the expectations of its designers. Computerized monitoring, control and coordination of colliery activities made possible such an increase in the rate of extraction that it was necessary to develop heavy-duty face equipment (HDFE) capable of sustained operation at high levels of output. Advanced technology mining (ATM) combines HDFE with MIDAS to give an anticipated daily face output of 4 kt [14]. The HDFE is installed first, but the return is only fully realised with the addition of the automation facilities. The preparation of faces with HDFE is approximately 60%, whereas MIDAS is installed on less than 30% of faces.

The parallel application of HDFE and microelectronic control systems brought a "huge reduction in down-time due to machine failure" [15]. Moreover, automation facilitated the restructuring which has reduced average colliery industrial manpower from 191,700 in the year ended March 1984 to 80,100 in the year ended March 1989

⌊16⌋. Over the same period, productivity grew from 2.43 tonnes per manshift to 4.14 tonnes per manshift, an increase of 70%. The continued application of ATM and associated industrial restructuring will result in further substantial improvements in productivity. By the early 1990s, output per manshift will be in excess of 6 tonnes ⌊17⌋.

The attainment of production objectives contrasts with the impact of new technologies upon the workforce in relation to health and safety and work organization. Computerized monitoring has contributed to improvements in the working environment since it is now possible to have continuous measurement of methane and dust. However, where systems designers appear not to have taken human factors into account, there is evidence of a deterioration in working conditions.

Higher rates of production have increased dust in proportion to machine running time. Even trusting that statutory levels are not being exceeded, face workers are inhaling a larger volume of dust and have a greater risk of pneumoconiosis or bronchitis. Nevertheless, it would be difficult to demonstrate empirically an increased incidence of dust-related illness using epidemiological techniques because of the rapid fall in the number and average age of faceworkers.

Reductions in manpower have been a major source of productivity gains and this has caused problems in the case of haulage workers. The MINOS coal clearance system has reduced elsewhere-below-ground employment, for example, by eliminating transfer-point patrol men on the belts. However, no monitoring system is as versatile as a human operator and neglect of human factors in this case may have contributed to an increase in the incidence of underground fires and suspected fires ⌊18⌋.

The conflict between production objectives and human factors is most marked in relation to work organization, particularly with the effects of new technology upon control and skill. In general, the design philosophy of MINOS reflects management ambitions of reducing workers' autonomy and limiting the system's dependence upon human skill. One study of staff in computerized colliery control rooms, for example, found that operators considered their skills and autonomy to have been reduced ⌊19⌋.

The effects of new technology upon maintenance work are equally profound. The introduction of power loading during the 1960s increased the strategic importance of maintenance workers and productivity became crucially dependent upon machine running time. Craft workers represented 20% of the workforce; by the 1980s more manshifts were spent on maintenance than on coal production ⌊20⌋ and the proportion was increasing ⌊21⌋. In reply to recommendation 9 of the Monopolies and Mergers Commission Report on the NCB that "priority . . . should be given to developing targets for reduction in the numbers of craftsmen" ⌊22⌋, the NCB argued that remote monitoring systems would substantially reduce the need for craftsmen. In an effort to reduce the time spent on maintenance, BC staff have developed a maintainability index for machinery in use [23].

Our analysis of MINOS, and particularly the elements of MIDAS and IMPACT designed to monitor machine health, concluded that the technology would facilitate a number of changes in the organization of craft work. First, computer diagnostics would enable maintenance work to be deskilled, with craft workers undertaking replacement rather than repair of faulty modules. Second, with improved machine reliability, the overall number of craft workers could be reduced; craftsmen would no longer be kept on 'standby' in a particular district but would be deployed centrally to replace parts identified by the system. Third, a small elite of craftworkers would be trained in condition monitoring techniques, undertaking highly-skilled maintenance of the system. The first two developments would both facilitate and necessitate two forms of functional flexibilility. A proportion of craft workers would be redeployed as machine drivers where they would also undertake nuisance breakdown repairs at the face. Remaining craft workers would eventually become 'multi-skilled', effectively deskilled but undertaking a wider range of maintenance repairs. These projected changes in the organization of craft work were made from an analysis of technical papers, especially evaluation reports, and extensive discussions since 1982 with miners experiencing prototype systems throughout Yorkshire and from every coalfield in the country. Such research established enormous regional variation both in the implementation of new technology and in management attempts to alter work organization.

DISCUSSION

Our analysis of changes in the organization of maintenance work in British coal mining has been subject to both theoretical and empirical critiques, which should be addressed before proceeding with the argument.

A theoretical critique held that we had neglected the process of reskilling which inevitably accompanies any technological change and assumed that a more likely scenario would entail "increasing numbers of maintenance workers with enhanced skills" [24]. In fact we had explicitly noted reskilling in relation to technician craftsmen, but pointed out that these were very few in number compared with the majority of craft workers being deskilled.

The empirical critique was based on research at two Lancashire collieries, neither of which could be considered a super-pit [25]. In one colliery reserves are almost exhausted, while the other is organized by the Union of Democratic

Mineworkers, making each an exceptional case. The effect of new technologies cannot be adequately examined in either colliery where neither IMPACT nor MIDAS has been installed. Moreover, there are serious questions over the interpretation of the evidence. While confirming that maintenance workers no longer undertake repairs but instead replace modules, it was claimed that this does not constitute deskilling and that maintenance work had become somehow more complex.

From the levels of craft deployment, descriptions of production systems and the levels of output, Agecroft is comparable with a medium-sized Yorkshire colliery before ATM. Their manpower statistics showed an increased proportion of craft workers, but the number of craftsmen had decreased slightly; the increased proportion is a result of significant job losses among surface and haulage workers, whose jobs have been eliminated by automated coal clearance and coal preparation, always the first phase of MINOS to be introduced. Their claims are not supported by their own evidence but it was nevertheless thought prudent to re-assess our earlier predictions of functional flexibility in the light of the restructuring of industrial relations in British Coal since the 1984-85 strike.

In May 1989 miners from every area were asked what forms of functional flexibility, if any, currently existed, and what they anticipated in the near future. In the Central and Yorkshire Areas, multi-skilling was reported in the form of fitters undertaking general purpose welding and as electro-mechanical craftsmen in face teams who undertake maintenance indicated by the diagnostic equipment on the shearer. Craft workers have also been redeployed as coal-face machine drivers, although there is no evidence of them undertaking machine repairs as yet. Craft workers have, however, negotiated their integration into the face bonus at many collieries, and in some pits there is agreement that craft workers will work in the face team if they are short of a miner. Some flexibility was written into the 1966 National Power Loading Agreement, but this only required craft workers to assist in an emergency, so was rarely invoked in practice. The incentive scheme is clearly being used to integrate craft workers into the face team, securing a greater degree of flexibility between craft and production functions. It is widely anticipated that the craft-trained machine drivers will soon be undertaking minor repairs, as British Coal have introduced a problem-solving 'quality circle' approach to coal-face repairs in selected high technology pits. These developments, like the increasing jurisdiction of craft workers in terms of their physical areas of responsibility, were associated with new technologies by all respondents.

These experiences contrasted with those of representatives from the peripheral areas, who reported no functional flexibility and claimed that craft-production and craft-craft demarcations were as clear as they had been a decade earlier. What all the peripheral areas have in common is a far lower application of MINOS than in the Central and Yorkshire Areas, again suggesting an association between the new technologies and functional flexibility. Such association should not be taken to constitute technological determinism because the new technologies merely offer the possibility of new forms of work organization.

CONCLUSIONS

In mining, productivity and reliability are crucially dependent upon human factors [26] and this is nowhere more apparent than in relation to maintainability and maintenance work [27]. The changes in the human factors of maintenance work in British coal mining are not an inevitable consequence of the new technology itself, but of a prevailing Taylorist perspective on work organization [28]. This approach will always create a sub-optimal solution because it stifles human ingenuity by reducing the scope for workers to exercise real skills and accept responsibility in return for autonomy. In Selby, the most modern mine complex to date, management have a target of reducing 'run of mine' stoppages to 20 hours per week. In August 1989 one week's stoppages amounted to almost 130 hours and the average for the last 10 weeks to end October 1989 was over 60 hours per week. The computerized monitoring systems have clearly not eliminated random breakdowns and in response management have established teams of 'Analytical Trouble Shooters', comprising ordinary miners, craft workers and charge hands, in an attempt to harness the skills and experience of the work force.

In the absence of 'Kaizen', the total commitment and loyalty on both sides of the idealized Japanese employment relationship, the ATS programme is unlikely to succeed, particularly given that the low-trust dynamic of Taylorist work organization has been exacerbated by poor industrial relations since the 1984-85 strike. Some Kepner-Tregoe ATS tutors have been exasperated when miners responded that production delays were management's problem and advocated strict observance of task demarcations in order to preserve jobs. Yet such attitudes are perfectly rational when work is organized according to Taylorist principles. In such an environment, the quality circle approach becomes just another attempt "to influence the attitudes of employees and their willingness to make greater efforts for the company, ie a new management philosophy aimed at introducing social control over the company work force" [29].

The need for ATS and its lack of success reinforce the argument that when systems are organized to reduce workers' autonomy and relieve management of dependence upon workers' skills, inferior solutions will result. A considerable body of research, particularly in Sweden, demonstrates that autonomous, democratic forms of

320

work organization not only create a more congenial psychosocial working environment, but also enhance efficiency ⌊30⌋. Rather than subordinating maintenance work to the computer system, the introduction of ATM provides a unique opportunity to utilise craft workers' skills and creativity through responsible autonomy to promote reliability. However, before maintenance work in coal mining can be organized to achieve such objectives, it is necessary to establish a high-trust environment in which innovative forms of work organization might flourish.

REFERENCES

1. Burns, A., Feickert, D., Newby, M., and Winterton, J., An Interim Assessment of MINOS, Working Environment Research Group Report No.2, University of Bradford, 1982.

2. Cleary, J., FIDO at Bold Colliery. The Mining Engineer, November 1981, 281-9.

3. Cooper, C.C., Improving machine utilization and reliability. COMMIT 82, Computer-based mine-management information technology exhibition and symposium, Harrogate, 8-10 December 1982.

4. Horton, E., Mining techniques in the 1980s. The Mining Engineer, February 1983, 451-5.

5. Goodrich, C., The Miners' Freedom, Marshall Jones, Boston, 1925.

6. Winterton, J., The trend of strikes in British coal mining, 1949-1979. Industrial Relations Journal, 1981, 12, 6, 10-19.

7. Winterton, J., and Winterton, R., Coal, Crisis and Conflict: the 1984-85 miners' strike in Yorkshire, Manchester University Press, Manchester, 1989, pp.9-12.

8. Winterton, J., The politics of new technology in British coal mining. Conference of Socialist Economists, Sheffield, 10-12 July 1987.

9. Chandler, K.W., MINOS - a computer system for central control at collieries. Second international conference on centralized control systems, London 20-23 March 1978.

10. Burns, A., et al., The miners and new technology. Industrial Relations Journal, 1983, 14, 4, 7-20; Burns, A., Newby, M., and Winterton, J., Second Report on MINOS, Working Environment Research Group Report No.6, University of Bradford, 1984; Burns, A., Newby, M., and Winterton, J., the restructuring of the British coal industry. Cambridge Journal of Economics, 1985, 9, 93-110; Winterton, J., Computerized coal : new technology in the mines. In Digging Deeper : Issues in the Miners' Strike, ed. H Beynon, Verso, London, 1985, pp. 231-43.

11. Hartley, D., and Wolfenden, J.R., Horizon control system designs for longwall face machines. International conference on remote control and monitoring in mining, NCB, 1977.

12. Fennelly, F., Coalface machine health. COMMIT 82, Computer-based mine management information technology exhibition and symposium, Harrogate 8-10 December 1982.

13. Bates, T.J., Inbuilt machine performance and condition testing - IMPACT The Mining Engineer, July 1981, 31-7.

14. Tregelles, P.G., The R and D history of computer-based control technology in the mining industry. The Mining Engineer, October 1986, 211-16.

15. Tregelles, P.G., The Coal Industry. Memorandum No.59, House of Commons Select Committee on Energy. The Coal Industry, HC 165-I, HMSO, London 1986, pp.312-14.

16. British Coal, Report and Accounts 1988/89, BC, London, 1989.

17. Winterton, J., The Effect of New Technologies on the Productivity and Production Costs of the British Coal Mining Industry, Working Environment Research Group Report No. 12, University of Bradford, 1988 [presented to House of Commons Select Committee on Asociated British Ports (No.2) Bill and North Killinghome Cargo Terminal Bill, 13 December 1988].

18. Feickert, D., Miners, safety and the technological revolution. Lancashire Area NUM Safety Conference, 6-8 November 1987.

19. Best, C.F., et al., The human aspect of computer-based monitoring and control of mining operations, Institute of Occupational Medicine, 1985.

20. Moses, K., Performance improvements in North Derbyshire, The Mining Engineer, May 1983, 581-5.

21. Horton, E., op. cit.

22. Monopolies and Mergers Commission, National Coal Board: A Report on the efficiency and costs in the development, production and supply of coal by the NCB, Cmnd. 8920. HMSO, London, 1983, p.377.

23. Mason, S., and Coleman, G.J., Designing for improved machinery maintainability. In Contemporary Ergonomics 1989, ed. E.D. Megaw, Taylor and Francis, London, 1989, pp.298-302

24. Penn, R., and Simpson, R., The development of skilled work in the British coal mining industry, 1870-1985. Industrial Relations Journal, 1986, 17, 4, 339-49.

25. Simpson, R., and Penn, R., The development of skilled work in the modern British coal mining industry : the case of maintenance work. Conference on Organization and Control of the Labour Process, UMIST, 22-24 April 1987.

26. Simpson, G.C., Promoting reliability : the human factors. Mineral Resources Engineering, 1988, 1, 1.

27. Rumani, R.V., Bhatterjee, A., and Pawlikowski, R.J., Reliability, maintainability and availability analysis of longwall mining systems. Mineral Resources Engineering, 1989, 2, 1.

28. Winterton, J., and Winterton, R., New Technology : the bargaining issues, Occasional Paper in Industrial Relations No. 7, Universities of Leeds and Nottingham/Institute of Personnel Management, 1985.

29. Swedish Work Environment Fund, Rewarding Work, Arbetsmiljofonden, Stockholm, 1987.

30. Gardell, B., Work Organization and Human Nature, Arbetsmiljofonden, Stockholm, 1987.

RELIABILITY ANALYSIS OF A CENTRAL COMPUTER SYSTEM POWER SUPPLY

B. B. W. OSTROM, C.F. PENSOM and D. J. WINFIELD

Atomic Energy of Canada Limited,
Chalk River Nuclear Laboratories,
Chalk River, Ontario, Canada KOJ 1JO

ABSTRACT

The Chalk River Nuclear Laboratories (CRNL) central computer facility provides mainframe computer service to the various functional groups on site. About 1500 local and remotely located people use this service for scientific research activities, inventory control, accounting and word processing and data transmission links off site.

The computer facility suffers from occasional unscheduled interruptions in service. This causes inconvenience to users and reduces their productivity. The unavailability of the computer causes activities dependent on it to be postponed or cancelled. These consequences can be minimized by reducing the frequency and duration of interruptions.

One major source of unscheduled computer service interruptions has been traced to power disturbances. This includes both short-duration power disturbances (< 5 s), and longer duration (> 5 s) power outages. This study is an analysis of the power-line disturbances at CRNL and their correlation to interruptions in computer service. The adequacy of the protection equipment for power-line disturbance related computer interruptions is also assessed.

PURPOSE

This study :

(1)   Characterizes the power-supply disturbances at CRNL.

(2) Develops a model to predict computer faults and downtime arising from power-supply disturbances.

(3) Assesses the cost effectiveness of additional power-conditioning equipment for the CRNL computer facility.

(4) Assesses the potential for disturbances in installations utilising disturbance-sensitive equipment in other on-site facilities.

SCOPE

This study is limited to examining power disturbances measurable by the Dranetz model 606 power disturbance monitor [1]. Performance specifications for this monitor are provided in Appendix 1. Data collection for 1984 to 1988 inclusive, has been analysed in the current study.

Types of Power Disturbances
The types of power supply disturbances experienced in practice have been previously documented and classified [2, 3] and are summarised below and shown schematically in Appendix 1. Apart from these studies, little is available in the open literature. With the growing use of solid-state devices, which are quite sensitive to such disturbances, this is surprising.

The disturbance classifications are :

(a) Voltage spikes/impulses: - overvoltage or undervoltage spikes (sometimes classed as impulses) superimposed on the ac 60 Hz waveform. Their duration is typically 0.5 to 800 µs and their magnitude, measured by the voltage above or below the ac waveform, is typically 50 V to a few kV.

(b) Undervoltages or sags: - voltage drops below the RMS average voltage level, having typical durations between one and several hundreds of cycles.

(c) Overvoltages or surges: - voltage increases above the RMS average voltage level, having typical durations between one and several hundreds of cycles.

(d) Outages: - total loss of the off-site power (power failures) for periods longer than the 5 s reclosure time of the off-site power supply switchgear. They have been separated into unscheduled and scheduled outages in the analysis.

(e) Frequency changes: - 1 s average deviations in frequency of more than ± 0.5 Hz from the site supply frequency standard of 60 ± 0.5 Hz.

DATA COLLECTION

Impulses
Impulse disturbance data were recorded on each phase and the highest phase voltage reached was used to represent the event. Measurements were made in volt-s (in 4 ranges from 0.0012 to 0.12 volt-s with 2 extra ranges covering values below and above these ranges), representing the integral of voltage with respect to time for which the voltage exceeds a pre-selected voltage threshold, taken as 208 V (43% of the RMS of 480 V).

Undervoltages and Overvoltages
Sets of undervoltages (sags) and overvoltages (surges), were recorded on each phase for each disturbance. The lowest/highest phase voltage reached and the associated duration in cycles were used to represent the under or overvoltage event. Threshold and incremental voltage increments of ± 4 V were recorded. The Dranetz records the under and overvoltages with respect to the current RMS value. Changes in the current RMS line voltage (designated the slow averaged steady-state voltage) of more than ± 4 V over a 10 s moving average, are noted by the Dranetz.

Sags which are effectively a complete loss of power (100 % sags and > 5 s) are classified separately as unscheduled outages (power failures). Where a number of disturbances occurred in rapid succession, without a new time stamp on the monitor, each record was treated as a separate disturbance, unless the sag/surge values on all three phases were unchanged.

Frequency Changes
These represent 1 s average deviations in frequency of more than ± 0.5 Hz from the previous steady-state frequency.

RESULTS

Undervoltages/Overvoltages
Undervoltage (sag) disturbances longer than one cycle or more, for 1984 to 1988, are shown as a function of the percent voltage sag, in the density distribution plot of Figure 1. The solid line represents the sag tolerance discussed below. The cluster of sags lasting around 50 to 90 cycles, and 5 to 8% sag, has been identified with the starting of the 3-phase, 2300 V, 200 kW ac motors which drive CRNL's, NRU reactor's eight main circulating pumps. Sags lasting 130-160 cycles and 3-5% sag are identified with startup of the computer building motor generator sets. These types of operations take place at essentially random times throughout the year, with the resulting power surge causing a power in-rush, depressing the line voltage for several cycles. No other clusters of disturbances could be identified with specific on-site root causes.

Figure 2 shows the equivalent surge data to Figure 1. Most surges were concluded to originate off-site as no on-site equipment events could be correlated to these incidents. The solid line represents the surge tolerance discussed below.

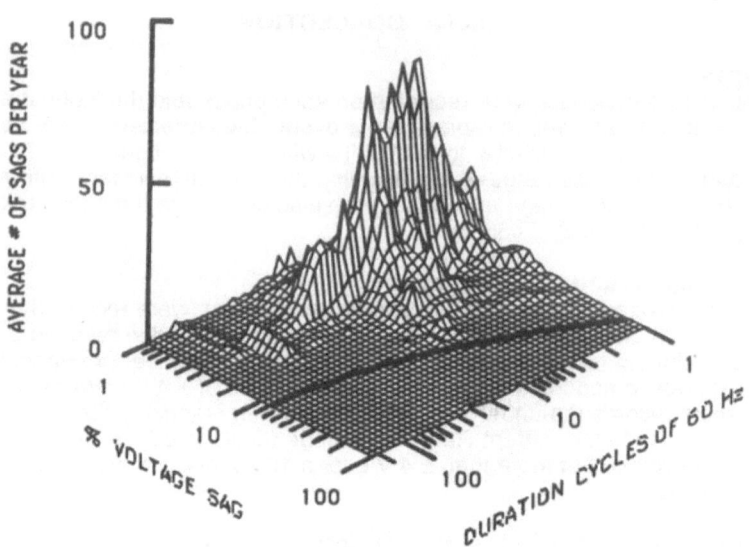

Figure 1. Undervoltage (Sag) Density Distribution as a Function of Cycle Duration, 1984-1988. Solid Line is Sag Tolerance/Undervoltage Ride Through.

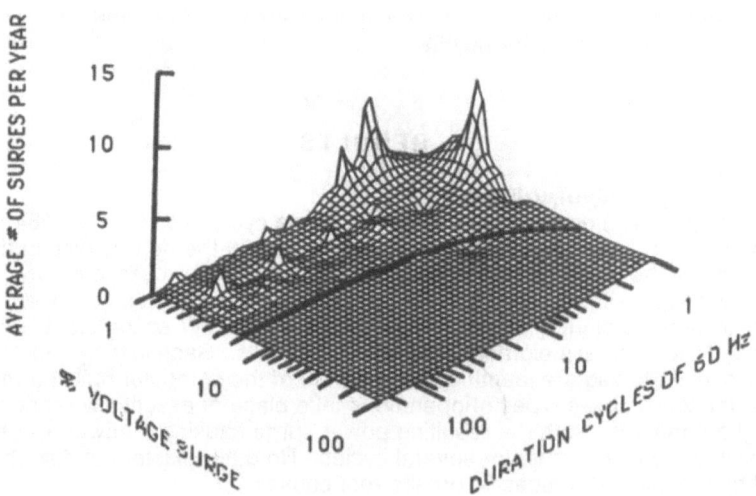

Figure 2. Overvoltage (Surge) Density Distribution as a Function of Cycle Duration, 1984-1988. Solid Line is Surge Tolerance/Overvoltage Ride Through.

A sag incident rate seasonal plot is shown in Figure 3. Surges and unscheduled outages, totalling 11% and 1% respectively, of the incidents in Figure 3, are indicated separately. The upper 95% confidence limit is shown for the maximum and minimum data points on this and subsequent similar histogram plots. A seasonal variation is indicated for all three incident types, illustrating the effect of summer storm activity. This variation has been noted in previous analysis of unscheduled outage data [2, 5].

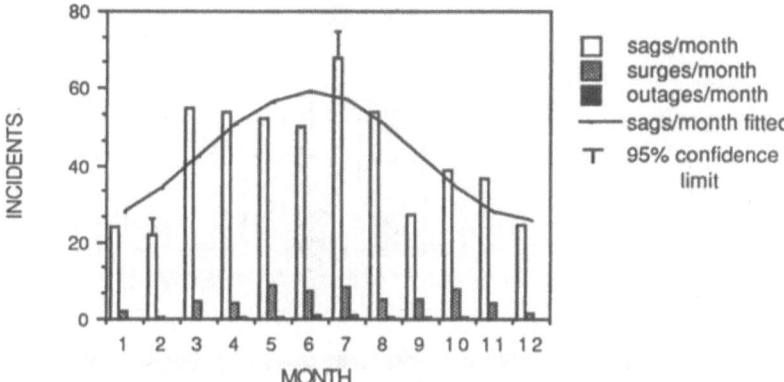

Figure 3. Annual Average Sag/Surge/Outage Monthly Incidents, 1984-1988.

A simple predictive model, based on an assumption of annual sinusoidal periodicity, was used to represent the sag disturbance data of Figure 3:

$$n_{sm}(t) = a_{sm} + b_{sm} \sin (0.524\, t + c_{sm}) \qquad (1)$$

where $n_{sm}(t)$ represents a 5-year average, number of sags per month (760 h) occurring at time t (months) from January 1st. The first term represents the time-independent contribution, and the second term, a time dependent sinusoidal contribution. A least squares fit gave the following values:

a_{sm} : 42.3 (per month)
b_{sm} : 16.7 (per month)
c_{sm} : - 1.3

Because of the sparsity of data, no model fit was made to the surge/outage disturbance seasonal data.

A 5-year averaged hourly dependence of the sag data is also demonstrated in Figure 4. The data clearly indicate an early morning peak between 07:00 and 11:00 h followed by a dominant mid-day peak between 12:00 and 16:00 h. Equation 1 was considered to provide an adequate model fit, without incorporating the two peaks, in view of the statistical accuracy. The least squares fit parameters, using $n_{sh}(t)$ as the number of sags per hour at t hours from 00:00 h, are given as:

$$n_{sh}(t) = 21.6 + 13.4 \sin (0.262 \, t - 2.1) \tag{2}$$

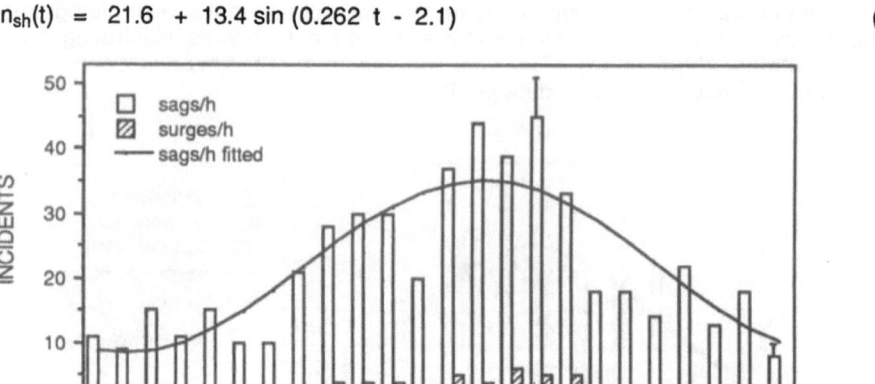

Figure 4. Annual Average Sag/Surge Incidents on Hourly Basis, 1984-88.

The morning and mid-day peaks are identified primarily with industrial activities (causing load changes) being transmitted to CRNL via the Ontario Hydro power system. From these types of distributions the reliability analyst can determine the effectiveness of disturbance protection equipment, given the tolerance specifications of the equipment.

Impulses
Results of the impulse disturbance volt-s distribution data, in Table 1, show that 94% of the impulses are within the 0.04 and 0.12 volt-s range.

TABLE 1
Impulse Disturbance volt-s Frequency Distribution Data,1984-1988.

Impulse volt-s	% of total events
0.0012 - 0.004	0.8
0.004 - 0.012	0.0
0.012 - 0.04	5
0.04 - 0.12	94
> 0.12	0.4

Figure 5 shows the seasonal variation of impulse data in impulses/month for 1984-1988, giving a similar profile to that of sag disturbances of Figure 3. Fitting an equation similar to (1), assuming annual periodicity, gives:

a_{im} :   7.62   (per month)
b_{im} :   5.06   (per month)
c_{im} :   - 1.4

as the least squares fit to the monthly impulse data.

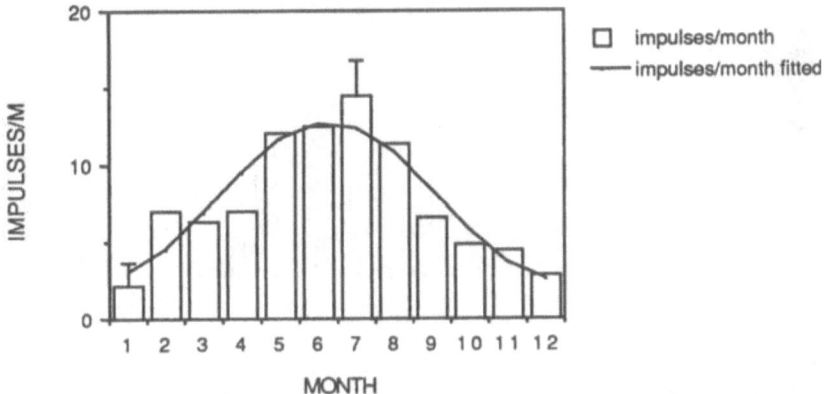

Figure 5. Annual Average Impulse Incidents on Monthly Basis, 1984-1988.

Daily periodicity is also indicated in the impulses/hour plot of Figure 6. For 1984-1987 a very pronounced peak at 07:00 h is indicated, with smaller peaks at 12:00 and 15:00 h. Data for 1988 are shown separately and the morning peak is no longer evident. No curve fitting was made for these data because of the recent profile change. Using only 1988 data, the values for a_{im} and b_{im} decrease by 36%, from the 5-year-average best-fit monthly impulse model above. Installation of a 0.9 MVAR capacitor on the incoming power lines of the site switchyard in late 1987 has reduced the prominence of this peak which is attributable primarily to off-site sources. The high-frequency nature of these transients enables them to propagate many kilometres from their source. They are initiated typically by lightning strikes, power network switching surges (e.g., from large transformers, capacitors and inductors) and operation of industrial equipment.

Frequency Changes
Analysis of the data indicated an average of only 8 disturbances per year where the frequency varied from the long-term 60 Hz average.

The sparsity of frequency disturbances is indicative of the expected good stability because of the large interconnected Ontario Hydro power system. The frequency disturbances experienced are related to either power distribution paths

330

being switched, resulting in momentary frequency changes, or plant synchronising problems bringing capacity on line. Although statistics are limited, no evidence of daily or seasonal periodicity is evident in the frequency disturbance data. This is not unexpected because of the types of incident causing these disturbances. There were no instances where frequency disturbances resulted in equipment problems.

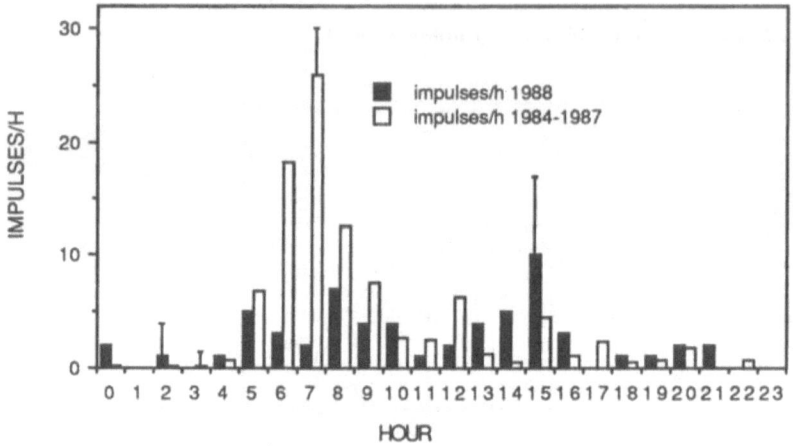

Figure 6. Annual Average Impulse Incidents on an Hourly Basis, 1984-1988

Outages
Figure 3 indicates the seasonal variation of unscheduled outages with a mean frequency of 4.4 per year. Figure 7 shows the corresponding computer power supply unscheduled-outage-time cumulative distribution function. The outage time has a mean value of 58 min, giving a mean unscheduled-outage-time total of 255 min per year. The computer equipment in general will be out of service longer, depending on the time of outage, user requirements and any associated re-start problems.

Overall Line Disturbances
Figure 8 shows the distribution of the various disturbance types as a percent of the total. Sags are predominant, followed by impulses, surges, frequency incidents and unscheduled outages. Low voltage conditions, sags and outages represent 76% of all disturbances. This behaviour appears typical of other large distribution systems and is very similar to that reported in [3].

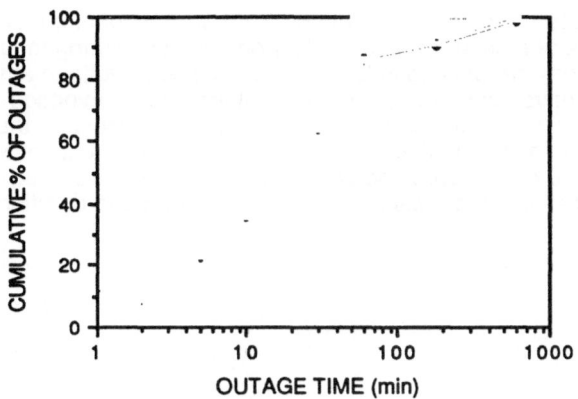

Figure 7. Unscheduled Power-Supply Outage-Time Cumulative Distribution Function

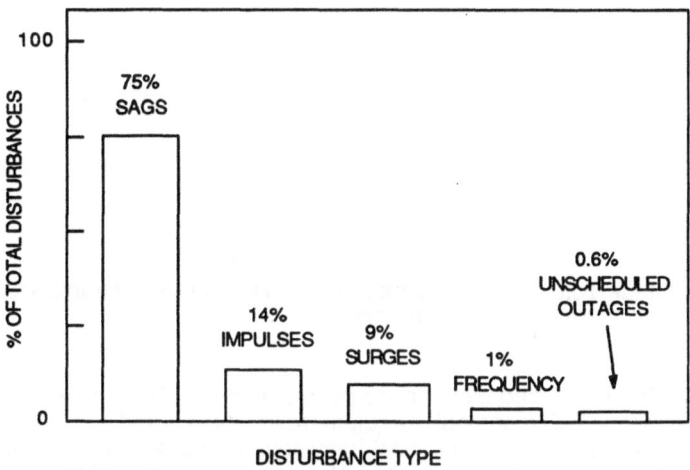

Figure 8. Distribution of Disturbance Types

COMPUTER VOLTAGE TOLERANCE ENVELOPE

Previous studies and test results [6] have generated computer voltage tolerance envelopes that provide design standards specifying the maximum voltage deviations within which computer equipment can operate without faults or outages.

Figure 9 illustrates a representative generalized design goal for disturbance-tolerance envelopes of a cross section of electronic manufacturing industry taken from [6]. Disturbances are defined in terms of percent rated voltage amplitude and duration. The envelopes represent undervoltage and overvoltage ride-through/tolerance limits. Three basic regions are used to represent the disturbance types as shown. Outages are defined as a power loss for > 5 s (300 cycles of 60 Hz). The curves are based on tests, computer power studies and manufacturer suggestions, the tests being of limited scope and duration, however.

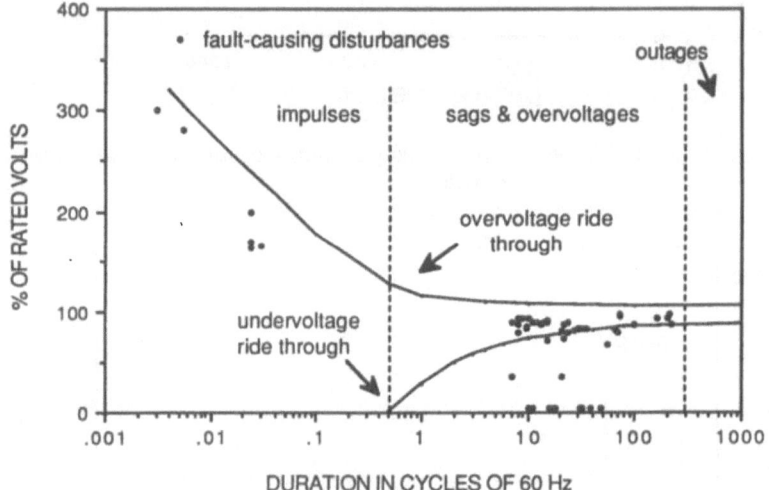

Figure 9. Voltage Tolerance Envelopes and Fault- Causing Disturbances, 1984-1988.

The 1984-1988 recorded disturbance events, not including unscheduled outages, which resulted in computer equipment immediate faults, are shown in Figure 9. They consist of 49 sags and 6 impulses. No surges or frequency disturbances resulted in computer faults. The average number of fault disturbances was 11 per year, which represents 1.6% of all recorded disturbance from the types shown in Figure 8. One assumption, however, is that a disturbance was completely responsible for the fault, rather than a contributing factor from previous 'hits'.

. The sag disturbances below the sag tolerance line, resulting in faults, represent, from the density data of Fig.1, 100% of all the sag disturbances experienced below the line. The generalized sag tolerance line thus provides a good indication of a facility-specific lower-range limit, beyond which faults are certain. An upper limit, facility-specific sag line, above which no faults are expected, can easily be established to encompass the remaining sag fault data.

Between the two lines it is then simple to assign fault probabilities for a given sag/cycle duration, although this is subject to the causal-relationship assumption noted above.

For the impulse fault data, a total of 6 events fall below the generalized overvoltage tolerance line. As the data are more sparse than for sags, it is more useful to specify a facility-specific single lower limit for the overvoltage line, set below the recorded fault data, than to specify a range between two lines.

In Table 2, statistical predictions at various confidence bounds are provided to predict the expected number of fault-causing disturbances. These predictions were made assuming a Poisson distribution of disturbances. For example, less than 17 fault-causing disturbances per year are expected to be experienced 95% of the time, assuming unchanged power system conditions.

TABLE 2
Expected Number of Sag/Impulse Disturbances per Year Resullting
in Computer Equipment Faults

year	observed # of disturbances/year causing faults	# of disturbances expected at indicated probability level		
		5%	50%	95%
1984	5	2	4.5	9
1985	12	6	12	17
1986	12	6	12	17
1987	10	5	9.5	15
1988	16	10	16	22
annual average	11	6	11	17

ADEQUACY OF EXISTING POWER-CONDITIONING EQUIPMENT

The reported study period 1984-88 was a period for which most variables were fixed. Data collection actually began in 1974 and is still a continuing activity. As a result of gaining experience in interpreting disturbance data, the computer operations section has been able to establish an advance notification system, involving on-site and also off-site power utility personnel, for those types of scheduled operations that would result in fault-causing disturbances.

The data presented indicated an average of 11 computer fault-causing disturbances per year and 4.4 unscheduled outages per year. Hardware faults affected by these disturbances have been disk storage units and communications equipment. The disk faults were linked primarily to sags and the communications equipment faults to impulses.

The existing power-conditioning for the computer mainframes consists of a two out of three ac to ac synchronous motor generator set with flywheel inertia. This method is estimated [3] to provide essentially complete protection against impulses and about 80% protection against sags. Disk storage units are not protected by the motor generators. Since mid-1988 a solid-state voltage regulator power conditioner has also been in service as an integral part of newly installed computer hardware. In addition, as noted above, since late 1987, capacitor banks have been used for overall site voltage and power factor correction, which has reduced the transient voltage impulse frequency.

The associated costs of the current level of faults and unscheduled outages are more complex to establish in a research site environment than, for example, in computer installations used by banks, air traffic control and stock exchanges. Estimates of all factors involved are 26 k$/year in firm costs (maintenance, computing charges, re-run costs) and 500-950 user-hours per year (lost time, re-run time). Typically, a fault-causing transient or unscheduled outage had an immediate impact on 30 to 80 interactive-user sessions, and 10 to 50 batch tasks up to the end of 1988.

From these dollar and time cost estimates, existing failure rates from unscheduled outages and fault disturbances do not currently justify the additional cost of installing an uninterruptable power supply, UPS, be it with either static or rotary devices, for transient and extended outage protection. Typical costs of these would result in a payback time of about 10 years. The protection provided by the existing MG sets and the recently installed power conditioner is, however, considered essential. Based on the protection factors quoted in [3], the sag disturbance fault-causing rate without these would not be acceptable.

The laboratory has recently installed a distributed computer-based financial information system, with users at sites some 3000 km apart, and has battery protected communications hardware. The impact of fault-causing disturbances on the use of this on-line data entry system is therefore similar to the financial and accounting type industries, affecting clerks, secretarial staff and managers directly. For a scientific research laboratory, this new, additional requirement could shorten the payback period for a UPS to less than 5 years, considering the cost-impact on the interactive and batch users, which has risen to 70-120 interactive users and 40-100 batch users at any time, since 1988.

As a purchaser of continually upgraded and expanding computer hardware the importance of continued disturbance monitoring is recognized by CRNL. Power disturbance data for 1989 is currently under review for cost-benefit decisions regarding future power conditioning requirements.

The data presented in this paper are also being used to assess the potential for disturbances in new and existing installations on-site, utilising disturbance sensitive solid-state equipment in critical applications.

REFERENCES

[1] Dranetz Engineering Laboratories Inc., 2385 South Clinton Avenue, South Plainfield, N.J.

[2] Allen, G.W. and Segall, D., Monitoring of Computer Installations for Power Line Disturbances, IEEE PES Winter Meeting, Conference Paper C74 199-6, New York, N.Y., Jan. 27, 1974 .

[3] Goldstein, M. and Speranza, P.D., The Quality of U.S. Commercial Power", IEEE Paper CH1818-4/82/0000-00028, 1982.

[4] Thomas, L. T. and Key, S., Diagnosing Power Quality-Related Computer Problems, Industrial and Commercial Power Systems Technical Conference, IEEE, Cincinnati, June 6-8, 1978.

[5] Winfield, D. J., Reliability Study of Loss of Off-Site Power Supply, IEEE Transactions on Reliability , Vol. R-36, No. 1, 1987.

[6] IEEE Recommended Practice for Emergency and Standby Power Systems for Industrial and Commercial Applications, IEEE Std. 446-1980, p. 62.

APPENDIX

Dranetz Power-Line Disturbance Analyser Specifications [1]

Input Monitored
3 wire 60 Hz, 480 V power supply.

Disturbance Modes Monitored
Slow averaging: Slow steady state change in average RMS voltage level, based on a 10 s moving average with increments of > 4 volts.

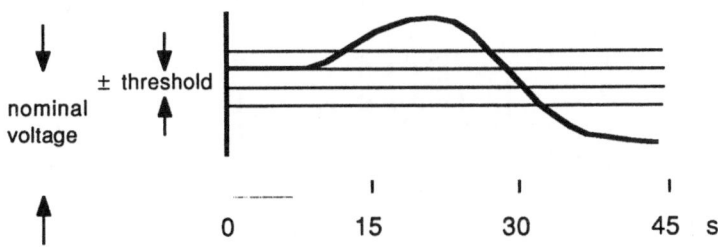

Figure A1.   Slow Averaged RMS Disturbance.

336

Sag/surge: Sudden changes in average RMS voltage level with threshold increments of 4 volts. Typical durations are between 1 cycle and 10 seconds.

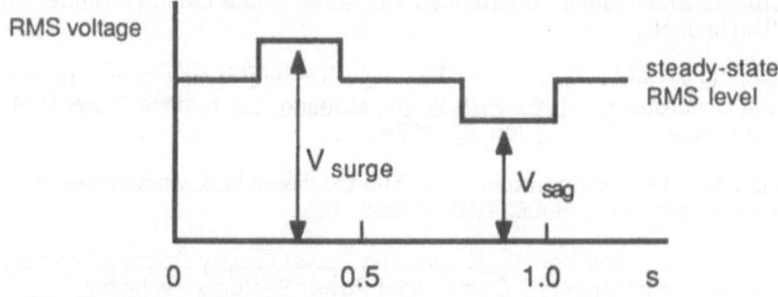

Figure A2. Sag/Surge Disturbances.

Impulse: Peak measurement of brief voltage excursion (impulse), after filtering out line waveform having a duration between 0.5 and 800 μs.

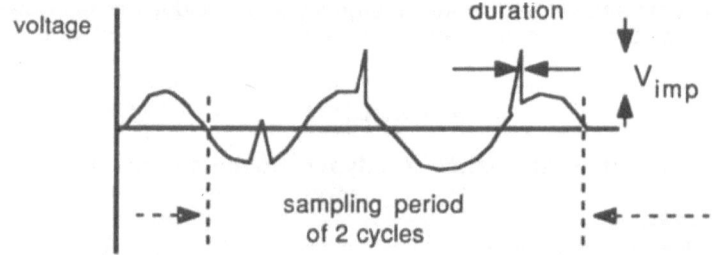

Figure A3. Impulse Disturbances.

Frequency: One-second average frequency deviations of more than 0.5 Hz.

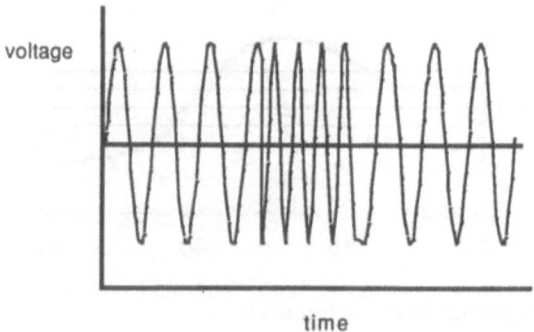

Figure A4. Frequency Disturbance.

AN EXPERT SYSTEM FOR FAULT DIAGNOSIS IN FIBRE PRODUCTION

PATRICK THORPE, ANDREW RUSHTON
Department of Chemical Engineering,
Loughborough University of Technology
Loughborough, Leicestershire LE11 3TU, UK

ABSTRACT

The maintenance of consistent product quality from an acrylic fibre manufacturing plant requires the rapid diagnosis of plant malfunctions. An expert system is being developed to aid process operators in this task. Various fibre quality parameters are monitored using control chart techniques. When predetermined tolerance limits are violated the system will attempt to identify the cause in terms of specific process equipment failures or mal-operations. In the proposed design, the diagnostic task is treated as a three stage process of data abstraction, hypothesis selection and solution refinement. Knowledge acquisition then involves eliciting the specific knowledge types associated with each of these tasks. Solution refinement is addressed by an hierarchical structure based on process objectives. Decomposition of the process, in terms of the fundamental tasks which it is designed to perform, aids the structuring of the knowledge base and facilitates efficient fault identification. The principal benefits of this approach are improved knowledge acquisition and knowledge base validation and modification.

INTRODUCTION

The maintenance of consistent product quality from a continuous production plant is dependent on the control of critical quality parameters. These parameters are often monitored through off-line laboratory testing. Techniques such as statistical process control [1] use suitable forms of control chart to plot quality data, this provides an efficient means of monitoring the production process and identifying process deviations. However, the ultimate maintenance of product quality is dependent on the identification of the root causes of deviations in terms of specific equipment failures or process mal-operations as well as the efficient rectification of these faults. This requires the process operator to draw on experience and knowledge of the process to interpret incoming data, gather additional information as required and take necessary corrective action.

The work describe here is the development of an expert system to assist process operators in the interpretation of quality data and to advise on the trouble-shooting of quality problems in the manufacture of acrylic fibre.

338

THE ACRYLIC FIBRE PRODUCTION PROCESS

The production of acrylic fibre involves, essentially, two main stages; the controlled reaction of monomers to form polymer dissolved in solvent which is termed dope, followed by the extrusion and processing of the polymer to form fibre. The process area addressed by this work is the spinning line which covers the process from extrusion onwards.

Polymer solution is extruded through fine holes into a bath of weak solvent where it coagulates into continuous filaments. The filaments are heated, stretched, cooled in a water bath and washed of residual solvent by counter-current flow of de-mineralised water. The stretching process aligns the molecular chains within the fibre giving it strength in the axial direction. After washing, the fibre is dried under controlled conditions to optimise fibre properties to a pre-defined moisture content. Next an anti-static solution is applied to the fibre before passing it through a crimping process which produces an axial undulation characteristic of natural woollen fibres. Finally, the fibre is plaited into cartons. The fibre can also be dyed on line when required.

Variables such as the bath temperatures, concentrations and flow rates, the speed of the fibre passing through various stages of the process as well as its alignment, tension and temperature profile through the dryer, will all have a critical effect on the final properties of the fibre, including its axial strength, brittleness, extensibility and dye quality.

Process operators are required to monitor various indicators of product quality and take action when quality standards are not met. This may require locating specific equipment faults within the process or adjusting various process parameters. If the latter course of action is taken they must be aware of the consequences of their actions; certain modes of operation will improve product quality in some areas while being detrimental to others.

A THREE STAGE DIAGNOSTIC PROCESS

The general architecture of the proposed expert system attempts to mimic the cognitive process of the operators in performing this diagnostic task. The process is essentially similar to Clancey's model of heuristic classification [2] and is broken down into three steps.

In the first step, described by Clancey as data abstraction, various quality variables are monitored, often with the use of control chart techniques. Raw numerical data is abstracted by the operator into a qualitative description of the state of the process.

The second stage, equivalent to Clancey's heuristic match, is triggered when the data, in its abstracted form, indicates a process abnormality. The operator uses experience of the process to form a number of hypotheses as to the cause of the deviation in product quality.

These fault hypotheses relate to general functions of the process and in the final refinement stage, each general hypothesis is explored in an attempt to locate the low level solution to the problem.

Hence, for example, in an attempt to explain a brittle fibre problem, the operator may propose poor spin bath temperature regulation as a possible cause, if the spin bath temperature is indeed abnormal, then, in the solution refinement phase, an attempt will be made to locate the fault to a component level such as a faulty cooling water circulation pump or clogged valve.

A DIAGNOSTIC HIERARCHY BASED ON PROCESS OBJECTIVES

The solution refinement task is essentially a search problem, and intelligent search entails efficient structuring of the search space. Rasmussen [3] has shown that human experts reason at successive levels of abstraction enabling them to reduce the complexity of the domain in question. This strategy suggests hierarchical modelling of the process; a hierarchical structure naturally supports the top-down solution refinement procedure.

In the proposed design, a method of hierarchical structuring is used to create a description of the process plant which forms the basis of the diagnostic system. Its construction forms the initial phase of the knowledge acquisition process in which the expert system builder, aided by an expert, experienced in the operation of the process, attempts to collect knowledge to be used in the development of the system.

Several researchers have made use of hierarchical structuring in the design of expert systems for process industry applications. The expert system shells Picon [4] and G2 [5] use sets of rules to focus diagnosis on successively smaller areas of the process plant. The systems described by Schum et al. [6] and Ramesh et al. [7] use a hierarchy based on successively refined malfunction hypotheses to act as a framework upon which the entire diagnostic system is based.

The method used to build the hierarchy here is a decomposition in terms of the elementary objectives of the process plant. That is, the fundamental tasks which it is designed to perform. This approach is useful in developing hierarchies which are similar to the process operators own conceptual view of the process. The approach is a generalisation of the work described by Shafaghi et al. [8] who used a process description based on control systems for the construction of fault trees. A similar method was used by Finch and Kramer [9] to develop an algorithm for fault diagnosis.

A section of the hierarchy, describing the function *fibre spinning*, is shown in Fig. 1. The relevant section of the process line diagram is shown in Fig. 2. At the highest level of

340

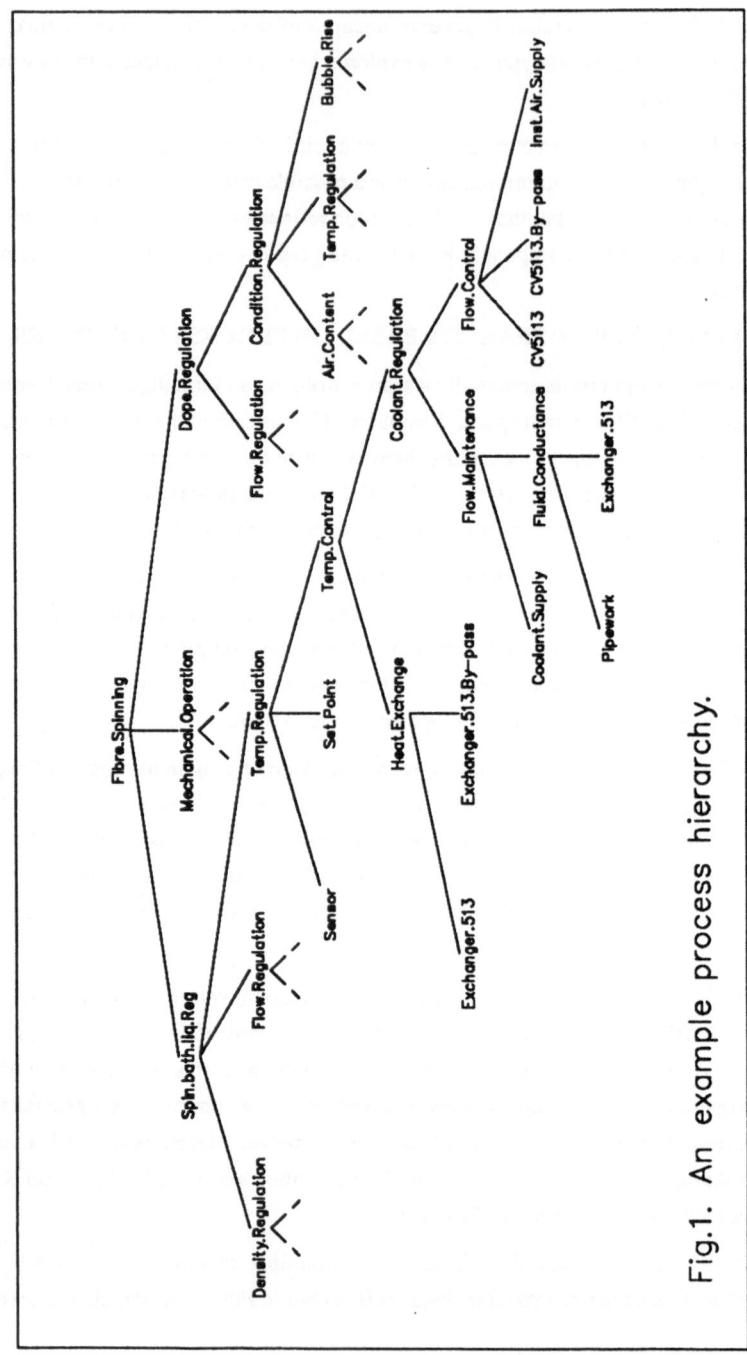

Fig.1. An example process hierarchy.

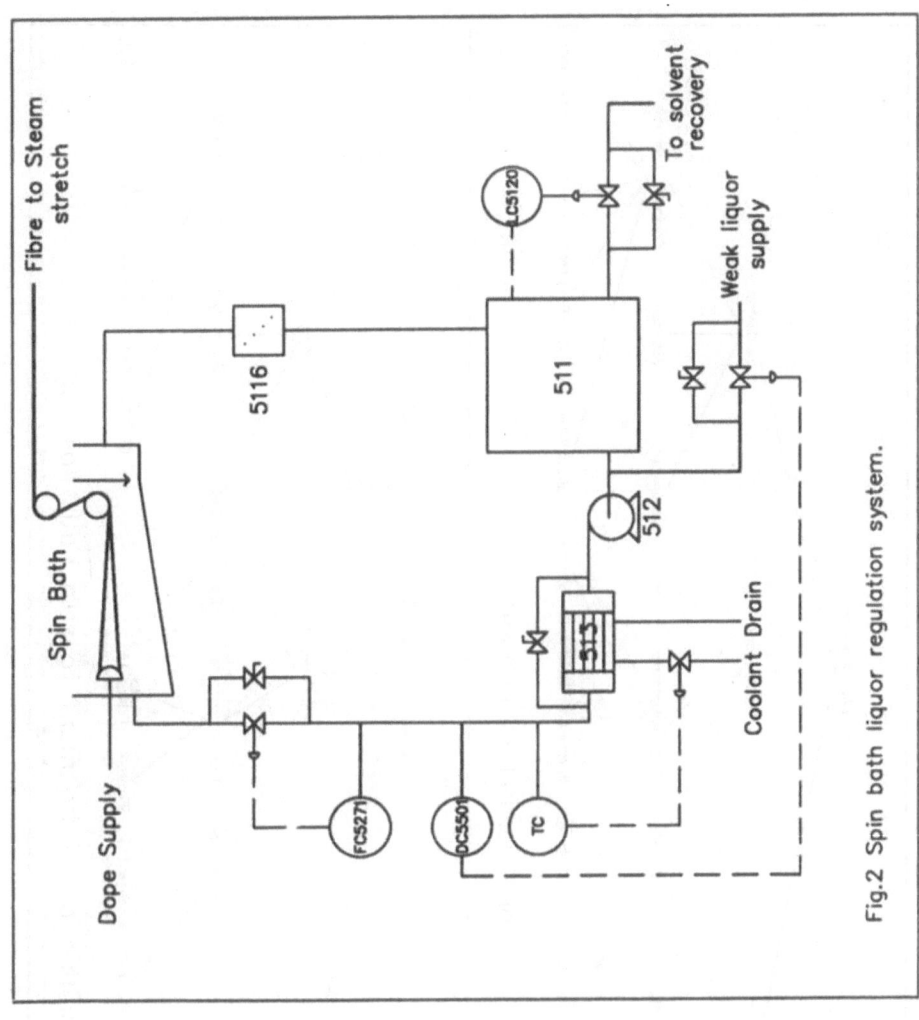

Fig.2 Spin bath liquor regulation system.

342

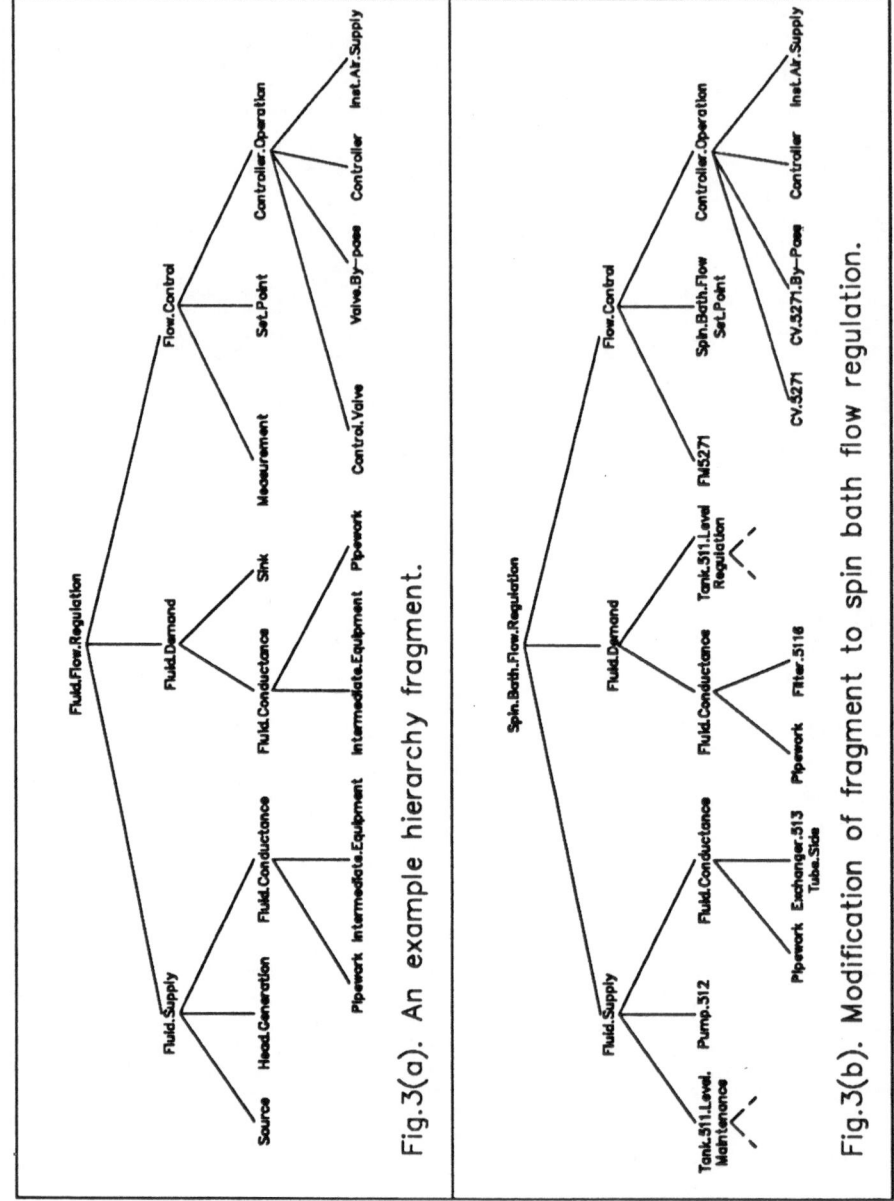

Fig.3(a). An example hierarchy fragment.

Fig.3(b). Modification of fragment to spin bath flow regulation.

abstraction within the hierarchy are functions such as *fibre spinning* and *fibre drying*, these may represent individual units of process equipment or collections of units performing a common general function.

The next level represents the sub-tasks which are required to be performed in order to achieve the general process objectives and, in turn, each of these may be dependent on further sub-tasks. Examples of these sub-tasks are spin-bath temperature and flow regulation, which may be dependent on further sub-tasks such as fluid containment or fluid conductance.

In the expert system each of the nodes within the hierarchy contains rules for testing whether the objective which the node represents is actually being fulfilled by the process. For example, the objective *spin bath temperature regulation* is only achieved if the spin bath temperature lies within specified tolerance limits. In this way, the hierarchical structure is used to guide the diagnostic search. The system successively locates those objectives which are not being fulfilled and eliminates those objectives that are being fulfilled from the diagnostic search.

Leaf nodes within the hierarchy will represent the elementary components of the process plant described at the required level of detail. These components will typically be items such as valves, pipes or pumps; failure of any one of these elementary components will represent a single, identifiable fault within the plant.

In order to aid the development of the hierarchical structure a set of hierarchy fragments was developed which describe typical process tasks. These tasks are based around control loop structures such as temperature or flow regulation, the hierarchy fragments are adapted to fit specific plant descriptions. Fig. 3 shows how the fragment relating to the general objective *fluid flow regulation* is adapted to describe *spin bath flow regulation*.

HYPOTHESIS SELECTION USING CAUSAL RELATIONSHIPS

The knowledge applied to the solution refinement task is rigourously structured, corresponding to the structure of the process plant. The knowledge associated with the task of hypothesis selection is somewhat less formalised, being essentially based on a set of causal relationships which map between abstracted data and general fault hypotheses. The general hypotheses formulated by the system relate to general process tasks at a high level of the structured hierarchy.

Sets of rules are used to imitate the way in which the process operator, faced with a product quality deviation, assesses the state of the plant and attempts to identify the cause. The knowledge relating to this stage of the diagnostic task was obtained using several techniques. These included protocol analysis [10], in which an operator is given a problem to solve and then monitored as the available information is analysed and solution hypotheses formed, as well as more common cause and effect analysis methods such as Pareto analysis [1].

SYSTEM OPERATION

The system monitors the various quality variables which are entered by the user as they become available. The data is compared with pre-defined tolerance limits and displayed in a control chart format. If these limits are violated for any particular variable then a rule base containing the set of rules specific to that particular variable is triggered.

Statistical data from the various quality control charts is abstracted, by the system, into qualitative descriptions of both the magnitude of the variables, with reference to their mean values, and the trend of data points. The rule sets triggered in the second stage of the diagnostic task use this abstracted data to generate an ordered list of solution hypotheses.

Each of the hypotheses is considered in turn by activating the appropriate section of the solution hierarchy. The solution refinement procedure will then attempt to accept or reject a particular hypothesis, if it is accepted then the relevant section of the hierarchy is searched until a leaf node is reached which is identified as the fault origin.

For example, if the sample data entered by the user indicates a *brittle fibre* problem then the set of rules specific to that particular quality variable will be activated. These rules will be used to consider the state of various other quality variables and generate a list of solution hypotheses. The first hypothesis on this list might be *faulty spin bath temperature regulation* and the system will activate the relevant node in the solution refinement hierarchy. Referring to Fig. 1, if this hypothesis is accepted then the objective *temperature control* may be considered next. This will have an associated test involving a comparison of the actual spin bath temperature with the set point value within specific tolerance limits. If this test fails, indicating that the objective is not being fulfilled, then the search space can be reduced to that area of the solution hierarchy. The heat exchange node will be activated if the bath temperature is below the set point value, this may lead to the identification of the fault location as the exchanger by-pass valve being open in error.

SUMMARY AND CONCLUSIONS

This paper has described an expert system which is being developed to aid process operators in the maintenance of consistent product quality from an acrylic fibre manufacturing plant. The system approaches diagnosis as a three stage process of data abstraction, hypothesis selection and solution refinement. Hypothesis selection is performed using sets of rules which relate symptomatic product quality information to operating faults in terms of general process objectives. Solution refinement uses a hierarchical structure based on process objectives to guide the diagnostic search.

The delivered system will couple this architecture with a graphical user interface and machine-machine interface facilities which will allow automatic access to plant data via existing process control hardware.

The method described here is generally applicable to the development of rule based expert systems for diagnosis in a chemical process plant environment. The particular benefits of this approach being :-

i) Improved Knowledge Acquisition

By analysing the diagnostic task before building the system and adapting the knowledge acquisition methods to the various types of applicable knowledge, this problematical stage of expert system development becomes more formal and rigourous. The construction of hierarchies based on process objectives provides a description of the plant which is similar to the operators' own conceptual view and is therefore more easily understood. The construction of the hierarchies is aided by the use of pre-defined hierarchy fragments.

ii) Improved Knowledge Base Validation and Modification

One major difficulty in the development of rule based expert systems for the process industry is efficient knowledge base validation. This includes both the location of errors in the reasoning process of the system, whereby a wrong conclusion is reached in the light of given evidence, and the location of missing or incomplete knowledge. Clearly, the more structured the architecture of the system, the easier this validation process will be.

The sectioning of rules into small rule bases, for hypothesis selection, should make their validation more manageable. In solution refinement, the use of a structured hierarchy corresponding to the process operators own conceptual view of the plant, should again be an aid to efficient validation.

The modular nature of the whole system will aid the process of modification and development. Alterations can be made to specific areas of the system without affecting the overall reasoning process.

FUTURE WORK

The structured hierarchies developed for this system provide a useful description of the process plant which could be adapted to a number of useful tasks and this may form the basis for future development work. One possible application is an intelligent alarm interpretation and display system, using process alarms to activate specific nodes within the hierarchy. The system could move down the hierarchy from the activated node to locate the possible cause of the alarm condition or up the hierarchy to display the possible consequences of the present situation. By creating graphic displays based on the hierarchy the user could be presented with plant mimics and alarm displays at various levels of detail which could aid in the task of alarm interpretation.

ACKNOWLEDGEMENTS

The authors gratefully acknowledge the assistance and financial support of the Science and Engineering Research Council, Swindon, and Courtaulds Research, Coventry.

REFERENCES

1. Juran, J.M., Juran's Quality Control Handbook, 4th.Edn., McGraw-Hill, 1988.

2. Clancey, W.J., Heuristic Classification. Artificial Intelligence, 1985, 25, 289.

3. Rasmussen, J., The Role of Hierarchical Knowledge Representation in Decisionmaking and System Management. IEEE Trans. Systems, Man, Cybernetics, SMC-15, 1985, 2, 234.

4. Moore, R.L., Hawkinson, L.B. Knickerbocker, C.G. and Churchman, L.M., A Real Time Expert System for Process Control. IEEE Proc. 1st Conf. Artificial Intelligence Applications, 1984, 569.

5. Moore, R.L., Hawkinson, L.B., Levin, M., Hofmann, A., Mathews, B.L. and David, M.H. The G2 Real Time Expert System. In Knowledge Based Expert Systems for Engineering, Classification , Education and Control, eds. D. Sriram and R.A. Adey, Computational Mechanics 1987.

6. Shum, S.K., Davis, J.F., Punch, W.F. and Chandrasekaran, B., An Expert System Approach to Malfunction Diagnosis in Chemical Plants. Comput. Chem. Engng, 1988, 12(1), 27.

7. Ramesh, T.S., Shum, S.K. and Davis, J.F., A Structured Framework For Efficient Problem Solving in Diagnostic Expert Systems. Comput. Chem. Engng, 1988, 12(9/10), 891.

8. Shafaghi, A., Andow, P.K. and Lees, F.P., Fault Tree Synthesis Based On Control Loop Structure, Chem. Engng Res. and Des., 1984, 62, 101.

9. Finch, F.E. And Kramer, M.A., Narrowing Diagnostic Focus Using Functional Decomposition. AIChE Journal, 1988, 34(1), 25.

10 Johnson, P.E., Zualkernan, I. and Garber, S., Specification of Expertise, Int. J. Man-Machine Studies, 1987, 26, 161.